# Springer Series on Polymer and Composite Materials

The "Springer Series on Polymer and Composite Materials" publishes monographs and edited works in the areas of Polymer Science and Composite Materials. These compound classes form the basis for the development of many new materials for various applications. The series covers biomaterials, nanomaterials, polymeric nanofibers, and electrospun materials, polymer hybrids, composite materials from macro- to nano-scale, and many more; from fundamentals, over the synthesis and development of the new materials, to their applications. The authored or edited books in this series address researchers and professionals, academic and industrial chemists involved in the areas of Polymer Science and the development of new Materials. They cover aspects such as the chemistry, physics, characterization, and material science of Polymers, and Polymer and Composite Materials. The books in this series can serve a growing demand for concise and comprehensive treatments of specific topics in this rapidly growing field. The series will be interesting for researchers working in this field and cover the latest advances in polymers and composite materials. Potential topics include, but are not limited to:

**Fibers and Polymers:**

- Lignocellulosic biomass and natural fibers
- Polymer nanofibers
- Polysaccharides and their derivatives
- Conducting polymers
- Surface functionalization of polymers
- Bio-inspired and stimuli-responsive polymers
- Shape-memory and self-healing polymers
- Hydrogels
- Rubber
- Polymeric foams
- Biodegradation and recycling of polymers

**Bio- and Nano- Composites:-**

- Fiber-reinforced composites including both long and short fibers
- Wood-based composites
- Polymer blends
- Hybrid materials (organic-inorganic)
- Nanocomposite hydrogels
- Mechanical behavior of composites
- The Interface and Interphase in polymer composites
- Biodegradation and recycling of polymer composites
- Applications of composite materials

Sapana Jadoun • Nirmala Kumari Jangid
Editors

# Tuning of Conducting Polymers

## Applications from Environmental to Biomedical Fields

Springer

*Editors*
Sapana Jadoun
Sol-ARIS, Departamento de Química
Universidad de Tarapacá
Arica, Chile

Nirmala Kumari Jangid
Department of Chemistry
Banasthali Vidyapith
Banasthali, Rajasthan, India

ISSN 2364-1878 ISSN 2364-1886 (electronic)
Springer Series on Polymer and Composite Materials
ISBN 978-3-032-00460-4 ISBN 978-3-032-00461-1 (eBook)
https://doi.org/10.1007/978-3-032-00461-1



This Springer imprint is published by the registered company Springer Nature Switzerland AG
The registered company address is: Gewerbestrasse 11, 6330 Cham, Switzerland

# Preface

Conducting polymers have emerged as a remarkable class of materials that combine the electrical properties of metals with the structural versatility and mechanical flexibility of polymers. Their unique ability to be chemically tuned and structurally modified for specific functions has attracted tremendous attention from researchers and industries alike. These materials now play a crucial role across a wide range of disciplines, including environmental remediation, energy storage, biomedical engineering, and wearable electronics.

The tuning of conducting polymers has opened new pathways for advancing applications in both traditional and emerging domains. Their properties can be precisely engineered through chemical modifications, post-synthesis treatments, and hybridization, allowing for enhanced conductivity, selectivity, biocompatibility, and environmental sustainability. This versatility has led to an explosion of interest in their use for photocatalytic degradation of pollutants, heavy metal adsorption, corrosion protection, smart drug delivery, flexible electronics, and architectural applications, among others.

This book, *Tuning of Conducting Polymers: Applications from Environmental to Biomedical Fields*, presents a comprehensive overview of the strategies used to modify and enhance conducting polymers, showcasing the latest developments and interdisciplinary applications. The chapters are contributed by experts from diverse fields, including chemistry, materials science, environmental engineering, healthcare, electronics, and architecture. Each chapter offers detailed insights, experimental data, case studies, and forward-looking perspectives that aim to bridge the gap between fundamental science and practical implementation.

We believe this book will be a valuable resource for students, researchers, scientists, engineers, and professionals across multiple disciplines. By offering a deep dive into the tunability and utility of conducting polymers, we aim to promote innovation, foster interdisciplinary collaboration, and encourage new explorations into the vast potential of these dynamic materials.

We would like to thank all the contributors and researchers whose work has shaped this growing field. Their dedication and expertise have been instrumental in assembling the knowledge presented in this volume.

Arica, Chile Sapana Jadoun
Banasthali, Rajasthan, India Nirmala Kumari Jangid

# Contents

# Chapter 1
# Introduction to Conducting Polymers

**Indrani, Riya Chugh, and Rajeev Kumar**

**Abstract** Conducting polymers (CPs) are organic materials that exhibit electrical and optical properties like metals and semiconductors. They combine the flexibility of traditional polymers with the ability to conduct electricity, making them ideal for applications in electronics, sensors, energy storage, and biomedical devices. The discovery of CPs dates back to 1862, with significant advancements in the 1960s and 1970s, including developing materials like polyacetylene and poly(sulphur nitride). The conductivity of CPs is primarily determined by doping processes involving the introduction of charge carriers, such as solitons and polarons. CPs offer various advantages, such as chemical tunability, lightweight nature, and corrosion resistance. However, their performance continues to evolve, showing significant potential to rival metals and semiconductors. CPs are often combined with carbon nanomaterials like graphene to enhance their properties. However, their conjugated backbone structure makes them insoluble, limiting their processability. Efforts to improve solubility, such as adding flexible side chains and using polar or nonpolar solvents, have been explored. Tailoring CPs through chemical modifications, copolymerization, and functionalization is key to enhancing their conductivity, flexibility, and chemical stability. This has expanded their applications in flexible electronics, organic photovoltaics, biosensors, and drug delivery systems. Characterization techniques such as FTIR, XRD, and AFM are crucial for assessing their properties. However, challenges like scalability, maintaining conductivity while enhancing flexibility and addressing environmental concerns remain. The ongoing development of CPs through innovative design approaches holds significant promise for next-generation technologies. Hence, tailored CPs offer vast potential, addressing current limitations and enabling advancements in various industries.

Indrani (✉) · R. Chugh · R. Kumar
Department of Applied Sciences, Manav Rachna International Institute of Research and Studies, Faridabad, Haryana, India
e-mail: indrani.set@mriu.edu.in; riyachugh.set@mriu.edu.in; rajeevkumar.set@mriu.edu.in



S. Jadoun, N. Kumari Jangid (eds.), *Tuning of Conducting Polymers*, Springer Series on Polymer and Composite Materials,
https://doi.org/10.1007/978-3-032-00461-1_1

**Keywords** Applications · Characterization · Conducting polymers · Modification techniques tailored conducting polymers

## 1.1 Introduction

A well-known fact about polymers is that they are created from long chains of molecules known as monomers [1]. In general, the use of polymers, whether natural or man-made, plays a significant role in our everyday lives due to their optical and mechanical features, ease of processing, heat resistance, and low production costs [2]. Polymers can be shaped into more complex forms than materials such as ceramics and metals. In general, polymers are typically considered insulators. Usually, we observe the coating of wires with plastics, which are also used for making switches because of their insulation properties. However, not all the available polymers are insulators. Polymers can show conductive properties in the same way as metals. Conductive polymers (CPs) are considered a category of organic materials possessing distinct electrical and optical characteristics on a trend like inorganic semiconductors and metals with the added advantage of basic, multifaceted, and budget-friendly approaches. Basically, an intriguing class of materials known as CPs combines the characteristics of traditional polymers with the capacity to conduct electricity. Scientists have devised many ways to tweak and adjust CPs to link them with various medical uses, like biosensors and biomaterials. These new findings are in high demand in bioengineering, regenerative medicine, and biosensors as they could lead to more progress. CPs have demonstrated the ability to initiate various biological processes, paving the way for new applications in the medical field. Also, they react to electrical signals from various body parts, like muscles, connective tissue, skin, and nerves, which makes them great for many biological uses. CPs have made many medical devices work better, faster, and more when they connect with body tissues. Before the discovery of CPs, most of these uses didn't involve electronics. Since then, experts from various fields have taken over to further the development of CPs. These polymers catch researchers' attention because they keep their polymer-like physical traits while showing a wide range of electrical conductivity similar to metals [3]. In the last three decades, CPs have caught the eye of many scientists. These materials are notable for their exceptional low mass, corrosion resistance, electrical conductivity, ease of usage, and many other qualities. CPs are used in various applications, including printed electronic circuits, batteries, sensors, and display devices, and that's just the beginning. These materials are impacting many other areas as well [4–6]. CPs are classified into nine families, as proposed by Otero [7]. These include substituted CPs like poly(3,4-ethylene dioxythiophene) (PEDOT), salts of pure CPs such as polythiophene doped with $ClO_4$, and self-ionized conjugated CPs like PEDOT-3-butanesulfonic acid. Blended forms also play a key role, including CPs combined with organic compounds (e.g., PEDOT: PSS), inorganic compounds (e.g., polypyrrole (PPy) with polyoxometalates), and organic-carbon electroactive composites like fullerene-charged poly(3-hexylthiophene). Additionally, copolymer-type CPs, such as poly (EDOT-pyrrole-EDOT) and other hybrid configurations, further showcase the versatility of these materials in various applications [8].

## 1.2 The Origins and the Development of Conducting Polymers

The history of CPs began in the year 1862 when Letheby et al. [9] synthesized polyaniline (PANI), initially known as "aniline black," through the oxidation of aniline under mild conditions. PANI exhibited conductivity and electrochromic properties, making it useful in dyes and printing [10, 11]. The field advanced significantly in the 1960s with the discovery of semiconducting polymers by Pohl, Katon, and colleagues [12, 13]. John Pople [14] and S.H. Walmsley [14] theoretically predicted CPs in 1962, introducing the concept of "solitons" in polyacetylene (PA), suggesting that mobile charge-induced defects accounted for its conductivity [14]. PA was later synthesized through acetylene reactions in the 1970s [15]. Shirakawa et al. [16] observed that its conductivity increased when exposed to halogen vapors. Weiss et al. [17] and Gill et al. [18] further enhanced PA's conductivity by doping it with oxidizing and reducing agents, marking a breakthrough in CP research. Literature suggests that the first conducting polymer, known as poly(sulphur nitride), was discovered in the year 1975 by Greene and Street [19]. They found that (sulphur nitride) was found to be electronically conductive at lower temperature as well as a super conduction transition occurs occurred at a temperature of 0.26 K [20]. Furthermore, chemist Alan MacDiarmid examined the electrical properties of poly(sulphur nitride) in collaboration with physicist Alan Heeger [21]. Hideki Shirikawa was studying the Ziegler-Natta-catalyzed polymerization of acetylene at the time. It was observed that a glossy coating was formed on the reaction vessel's surface when the catalyst concentration approached molar levels [19]. Shirikawa and MacDiarmid Heeger were drawn to work together on poly(acetylene) because of this "metallic" polymer. The conductivity and structure of poly(acetylene) were investigated after donors and acceptors were doped on it [19]. For their role in discovering and advancing CPs, Heeger, MacDiarmid, and Shirakawa were granted the Nobel Prize in 2000. There are two ways of doping in which the CP backbone gains an electron through the process of reduction, which is popularly known as n-type doping, while the removal of an electron, which is commonly known as p-type of doping, results in such a structural arrangement that assists the movement of charges. The range of electrical conductivity from insulator to semiconductor to metal is determined by the doping degree during the chemical and electrochemical doping processes. Since the anisotropic alignment of the doped polymer chains increases the conductivity by the orders of magnitude of two, the electrical conduction is highly reliant on the type of dopants, the polymeric nature, and the processing technology utilized to manufacture the polymers [22]. CPs usually show a remarkable range of optical, electrical, and mechanical properties, especially when subjected to methods like redox doping, reversible doping, and dedoping processes. CPs found numerous ranges of applications in various technological areas, which is basically due to unique structural features [23].

## 1.3 Essential Characteristics of Conducting Polymers

Compared to the inorganic materials used nowadays, CPs offer various advantages such as varied chemical composition, lightweight nature, adaptability, resistance to corrosion, customizable shape and structure, and adjustable conductivity. However, CPs haven't caught up to metals and inorganic semiconductors regarding their properties. So, to overcome their built-in limitations in how well they dissolve, conduct electricity, and last over time, scientists have changed CPs or mixed them with other materials [24, 25]. If conjugated polymers CPs are mixed with other materials in a controlled manner, it is possible to fabricate materials with appealing properties and great application prospects across a range of industries, such as electronics and energy devices. In this sense, researchers in this field have reported several ways to fabricate CP-based composites and hybrids with complex structures and enhanced functional properties. One representative case is the synthesis of CP nanocomposites using carbon nano-species such as graphene, carbon nanofibers, and carbon nanofibers [26–29]. These carbon nano-species enhance conductivity by facilitating the charge carriers' delocalization, thereby improving the CP chains' structural ordering. Conductivities have been achieved from metallic to insulator among others [25, 30, 31].

### *1.3.1 Electrical Conducting Properties*

The electrical conductivity of materials is majorly determined by their electronic structure, with the energy band theory providing insights into the behavior of various conductors, insulators, and semiconductors. In general, in the case of conductors, conduction and valence bands overlap, allowing free movement of electrons. However, semiconductors have little band gaps that allow electrons to be excited into the conduction band, facilitating current flow. Insulators have significant band gaps that prevent electron movement, thus hindering conductivity. However, the energy band theory does not fully explain the conductivity of CPs, organic materials with conjugated single and double bonds along their polymer chains. While these conjugated bonds allow some charge flow, they do not inherently make CPs highly conductive. A breakthrough study demonstrated that doping CPs with halogens could remove electrons from the polymer, creating holes [32]. Electrons from neighboring positions can then jump to fill these holes, enabling charge flow through the polymer chain. Following this work, theories on the conductivity of CPs have focused on forming nonlinear local excitations, such as solitons, polarons, and bipolarons, which act as charge carriers and enhance conductivity [32]. The electrical characteristics of polyconjugated materials are greatly governed by the behavior of π-electrons in the material, especially the degree of π-electrons delocalization and polarization. The electrical conductivity of first-generation CPs remained low due to the high degree of disorder, both structural and morphological, within the polymer matrix. As a result of this disorder, the polyconjugated structures did not allow considerable delocalization of π electrons and thereby do not promote charge transport to levels more than weak metallic conductivity range. Doping these CPs, however, leads to a reduction in the structural and morphological disorder; hence, the new type of CPs exhibited much higher CPs. Generally, doping gives rise to charge carriers with lattice distortions like solitons, polarons, and bipolarons. Undoped CPs have very low conductivity, on the order of $10^{-6}$ to $10^{-10}$ S/cm, which puts them in the interfacial region of semiconductors and insulators. Conversely, this conductivity can increase over 10 or more orders of magnitude when doped, making them near or fully metallic.

### *1.3.2 Magnetic Properties*

The magnetic properties of CPs have been comprehensively explored to understand charge carriers and unpaired spins [33–36]. Studies done by Lu et al. [37] and Long et al. [38] investigated the magnetic properties of PANI-$Fe_3O_4$ composites in different nanostructures. Lu et al. [37] synthesized PANI-$Fe_3O_4$ composite nanotubes using an ultrasonic irradiation technique, while Long et al. [38] used a self-assembly method to create PANI-$Fe_3O_4$ nanorods. The ultrasonic irradiation technique enhanced the dispersion of $Fe_3O_4$ particles and resulted in superparamagnetic behavior in the composite nanotubes. Additionally, the magnetic properties of other 1D conducting polymer-metal (Fe, Co, Ni) nanocomposites have also been studied, contributing further insights into the interaction between CPs and magnetic materials [39–41].

### *1.3.3 Mechanical Properties*

Sulong et al. [42] investigated the mechanical properties of carbon nanotube/graphite/polypropylene nanocomposites for conducting polymer composites (CPCs). Furthermore, it was identified that chemical functionalization and filler concentrations significantly influenced the mechanical characteristics of the CPCs. After functionalization, the CPC exhibited improved strength, tensile strengths of 80 MPa and 35 MPa, and greater hardness than the non-functionalized CPC. The mechanical properties of uniaxially oriented PPV thin films were studied by synthesizing them through simultaneous thermal exclusion and uniaxial extension of the poly sulphonium salt precursor [43]. The properties were anisotropic, dependent on molecular orientation, with Young's modulus ranging from 2.3 to 37 GPa in the machine direction and from 2.3 to 0.5 GPa in the transverse direction. Somanathan [44] studied the tensile properties of conducting poly(3-cyclohexyl thiophene) films at different temperatures. Additionally, Sgreccia et al. [45] examined the mechanical properties of hybrid proton conducting polymer composites based on sulfonated polyetherketones.

### *1.3.4 Thermal Properties*

Thermal tunability and stability are essential for the processibility of CPs. CPs are relatively stiff, with higher melting points and glass transition temperatures (Tg) compared when compared to non-conjugated systems [46]. Various research studies by different groups have proposed that elevated temperatures can enhance the alignment between the polymer and doping agent inside a CP, strengthening intermolecular connections and improving conductivity. However, this is also concerned with a reduction in thermal stability, which materializes due to reorientations of the molecules that expose polymer side groups for reactions with moisture or air. CPs don't have a distinct melting point like metals and other things made of polymers do. This has severely hampered their application in temperature-mediated manufacturing.

### *1.3.5 Solubility*

In general, CPs are usually insoluble and infusible due to the stiff conjugated polymer chain, which restricts their direct use. Researchers have worked on various strategies to solve this difficulty and make them more useable. The successful synthesis of soluble Polyaniline (PANI) by Cao et al. [47], for example, resolved the polymer's intrinsic solubility problems and opened the door for large-scale manufacture. One of the popular and successful strategies for increasing the solubility of conjugated CPs is the addition of flexible side-chain replacements. It has been observed that even though the monomer unit of the majority of CP is usually dissolvable in various commonly available solvents, the polymerized structure usually results in limited solubility. A fundamental principle in chemistry, "like dissolves like," indicates that for a conducting polymer (CP) to dissolve, the solvent must have similar physicochemical properties [48]. The unique polarity of most CPs, stemming from their heterocyclic structure, makes it very difficult to find appropriate solvents capable of dissolving or dispersing them. Solvents are broadly classified into two categories: polar and nonpolar. Polar solvents interact through dipole moments arising from differences in electronegativity between adjacent atoms, as seen in hydrogen bonding with water molecules. These solvents are also categorized as protic and aprotic. In contrast, nonpolar solvents lack such interactions and do not support charge-driven mechanisms such as dipoles. In addition to solubility, selecting solvent also influences reaction kinetics and chemical stability, which may lead to unwanted reactions or the breakdown of the CP structure. Due to their conjugated nature and varying localized charge states, most CPs cannot be fully dissolved by polar or nonpolar solvents alone. For solubilization to occur, the interaction forces between the polymer and solvent molecules must exceed the intermolecular forces holding the polymer chains together. Longer molecular chains increase these intermolecular interactions, further reducing solubility. CP chains are polymerized to maintain conductivity and ensure uninterrupted conjugation along their backbone, which limits their solubility.

## 1.4 Applications of Conducting Polymers Across Industries

CPs are extraordinarily versatile materials with applications extending across numerous domains. They result from their incredible electric conductivity, chemical tunability, and mechanical flexibility. In electronics, CPs are used to fabricate organic light emitting diodes (OLEDs), natural photovoltaics (OPVs), and transistors. Their light weight and flexibility make them additionally suitable for electromagnetic protection. Moreover, CPs are crucial in power storage as nanostructures have excessive capacitance, making them suitable for supercapacitors, lithium-ion batteries, gasoline cells, and solar cells, which might be an alternative way to resolve strength troubles. In the biomedical field, CPs are used to develop biosensors, drug delivery systems, and scaffolds for tissue engineering to discover organic signs and enhance cutting-edge medical technologies by using their biocompatibility. CPs are also implemented in environmental-friendly sensors for monitoring and detecting flashes of chemical or organic materials, which are characterized through reversible phenomena like drops or changes in electrical conductivity, color, and quantity. The performance of sensors has also accelerated with the incorporation of carbon nanotube, graphene, and metallic oxide substances. CP-based biosensors featuring enzyme-functionalized nanotubes offer remarkable sensitivity and stability. Thanks to their novel structural design and unique bonding mechanisms enabled by organic polymers, they excel in detecting biomolecules such as glucose, proteins, and hormones. CP composites are also effective in catalytic applications, with PANI-Pd nanofibers excelling in chemical catalysis and $TiO_2$-based composites showcasing enhanced photocatalytic properties for environmental remediation. Additionally, CP composites such as PANI-Au nanotubes and PPy-cobalt porphyrin nanorods exhibit remarkable electrocatalytic activity for reactions like NADH oxidation and oxygen reduction in fuel cells. Their incorporation into smart textiles and wearable devices has led to innovative solutions combining flexibility with conductivity. These multifaceted applications underscore the transformative potential of CPs in driving advancements across energy, healthcare, environmental, and electronic technologies.

## 1.5 Techniques for Tailoring Conducting Polymers

### *1.5.1 Doping*

The synthesis of the attached CPs can be significantly optimized by chemical modifications of solubility, thermal stability, or flexibility by enhancing their physicochemical properties. These modifications often occur in precursor monomers but can also be introduced into conducting polymer materials after synthesizing the products. Special efforts were identified to increase CP's approachability and include additional functionality through chemical modification [49, 50]. Doping has been a known method for modifying the physiochemical properties of CPs. One approach for controlling the impact of doping agent on the structure and properties of CPs is through their covalent incorporation as a functional group attached to the monomer unit. On the same line, a water-soluble CP was engineered by Patil et al. [51] by modification of monomer followed by polymerization of these monomers. Bu et al. [52] further utilized the same technique to synthesize solvent-soluble polymer. The conductivity, in general, depends on the charge carrier density as well as the mobility of the charge carriers. Both factors are affected by doping; hence, doping can directly increase electrical conductivity. Essentially, the dopant concentration primarily increases the host's molecular anions or cations, which serve as charge carriers. The electrical properties of CPs are further dependent on the degree of doping. This increased conductivity is facilitated by redox reactions, which involve the transfer of charges and the creation of charge carriers [53]. Dopants function as electron acceptors, which promotes the influx of additional electrons into the system. From the viewpoint of electronic structure, doping is a process which entails electron removal from the highest occupied molecular orbital (HOMO) present in the valence band and their transfer to the lowest unoccupied molecular orbital (LUMO) present in the conduction band. This process of oxidation and reduction takes place in the polymer's backbone and results in generating various charge carriers popularly known as polarons, bipolarons, and solitons [54]. The movement of these charge carriers is the primary mechanism that enables electrical conduction in polymer structures.

### *1.5.2 Copolymerization*

In the recent years, modifying CPs through copolymerization has garnered significant interest due to their exceptional properties and wide range of applications. The copolymers can be synthesized in various forms, such as thin films and powders, through the oxidation and reduction of different monomers. These copolymers synthesized via copolymerization are intriguing materials with high demand, especially in biomedical, optoelectronics, and energy applications. Copolymers are created by combining two or more monomers, offering enhanced properties compared to their components. Polymers with three or more monomers, often "multicomponent polymers," can form as blends, blocks, or grafted structures [55]. Copolymerization allows the use of monomers that cannot form homopolymers and provides precise control over their chemical and physical properties, such as composition, polarity, solubility, and hydrophobic or hydrophilic balance. Additionally, copolymers' optical and electronic properties can be adjusted by varying the proportion of the monomers. Copolymer formation is typically random, with the relative amounts of each monomer having a more significant influence than their arrangement in the polymer chain [56]. Copolymerization offers various applications in diverse fields. In photovoltaic devices, copolymerization is essential for the design of donor–acceptor systems that enhance photoabsorption, exciton dissociation, charge transport, and electrical exchange. Improved performance in photovoltaic devices is achieved through side-chain engineering and other modifications, enhancing water solubility, electronic properties, and overall device efficiency without compromising electron transfer. Innovations like random block copolymers, including diketopyrrolopyrrole (DPP), dithienothiophene (DTT), and thiophene-based systems, enable efficient charge transport and energy conversion. Tailor-made interactions, through the strategic placement of functional groups, refine backbone structure and charge dynamics, further optimizing device performance. Thus, copolymerization facilitates precise tuning of material properties, driving advancements in high-performance solar cells. Copolymerization offers various applications in diverse fields. In photovoltaic devices, copolymerization is essential for the design of donor–acceptor systems that enhance photoabsorption, exciton dissociation, charge transport, and electrical exchange. Improved performance in photovoltaic devices is achieved through side-chain engineering and other modifications, enhancing water solubility, electronic properties, and overall device efficiency without compromising electron transfer. Innovations such as random block copolymers, including diketopyrrolopyrrole (DPP), dithienothiophene (DTT), and thiophene-based systems, enable efficient charge transport and energy conversion and tailor-made interactions through the strategic placement of functional groups, refining backbone structure and charge dynamics, and further optimizing device performance. Thus, copolymerization facilitates precise tuning of material properties, driving advancements in high-performance solar cells. Electrochromic copolymers have shown promising performance in devices, offering good optical contrast, fast response times, and high coloration efficiency. These materials also exhibit excellent optical memory and redox stability, making them suitable for commercial applications. Additionally, copolymers like poly[4-(5-(6-(2-decyltetradecyl)-6H-phenanthro[1,10,9,8-c,d,e,f,g]carbazol-4-yl)thiophen-2-yl)-5,6-bis(octyloxy)-7-(thiophen-2 yl)benzo[c][1,2,5]thiadiazole] (PPTBTT) demonstrate enhanced solubility, molecular packing, and structural order, making them ideal for cost-effective solution printing in electronic applications. Copolymers like poly(aniline-co-fluorene) also show good cycling performance, despite some capacity loss after extended use [57]. Supercapacitors, which are energy storage devices, have seen significant research into using CPs and copolymers to improve their performance. Copolymers like ethylenedioxythiophene (EDOT) and MPy (Methylpyrrole), along with functionalized

MWCNTs, have shown enhanced capacitance, rate capability, and excellent cycling stability, with some copolymers maintaining performance after 20,000 cycles. Poly(ANI-co-PPDA) and other copolymers have also demonstrated high capacitance and stability, making them ideal for supercapacitors. Additionally, copolymers are explored for chemical sensors, such as those detecting nitrogen dioxide, and drug delivery systems, showing controlled release properties. They are also applied in wastewater treatment, particularly in the adsorption and degradation of pollutants like dyes and toxic metals.

### *1.5.3 Functionalization of Conducting Polymers*

CPs like polyaniline (PANI), polypyrrole (PPy), and polythiophene (PT) have been extensively studied and functionalized using organic chemical reactions. This functionalization primarily follows two approaches: Direct modification of CP chains and Reactions involving substituted CPs with reactive groups (e.g., esters). In the direct approach, reactions such as electrophilic aromatic substitution (SEAr) are performed on the non-conductive or reduced form of CPs. For instance, in PANI and PPy, the N-H group can undergo electrophilic substitution. Similarly, nucleophilic substitution (SN) reactions can occur where nitrogen nucleophiles replace alkyl or acyl groups. Oxidized forms of these polymers can also undergo nucleophilic conjugate addition. However, functionalization is usually indirect for PT and other CPs like poly(fluorene). Functional groups are incorporated by attaching them to reactive groups already in the monomer during polymer synthesis. The primary goal of these functionalization processes is to enhance specific properties of the CPs, making them more suitable for targeted applications [58].

### *1.5.4 Advancements in Biofunctionalization of Conducting Polymers*

Polymers' electrical and mechanical attributes can be altered to complement the biologics and, thus, improve bio-interface interactions. The unintelligent coverage of the CPs with biologics through physical or chemical means is one of the ways of improving the compatibility. The reactive functional groups can covalently bind biomolecules to either the monomer (pre-functionalization) or the polymer itself (post-functionalization) [59]. Moreover, functionalized dopants can also incorporate the biomolecules. The pre-functionalization process involves the alteration of the monomers before the actual polymerization, which makes it easy to characterize but may produce steric hindrance during polymerization. These problems are addressed through copolymerization, and it is now possible to create CPs that can easily interact with biological systems with minimal foreign substance reaction. Post-functionalization allows the immobilization of large biofunctionals, which can greatly help tailor CPs and even create certain structures, such as hydrogels. But there are still some processing issues. Functionalization of the dopant makes the synthesis process easier and improves the interactions of the biointerfaces, especially when polymeric dopants are used.

### *1.5.5 Nanostructuring*

In the recent years, nanostructured CPs have received considerable attention due to their superior properties compared to many properties, such as increased surface area, reduced charge, and mass transport mechanisms, and these advantages make them ideal candidates for a variety of applications, including energy storage and conversion, sensors, actuators, and biomedical devices. Various new synthetic techniques have been developed for nanostructured CPs, resulting in high-performance devices using these materials.

## 1.6 Characterization and Evaluation of Tailored Conducting Polymers

These polymers' structural, electrical, optical, and chemical properties must be carefully characterized and assessed to customize them for uses [60]. An organized summary of their description and assessment of the attributes is provided below.

A crucial component of materials science is material characterization, which helps scientists comprehend materials' composition, characteristics, and functionality. Understanding polymers' mechanical, thermal, chemical, and physical characteristics and their molecular structure requires characterization [61]. The particular feature or trait being examined determines the methods for polymer characterization. There are five primary characteristics of polymers. The five primary facets of polymer characterization are depicted in Fig. 1.1.

### *1.6.1 Structural Characterization*

Structural characterization entails examining polymer molecular structures to comprehend their physical, chemical, and mechanical properties. The key characterization methods listed below have been applied to identify different polymer groups, forms, sizes, and crystallinity.

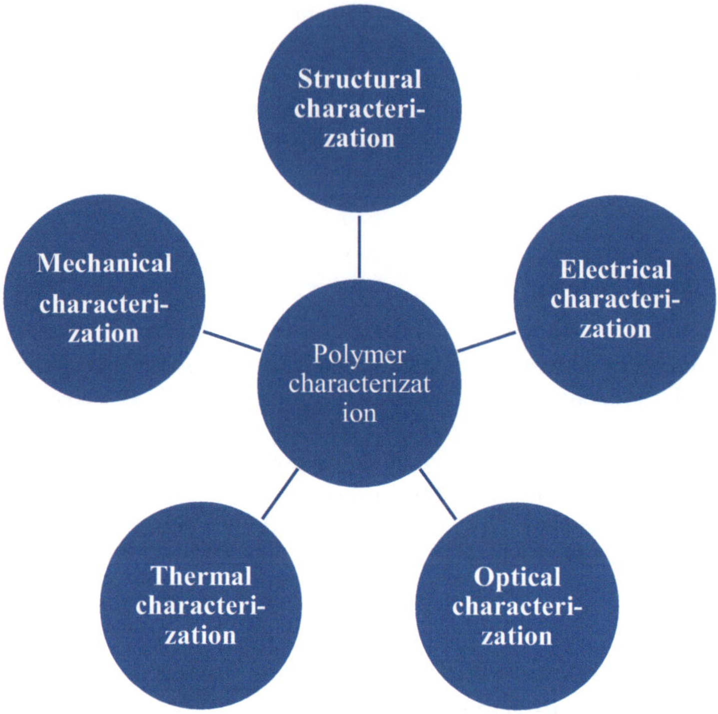


**Fig. 1.1** Characterization aspects of polymers

### *1.6.2 Fourier Transform Infrared Spectroscopy (FTIR)*

By examining the molecular vibrations of polymers, FTIR is a potent analytical tool for identifying and characterizing them. It details polymeric materials' molecular structure, functional groups, and chemical bonding. A polymer sample absorbs particular wavelengths corresponding to the vibrational modes of its chemical bonds when subjected to infrared (IR) radiation. After gathering the raw interferogram data, the device plots intensity versus wave-number to create an IR spectrum using a mathematical Fourier transform. The distinct absorption pattern is found between 400 and 1500 $cm^{-1}$ for identifying particular polymers. Characteristic peaks for functional groups such as -OH, -CH, -C=O, and so on are visible at 1500–4000 $cm^{-1}$ [62–65].

### *1.6.3 X-Ray Diffraction (XRD)*

The degree of crystallinity in polymers is evaluated using XRD. A large number of polymers have both crystalline and amorphous areas, making them semi-crystalline. Sharp peaks (crystalline regions) are placed on a broad hump (amorphous regions) in the diffraction pattern. Lattice parameters, crystal size, and the arrangement of molecules in the crystalline phase can be ascertained using XRD. XRD can show the preferred orientation of crystallites, commonly known as texture, in stretched or orientated polymers. A polymer material's various polymorphic forms or phases can be identified using XRD. In XRD, three primary peak types can be seen. When compared to metals and ceramics, polymers' comparatively poor atomic order was demonstrated by their broad peaks. Polymers with diffuse haloes showed amorphous areas with disordered molecules. Sharp peaks in polymers indicated crystalline areas and regular molecular packing [62–65].

### *1.6.4 Surface Topography*

The surface topography of polymers is the in-depth examination of the surface characteristics, textures, and structures of polymer materials. This involves analyzing the material's surface roughness, patterns, and other physical attributes that could affect its mechanical behavior, wettability, and adhesion. The surface topography is essential in many applications, such as coatings, biomaterials, microfluidics, and sensors. Techniques like atomic force microscopy (AFM), transmission electron microscopy (TEM), and scanning electron microscopy (SEM) are frequently employed to determine the surface topography of polymers. These methods assist in exposing surface alterations, cracks, textures, cross-sections for layered structures or flaws, phase separations or miscibility in composite systems, and the diameter, alignment, and deficiencies in fibers or thin films. The elemental composition and percentage in the polymers are provided using SEM in conjunction with energy dispersive X-ray (EDS). For polymers, AFM is utilized for mapping mechanical properties and detailed surface topography, TEM is the preferred method for internal micro/nanostructure investigation, and SEM is best for surface morphology [66, 67].

### *1.6.5 Electrical Characterization*

Analyzing the electrical characteristics of polymers is necessary to ascertain their possible uses in energy storage, electronics, and other domains. Various methods have been used to characterize the polymers electrically [68]. The diverse characterizations of polymers for various applications are shown in Fig. 1.2.

### *1.6.6 Optical Characterization*

Optical characterization of polymers is the study of how they interact with light to determine their optical characteristics, which can reveal information about their composition, structure, and uses. The methods frequently employed for

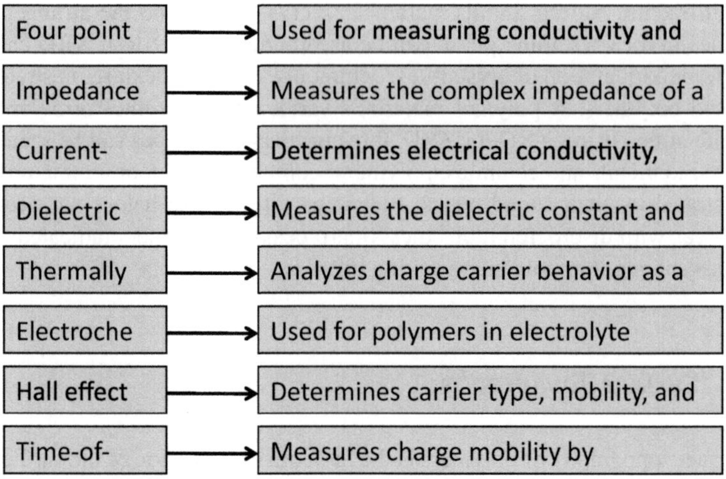


**Fig. 1.2** Electrical characterization techniques for polymers

optical characterization of the polymers include UV-Vis spectroscopy, FTIR, ellipsometry, photoluminescence, Raman spectroscopy, fluorescence spectroscopy, laser scattering techniques, and time-resolved spectroscopy. Using the methods above, the following characteristics have been found and ascertained: refractive index, absorbance and transmittance, photoluminescence, optical band gap, optical density, birefringence, scattering properties, and photoacoustic properties [60, 62, 64]. Understanding these characteristics and properties through various methods is useful for various applications of these polymers in different fields. Electrical characterizations also aid in the application of these polymers in particular fields, such as biomedical polymers (light-responsive drug delivery systems), coatings and films (transparency, reflectivity, or anti-reflective properties), optoelectronics (for the production of LEDs and solar cells), and polymer composites.

### *1.6.7 Thermal Characterization*

Understanding the thermal behavior of polymers and assessing their appropriateness for different applications requires the process of thermal characterization. Thermal gravimetric analysis (TGA) measures the weight loss of desired polymers as a function of temperature or time in a controlled atmosphere (air or inert gas). This study aids in the determination of residual solvents, moisture content, decomposition temperatures, and thermal stability. Differential scanning calorimetry (DSC) measures the heat flow related to polymer transitions as a function of the desired polymer's temperature. The glass transition temperature, melting point, crystallization temperature, heat capacity, and enthalpy of transitions are all determined by this study [69].

Dynamic mechanical analysis (DMA) is used to study the viscoelastic characteristics of desirable polymers under oscillatory stress. This research aids in measuring the damping factor, glass transition, relaxation phenomena, storage modulus, loss modulus, and time-temperature superposition behavior. Thermo-mechanical analysis determines how a polymer's dimensions vary as a function of temperature or time for desired polymers. This study aids in the measurement of softening points, shrinkage, and the coefficient of thermal expansion. Vicat softening point and heat deflection temperature (HDT) define the temperature at which a polymer deforms under a given load. This study compares the thermal performance of materials for structural purposes and evaluates the heat resistance of polymers. Thermal cycling tests determine how well a polymer performs during repeated heating and cooling cycles. This study helps to identify fatigue resistance and investigate thermal aging and stability [68].

### *1.6.8 Mechanical Characterization*

The process of measuring and examining a polymer's mechanical characteristics to comprehend how it behaves under different loads and environmental circumstances is known as mechanical characterization. This is essential for assessing their fit for certain uses and maximizing their functionality. Important characteristics of the polymers that require mechanical characterization include tensile strength, elastic modulus, elongation at break, toughness, hardness, impact strength, creep, stress relaxation, fatigue resistance, and viscoelastic behavior.

Stress–strain curves are produced by tensile testing to assess modulus, elongation at break, and tensile strength. Behavior under compressive loads is measured by compression testing. Three- or four-point bending tests are used in flexural testing to evaluate bending behavior. Common techniques for determining hardness include Rockwell or Vickers testing for tougher materials and Shore (durometer) testing for soft polymers. Impact testing measures resistance to abrupt impact loads, much like Charpy and Izod tests. The glass transition temperature and damping behavior are determined using dynamic mechanical analysis. Creep testing assesses long-term deformation at high temperatures and steady loads. To estimate longevity, fatigue testing simulates repeated loading and unloading cycles [60, 66, 68, 69].

## 1.7 Challenges and Future Directions and Emerging Trends in Tailoring Conducting Polymers

CPs must satisfy specifications about electrical, mechanical, and chemical properties, which presents several difficulties when customizing them for particular uses. Managing conductivity and process ability, stability issues, morphological control, electrical and mechanical property tuning, biocompatibility, scalability, functionalization issues, device integration, and raw material costs are the primary challenges in customizing CPs (Fig. 1.3) [70]. Because CPs in their doped forms are usually brittle and insoluble in most solvents, high conductivity sometimes comes at the sacrifice of process ability. Since CPs can break down at high temperatures, their use in high-temperature settings is restricted.

For certain applications like sensors, batteries, and organic electronics, achieving exact control over morphology, that is, shape and size, is essential. Doping levels in CPs determine their conductivity, and these levels must be carefully regulated for various uses. Because CPs are frequently brittle, they are inappropriate for uses requiring stretchability or mechanical flexibility [71]. Biocompatible and nontoxic CPs, including biosensors and neural interfaces, are essential for biomedical applications. It is challenging and expensive to synthesize CPs on a large scale with reliable quality and performance. Functional groups can be challenged to add to CPs without changing their π-conjugated structure, which is essential for conductivity. Complementary fabrication approaches are needed to include CPs in devices without compromising durability or performance. Because they are frequently costly, monomers utilized in polymer synthesis have limited use [72].

CP tailoring is a vibrant field with lots of room for development. Create polymers with conjugated backbones tailored to improve mechanical flexibility and electrical conductivity. Combine conductive polymers with nanomaterials such as metallic nanoparticles, graphene, or carbon nanotubes to increase conductivity and

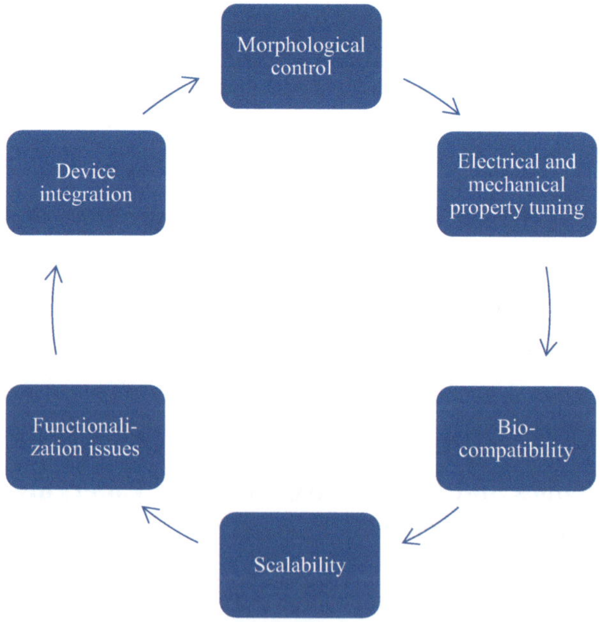


**Fig. 1.3** Challenges for tailoring conducting polymers

stability. Create environmental-friendly polymerization methods, including solvent-free or aqueous-phase polymerization, to reduce your influence on the environment. Create biodegradable CPs for uses like transitory electronics when environmental degradation is required. Add functional groups to adjust qualities like solubility, flexibility, or biocompatibility. To increase the lifespan of CP devices, include self-healing capabilities [73].

Create lightweight, stretchy, and flexible polymers for wearable and implantable technology. Enhance charge mobility and chemical stability in polymers to make them more suitable for fuel cells, batteries, and supercapacitors. Pay particular attention to materials highly compatible with drug delivery systems, biosensors, and neural interfaces. Enhance electrical and ionic conductivity in physiological settings for biological and medicinal uses. Create polymers for sensors and actuators that react to environmental stimuli such as pH, light, or temperature changes. Before synthesis, forecast the characteristics of novel CPs using machine learning and computational chemistry. Pay attention to polymers resistant to harsh environments, such as oxidative ones and high temperatures. For large-scale production, develop scalable manufacturing techniques like 3D printing or roll-to-roll processing [74].

CPs are multipurpose materials that find use in bioelectronics, sensors, and energy storage. New developments in the customization of these polymers center on improving their characteristics and capabilities for particular uses. The following are some new developments in CP customization. Because of their increased surface area and decreased dimensionality, nanofibers, nanowires, and nanotubes have improved charge transport. Combining characteristics at the micro and nano-scales is a new development to increase conductivity and mechanical flexibility [75]. Combining conductive polymers with metal oxides, graphene, or carbon nanotubes is also a new strategy to improve functionality, conductivity, and mechanical robustness. Molecular engineering is an advanced technology for adding functional groups to promote self-assembly, biocompatibility, or solubility. Green synthesis is the best emerging trend of CPs to avoid chemicals' harmful and toxic impact.

## References

1. Henry A (2013) Thermal transport in polymers. Ann Rev Heat Transf 17:485–520
2. Jadoun, S. Riaz, U. (2020) Biodegradable conducting polymeric materials for biomedical applications: a review. *MDPI Medical Devices & Sensors, 4*(1), e10141.
3. Tourillon G, Garnier F (1983) Effect of dopant on physiochemical and electrical properties of organic conducting polymers. J Phys Chemistry 87(13):2289–2292
4. Inzelt G (2017) J Solid State Electrochem 21:1965
5. Kar KK (2017) Composite materials: processing, applications, characterization. Springer
6. Singh KK, Prakash H, Akhtar M, Kar KK (2018) ACS Sustain Chem Eng:5381
7. Otero T, Martinez J (2016) J Mater Chem B 4:2069
8. Kar KK, Rana S, Pandey J (2015) Handbook of polymer nanocomposites processing, performance and application. Springer
9. Letheby H (1862) XXIX-On the production of a blue substance by electrolysis of sulphate of alanine. J Chem Soc 15:161–163

10. Inzelt G (2012) Conducting polymers: a new era in electrochemistry. Springer Science and Business Media
11. Noelting E (1889) Scientific and industrial history of aniline black. Matheson
12. Chilton JA, Goosey MT (1996) Special polymers for electronics and optoelectronics. Adv Mater:8.453–8.454
13. Harun MH, Saion E, Kassim A, Yahya N, Mahmud E (2007) Conjugated conducting polymers: a brief overview. UCSI Acad J J Adv Sci Arts 2:63–68
14. Pople JA, Walmsley SH (1962) Bond alternation defects in long polyene molecule. Mol Phys 5:15–20
15. Inzelt G (2008) Historical background (or: there is nothing new under the sun). Conduct Polym A New Era Electrochem:265–269
16. Shirakawa H, Loius EJ, MacDiarmid AG, Chiamg CK, Heeger AJ (1977) Synthesis of electrically conducting organic polymers: halogen derivatives of polyacetylene, $(CH)_x$. J Chem Soc Chem Commun:578–580
17. Bolto BA, McNeil R, Weiss DE (1963) Electronic properties of polypyrrole. Aust J Chem 16:1090–1103
18. Gill WD, Bludau W, Geiss RH, Grant PM, Greene RL, Mayerle JJ, Street GB (1977) Structure and electronic properties of polymeric sulfur nitride $(SN)_x$ modified by bromine. Phys Rev Lett 38:1305
19. Scott JC (2000) History of conductive polymers: nanostructured conductive polymers, p 1e17
20. Greene R, Street G, Suter L (1975) Superconductivity in polysulfur nitride $(SN)_x$. Phys Rev Lett 34(10):8e10
21. Yi N, Abidian MR (2016) Conducting polymers and their bio medical application. In: Biosynthetic polymers for medical applications. Chapter 10
22. Hundley M, Adams P, Mattes B (2002) Met 129:291
23. Pal T, Banerjee S, Manna PK, Kar KK Characteristics of conducting polymers. Springer series in material science, vol 300.
24. Huang WS, Humphrey BD, MacDiarmid AG (1986) Polyaniline, a novel conducting polymer. Morphology and chemistry of its oxidation and reduction in aqueous electrolytes. J Chem Soc Faraday Trans 82:2385–2400
25. McCullough RD, Lowe RD, Jayaraman M, Anderson DL (1993) Design, synthesis, and control of conducting polymer architectures: structurally homogeneous poly(3-alkylthiophenes). J Org Chem 58:904–912
26. Wang J, Dai J, Yarlagadda T (2005) Carbon nanotube conducting polymer composite nanowires. Langmuir 21:9–12
27. Lee BH, Lee JH, Kahng YH, Kim N, Kim YJ, Lee J, Lee T, Lee K (2014) Graphene-conducting polymer, hybrid transparent electrodes for efficient organic optoelectronic devices. Adv Funct Mater 24:1847–1856
28. Gupta S, McDonald B, Carrizosa SB, Price C (2016) Microstructure, residual stress, and intermolecular force distribution maps of graphene/polymer hybrid composites: nanoscale morphology-promoted synergistic effects. Compos Part B 92:175–192
29. Gupta S, Price C, Heintzman E (2016) Conducting polymer nanostructures and nanocomposites with carbon nanotubes: hierarchical assembly by molecular electrochemistry, growth aspects and property characterization. J Nanosci Nanotechnol 16:374–391
30. Huang WS, Humphrey BD, MacDiarmid AG (1986) Polyaniline, a novel conducting polymer. Morphology and chemistry of its oxidation and reduction in aqueous electrolytes. J Chem Soc Faraday Trans 1(82):2385–2400
31. Cantwell WJ, Morton J (1991) The impact resistance of composite materials—a review. Composites 22:347–362
32. Le T-H, Kima Y, Yoon H (2017) Electrical and electrochemical properties of conducting polymers. Polymers 9:150
33. Weinberger BR, Kaufer J, Heeger AJ, Pron A, MacDiarmid AG (1979) Magnetic susceptibility of doped polyacetylene. Phys Rev B 20:223–230

34. Ginder JM, Richter AF, MacDiarmid AG, Epstein AJ (1987) Insulatorto- metal transition in polyaniline. Solid State Commun 63:97–101
35. Nalwa HS (1989) Phase transitions in polypyrrole and polythiophene conducting polymers demonstrated by magnetic susceptibility measurements. Phys Rev B 39:5964–5974
36. Long Y, Chen Z, Shen J, Zhang Z, Zhang L, Xiao H, Wan M, Duvail JL (2006) Magnetic properties of conducting polymer nanostructures. J Phys Chem B 110:23228–23233
37. Lu X, Mao H, Chao D, Zhang W, Wei Y (2006) Ultrasonic synthesis of polyaniline nanotubes containing Fe3O4 nanoparticles. J Solid State Chem 179:2609–2615
38. Long Y, Chen Z, Duvail JL, Zhang Z, Wan M (2005) Electrical and magnetic properties of polyaniline=Fe3O4 nanostructures. Phys B Condens Matter 370:121–130
39. Cao H, Xu Z, Sang H, Sheng D, Tie C (2001) Template synthesis and magnetic behavior of an array of cobalt nanowires encapsulated in polyaniline nanotubules. Adv Mater 13:121–123
40. Cao H, Xu Z, Sheng D, Hong J, Sang H, Du Y (2001) An array of iron nanowires encapsulated in polyaniline nanotubules and its magnetic behavior. J Mater Chem 11:958–960
41. Cao H, Tie C, Xu Z, Hong J, Sang H (2001) Array of nickel nanowires enveloped in polyaniline nanotubules and its magnetic behavior. Appl Phys Lett 78:1592–1594
42. Sulong AB, Ramli MI, Hau SL, Sahari J, Muhamad N, Suherman H (2013) Rheological and mechanical properties of carbon nanotube/Graphite/SS316L/polypropylene nanocomposite for a conductive polymer composite. Composites 50:54–61
43. Machado M, Masse MA, Karasz FE (1989) Anisotropic mechanical properties of uniaxially oriented electrically conducting poly(p-phenylene vinylene). Polymer 30:1992–1996
44. Somanathan N, Wegner G (1996) Mechanical properties of conducting poly (3-cyclohexyl thiophene) films. Polymer 37:1891–1895
45. Sgreccia E, Khadhraoui M, Bonis C, Licoccia S, Vona MLD, Knauth P (2008) Mechanical properties of hybrid proton conducting polymer blends based on sulfonated polyetheretherketones. Sci Direct 178:667–670
46. Tieke B (2005) Chemie - Eine Einfuhrung M., 2 Auflage. Wiley-VCH Verlag GmbH & Co. KGaA, Weinheim
47. Cao Y, Smith P, Heeger AJ (1992) Counter-ion induced processibility of conducting polyaniline and of conducting polyblends of polyaniline in bulk polymers. Synth Met 48:91–97
48. Montes I, Lai C, Sanabria D (2003) Like dissolves like: a guided inquiry experiment for organic chemistry. J Chem Educ 80:447
49. Strakosas X, Wei B, Martin DC, Owens RM (2016) Biofunctionalization of polydioxythiophene derivatives for biomedical applications. J Mater Chem B 4:4952–4968
50. Mantione FD, del Agua I, Sanchez-Sanchez A, Mecerreyes D (2017) Poly(3, 4-ethylenedioxythiophene) (PEDOT) derivatives: innovative conductive polymers for bioelectronics. Polymers (Basel) 9:354
51. Patil AO, Ikenoue Y, Wudl F, Heeger AJ (1987) Water soluble conducting polymers. J Am Chem Soc 109:1858–1859
52. Bu H-B, Gotz G, Reinold E, Vogt A, Azumi R, Segura JL, Bauerle P (2012) "Click"-modification of a functionalized poly(3,4-ethylenedioxythiophene) (PEDOT) soluble in organic solvents. Chem Commun 48:2677
53. Trivedi DC, "Handbook of Organic Conductive Molecules and Polymers," Wiley, Chichester, 1997, pp. 505–572.
54. Tanvi P, Banerjee S, Manna PK, Kar KK (2020) Characteristics of conducting polymers. In: Kar KK (ed) Handbook of nanocomposite supercapacitor materials I. Springer series in materials science, vol 300. Springer Nature Switzerland AG
55. Smets G, Hart R (1960) Block and graft copolymers. Fortschritte Der Hochpolym 2:173–220
56. Jadoun S, Ashraf SM, Riaz U (2017) Tuning the spectral, thermal and fluorescent properties of conjugated polymers: via random copolymerization of hole transporting monomers. RSC Adv 7:32757–32768
57. Jadoun S, Rathore DS, Riaz UC, N. P. (2021) Singh: tailoring of conducting polymers via copolymerization-a review. Eur Polym J 155:110561

58. Abel SB, Frontera E, Acevedo D, Barbero CA (2023) Functionalization of conductive polymers through covalent postmodification. Polymers 15:205
59. Baker C, Wagner K, Wagner P, Officer DL, Mawad D (2021) Biofunctional conducting polymers: synthetic advances, challenges, and perspectives towards their use in implantable bioelectronic devices. Adv Phys X 6
60. Ding H, Hussein AM, Ahmad I, Latef R, Abbas JK, Abd Ali AT, Saeed SM, Abdulwahid AS, Ramadan MF, Rasool HA, Elawady A (2024) Conducting polymers in industry: a comprehensive review on the characterization, synthesis and application. Alex Eng J 88:253–267
61. Vinayak VV, Deshmukh K, Murthy VRK, Pasha SK (2024) Conducting polymer based nanocomposites for supercapacitor applications: a review of recent advances, challenges and future prospects. J Energy Storage 100:113551
62. Tripathi N, Bag DS, Dwived IM (2023) A review on Auxetic polymeric materials: synthetic methodology, characterization and their applications. J Polym Mater 40
63. Cuthbertson AA, Lincoln C, Miscall J, Stanley LM, Maurya AK, Asundi AS, Tassone CJ, Rorrer NA, Beckham GT (2024) Characterization of polymer properties and identification of additives in commercially available research plastics. Green Chem 26:7067–7090
64. Siddiqui MS, Rabbi MS, Ahmed RU, Billah MM (2024) Biodegradable natural polymers and fibers for 3D printing: a holistic perspective on processing, characterization, and advanced applications. Clean Mater:100275
65. Rahim AA, Shamsuri NA, Adam AA, Aziz MF, Hamsan MH, Rusdi H, Siong SOJ, Noor IM, Kadir MFZ, Shukur MF (2024) Characterization of nanocomposite polyvinyl alcohol/cellulose acetate blend gel polymer electrolytes for supercapacitor application. J Energy Storage 97:112964
66. Yuan X, Zhang Z, Mu X, Shan C, Gao X, Zhu B (2024) Recent progress on interface characterization methods of carbon fiber reinforced polymer composites. Chem Eng J 499:156220
67. Verma A, Bhushan K, Singh H (2024) Nanocomposites for extrinsic self-healing polymer materials: a comprehensive review of their repair behaviour. Results Chem 13:101973
68. Ali Z, Yaqoob S, Yu J, D'Amore A (2024) Critical review on the characterization, preparation, and enhanced mechanical, thermal, and electrical properties of carbon nanotubes and their hybrid filler polymer composites for various applications. Compos Part C Open Access 13:100434
69. Muna II, Mieloszyk M, Rimasauskiene R (2024) Characterization of the thermal and mechanical properties of additively manufactured carbon fiber reinforced polymer exposed to above-zero and sub-zero temperatures. J Mater Res Technol 33:9832–9842
70. Kangarshahi BM, Sojdeh S, Daneshgar H, Bagherzadeh M, Naghib SM, Rabiee N (2025) Electroconductive polymer-based biosensors for early cancer detection via liquid biopsy: advances, challenges, and future prospects. TrAC Trends Anal Chem 183:118062
71. Lee DE, Ali A, Kang KT, Danish M, Jo WK (2024) Advancing the integration of covalent-organic-framework with organic, inorganic, and polymeric materials for light-assisted green H2 generation: a review of emerging trends. Mater Sci Eng R Rep 161:10085
72. Kumar A, Indhur R, Bux F, Kumari S (2024) Recent advances in mechanistic insights into microplastics mitigation strategies via emerging advanced oxidation processes: legislation, challenges, and future direction. Sci Total Environ 957:177150
73. Zahra T, Javeria U, Jamal H, Baig MM, Akhtar F, Kamran U (2024) A review of biocompatible polymer-functionalized two-dimensional materials: emerging contenders for biosensors and bioelectronics applications. Anal Chim Acta 1316:342880
74. Dananjaya SAV, Chevali VS, Dear JP, Potluri P, Abeykoon C (2024) 3D printing of biodegradable polymers and their composites–current state-of-the-art, properties, applications, and machine learning for potential future applications. Prog Mater Sci:101336
75. Siddamallappa NG, Basavaju M, Sionkowska A, Gowda CGD (2024) A review on synthetic polypeptide-based blends with other polymers: emerging trends and advances. Eur Polym J:113225

# Chapter 2
# Tuning of Conducting Polymers According to Applications

**Manjinder Kour, Nirvair Singh Khosa, Diwya Kumar Lal, and Sapana Jadoun**

**Abstract** Conducting polymers (CPs) are cost-effective and lightweight and offer tunable electrical, optical, and mechanical properties that outperform conventional materials. Their areas of synthesis, fabrication, and high environmental stability make them highly attractive for several applications. CPs exhibit strongly reversible redox behaviors, combining desirable features of plastics and metals and allowing for versatile use across multiple fields. CPs are widely applied in sensing technologies, including air quality monitoring, gas sensing, and the detection of hazardous chemicals in the environment. In biosensing, CP-based devices study biological materials such as proteins and DNA, with applications spanning medical diagnostics and environmental monitoring. Advances in nanotechnology have enhanced CP nanomaterials, improving their performance in electronics, optoelectronics, sensing, and energy devices. Surface modifications can introduce physicochemical and biological cues, promoting cell adhesion at the polymer-tissue interface, making CPs ideal for biomedical applications. To be suitable for various technologies, CPs must overcome limitations like poor mechanical properties and instability, which are addressed through tuning. They allow the flexible tuning of structures and electrical properties while maintaining durability and are solution-processable for the large-scale production of flexible devices. This chapter discusses the tailoring of

M. Kour (✉)
Department of Chemistry and Biochemistry, Montana State University, Bozeman, MT, USA

N. S. Khosa
Mechanical & Industrial Engineering Department, Montana State University, Bozeman, MT, USA

D. K. Lal
Deurali Janta Pharmaceuticals Pvt. Ltd., Kathmandu, Nepal

S. Jadoun
Sol-ARIS, Departamento de Química, Facultad de Ciencias, Universidad de Tarapacá, Arica, Chile
e-mail: sjadoun@academicos.uta.cl



S. Jadoun, N. Kumari Jangid (eds.), *Tuning of Conducting Polymers*, Springer Series on Polymer and Composite Materials,
https://doi.org/10.1007/978-3-032-00461-1_2

CPs for specific applications, focusing on their role in environmental, biomedical, and energy-related fields.

**Keywords** Conducting polymers (CPs) · Tunable properties · Sensing technologies · Biosensing applications · Nanotechnology · Biomedical applications · Environmental stability

## 2.1 Introducing Conducting Polymers and Tuning Strategies

Conducting polymers (CPs) are unique materials with a blend of crystalline and amorphous characteristics. Structurally, CPs are synthetic macromolecules characterized by a delocalized π-conjugated backbone and adjustable side chains, which make them highly tunable. The backbone consists of alternating C–C, C=C, or C≡C bonds, in which pai (π) bonds weakly bind delocalized electrons, but the overall chain strength is governed by strong sigma (σ) bonds (Fig. 2.1). It is the conjugated double or triple bonds along the backbone that primarily contribute to CPs' excellent electrical conductivity [1]. Their electronic properties arise mutually from localized and delocalized states. The π bonds delocalization is critical in generating charge carriers such as polarons (radical ions), bipolarons (dications or dianions), and solitons. These charge carriers facilitate the transition from an insulating to a conducting state [2]. CPs behave as semiconductors in their pure form with a moderate bandgap (2–3 eV), combining the electronic and optical properties of semiconductors with the mechanical flexibility of polymers. CPs offer the benefits of low density, chemical diversity, controllable shape, modifiable morphology, corrosion confrontation, and tunable conductivity over their existing inorganic counterparts [2].

Doping enhances their conduction property by introducing nonlinear excitations, transforming them into conductor-like metals. This tunable conductivity makes them extremely useful for a range of applications. The dopants showcase redox reactions by adding and withdrawing electrons from the conducting polymers. Doping in conducting polymers is categorized into p-type and n-type, and both

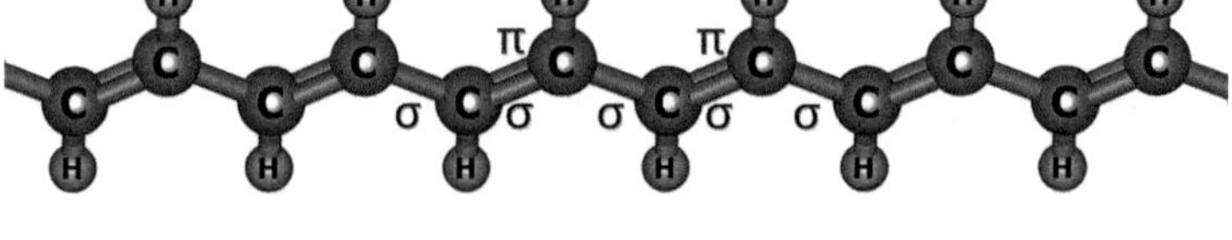


**Fig. 2.1** Chemical structure of polyacetylene (PAc), showing the backbone with alternating single and double bonds

involve the generation of charge carriers. These carriers are delocalized along the polymer chains, facilitating enhanced conductivity. In p-type doping, electrons are removed from the highest occupied molecular orbital of the polymer and relocated to the dopant, resulting in the formation of a hole, a localized electron deficiency within the polymer backbone. However, n-type doping involves the addition of electrons to the lowest unoccupied molecular orbit of the polymer, increasing electron density. Doping increases the densities of both hole and electron carriers, which alters the carrier mobility ratio, enhancing the material's overall conductivity. P-type doping is particularly significant in research due to the superior stability of positively charged carriers in the polymer structure [3, 4]. Several approaches have been established for doping CPs, including electrochemical, chemical, photo, non-redox, and charge-injection doping. Electrochemical and chemical doping are the most used among these due to their cost-effectiveness and practicality [2].

Customizing CPs is crucial for improving their suitability for various applications. The tuning process uses different processes to adjust key properties, including electrical conductivity, flexibility, thermal stability, and chemical resistance to meet specific requirements. One example is the environmental applications that benefit from CPs with improved sensitivity and selectivity for pollutant detection. Similarly, biocompatibility and chemical stability are crucial in biomedical fields, requiring precise molecular alterations to ensure safe use with biological tissues. Energy applications, including batteries and supercapacitors, demand optimized conductivity and thermal stability for efficient energy storage and conversion. Likewise, smart textiles and wearables require CPs that are flexible and durable to withstand mechanical stress. By tailoring these attributes, CPs can be optimized for usage in a wide range of environments, thereby broadening their applicability and improving performance.

The mechanical properties of CPs can be tailored through structural tuning, enabling modifications to suit specific applications. Their behavior can be significantly impacted by adjusting features such as degree of polymerization, regioregularity, length of side chain, and incorporation of non-conjugated monomers [5]. A well-known approach is increasing the length of alkyl side chains, which reduces the glass transition temperature, tensile strength, and elastic modulus as observed for polydiketopyrrolopyrrole [6]. Hydrogen bonding within side chains can further influence mechanical and electronic properties. Similarly, increasing regioregularity in polythiophene (PT) derivatives enhances tensile strength but reduces elasticity and toughness [7]. Similarly, chain entanglement can be increased by tuning the degree of polymerization, which improves resistance to cavitation, crack formation, and chain breakage [8]. Electrochemical and chemical oxidations are conventional techniques for CPs polymerization. Polyaniline (PANI), polypyrrole (PPy), and PT have attracted special interest due to their high processability, allowing straightforward customization and functionalization [9]. CPs and their derivatives have been vastly used across numerous fields (Fig. 2.2).

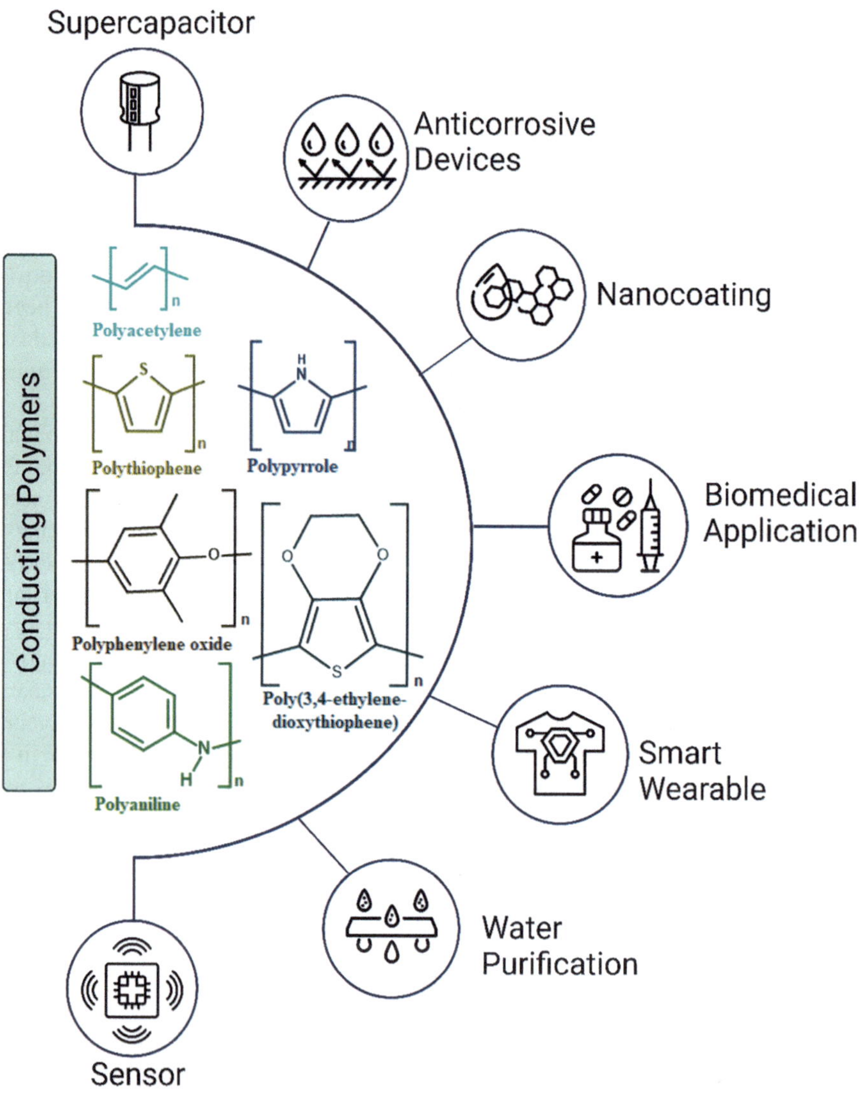


**Fig. 2.2** Applications of conducting polymers (CPs) in diverse fields

## 2.2 Application-based Tuning of Conducting Polymers

CPs are unique materials that unite metals' electrical properties with polymers' mechanical properties. This makes them highly versatile for various applications, including environmental, biomedical, energy, smart wearables, and textiles. Researchers could optimize CPs for specific functions by fine-tuning their chemical structures and doping levels, enhancing their performance and utility in various fields. One emerging area is environmental applications, where CPs are utilized in highly sensitive sensors for detecting pollutants [10]. Similarly, in the biomedical field, these are widely employed in sensors, drug delivery systems, and tissue engineering scaffolds, owing to their unique biocompatibility and electrical conductivity combination [11]. CPs' efficient charge storage and transfer capabilities benefit energy applications, making them ideal for batteries, supercapacitors, and solar cells [12]. Additionally, integrating CPs into fabrics for smart wearables and textiles enables the creation of flexible, conductive materials for wearable electronics and responsive clothing [13].

### 2.2.1 Tuning for Environmental Applications

CPs contribute to environmental remediation by facilitating contaminant detection, quantification, and removal. In impurity abstraction, CPs act as sorbents, ion exchangers, filters, membranes, and chemical reactors through complexation and redox reactions. When combined with materials like zinc oxide, CP-based composites exhibit enhanced photocatalytic capabilities, particularly under visible light. These composites rely on mechanisms such as pollutant uptake/release, doping/de-doping cycles, and electrocatalytic reduction/oxidation, often under applied potential. Their unique properties make CP-based materials highly effective for addressing various environmental challenges [14].

CPs can be tailored to act as efficient adsorbers for pollutants, with their performance highly dependent on their composition, surface morphology, and functional groups. For example, eco-friendly composites like PANI with biopolymers such as chitosan have demonstrated the ability to adsorb significant amounts of dyes from aqueous solutions [15]. Researchers have shown that the specific CP composites exhibit unique adsorption mechanisms. In an interesting finding by Hezarkhani et al., they have demonstrated the removal of methyl orange dyes using PANI/clinoptilolite composites. They exchanged the cations ($Na^+$, $K^+$, $Mg^{2+}$, $Ca^{2+}$) located in the clinoptilolite structure with anilinium cations, and a subsequent polymerization of these cations in and outside of the clinoptilolite networks resulted in PANI/clinoptilolite nanocomposite formation. This tuned CP composite exhibited unique adsorption mechanisms and is selective to a particular type of pollutant [16]. Polythiophene can efficiently capture toxic As(III) ions through interactions between the $\pi$-electrons in the polymer backbone and arsenic atoms [17]. Even more interesting are the multifunctional materials, such as PANI-Zr(IV) nanocomposites, which combine sorption and photocatalytic remediation, broadening the scope of CP applications [18]. These examples highlight the significance of fine-tuning CP structures and composites to optimize their adsorption capabilities, further enabling targeted environmental applications, including dye and heavy metal removal.

CPs can excellently remove charged pollutants from the liquid waste through incorporation into CP thin films deposited on conductive substrates such as gold, platinum, titanium, vitreous carbon, or graphite felts. For uncharged CPs, pollutant removal often involves metal ion complexation facilitated by nitrogen-containing functional groups. For example, nitrogen functionalities in PANI matrices or the pyrrole rings in PPy composites and nanofibers are key to capturing metal ions like $Hg^{2+}$, $Pb^{2+}$, $Ni^{2+}$, $Zn^{2+}$, $Cd^{2+}$, and $Cr^{6+}$. PANI and Ppy have shown an excellent profile as an adsorbent for dye-contaminated aqueous solution due to their excellent chemical features [19].

In environmental remediation, nanohybrids of CPs and zinc oxide (ZnO) have also gained significant attention for environmental photocatalysis, particularly in water purification. These nanohybrid materials combine the photocatalytic properties of ZnO with the unique characteristics of conducting polymers, which include extended light absorption, improved charge separation, and enhanced photostability. Combining these properties makes it an ideal application. Synthesis and optimization of structural and functional properties of the nanohybrids involve methods like in situ polymerization, sol–gel techniques, and semiconductor blending. Besides, incorporating CPs into ZnO systems addresses key challenges in photocatalysis, such as reducing charge recombination and increasing visible light absorption. This combination allows the bandgap of the hybrid to align within the visible light region, enabling efficient utilization of solar energy. Additionally, CPs inhibit aggregation during synthesis and stabilize photoinduced charge carriers by trapping holes within their $\pi$-conjugated chains, preventing self-oxidation and photocorrosion. These synergistic effects make conducting polymer-zinc oxide nanohybrids highly efficient for degrading toxic pollutants, harmful chemicals, and insecticides, highlighting their potential as advanced photocatalysts for environmental applications [20].

Treating wastewater contaminated with organic dyes remains a big challenge. This is also a growing environmental concern. CPs are not soluble in aqueous media, and when they are added to an organic dye solution, a decreasing dye concentration is reported. Dye adsorption and photocatalytic decomposition are the two main mechanisms reported in this process [21]. CP-based materials have received substantial attention in water purification due to their exceptional properties and versatility. Mechanical and thermal stability, chemical adaptability, ease of functionalization, and high active surface area of tuned CPs make them ideal for applications such as adsorbents, photocatalysts, and membranes. These features enable the efficient removal of contaminants and support advanced filtration and catalytic processes, positioning conductive polymers as promising materials for sustainable water treatment solutions [22]. PANI-based nanocomposites, incorporating metals, carbon nanostructures, or clay, have demonstrated significant efficiency in purifying contaminated water by leveraging their catalytic and adsorption capabilities [23]. Some selected findings are presented in Table 2.1.

**Table 2.1** Applications of some nanocomposites in water remediation

| Type of pollutant | Pollutant | Composites | References |
|---|---|---|---|
| Dyes | Methyl orange (MO) | PT/TiO2 | [24] |
| | Rhodamine B | PT/TiO2 | [25] |
| | MO, methylene blue (MB) | Porphyrin/$ZnFe_2O_4$@PT | [26] |
| | Reactive black 5 | PANI/ligno-cellulose | [27] |
| Cations | $Ca^{2+}$ and $Mg^{2+}$ | Ppy/Polystyrenesulfonate | [28] |
| | $Ag^+$ | Poly 3-methylthiophene/ saw dust | [29] |
| | $As^{3+}$ and $As^{5+}$ | PANI/$Fe^0$ | [30] |
| | $Hg^{2+}$ | PANI/reduced graphene oxide | [31] |
| | $Pb^{2+}$ | PANI/ iron oxide coated natural soil | [32] |
| Polycyclic aromatic compounds | Benzo(a)anthracene (BaA), benzo(b)fluoranthene (BbF), and benzo(a)pyrene (BaP) | PANI/alginate/$Fe_3O_4$ | [33] |

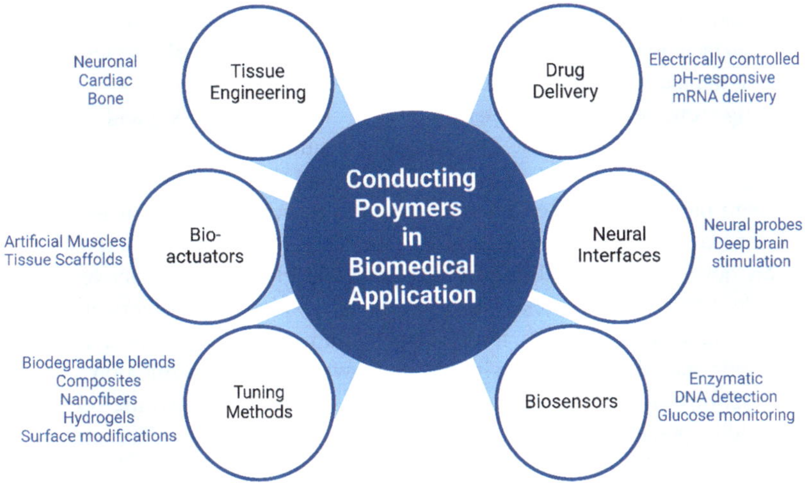


**Fig. 2.3** Overview of biomedical applications of conducting polymers (CPs)

## 2.2.2 *Tuning for Biomedical Applications*

The exploration of CPs in biomedical applications has grown significantly since the 1980s (Fig. 2.3). The compatibility of CPs with biological molecules has gathered much attention. These materials exhibit unique features, such as biocompatibility, reversible doping for controlled biomolecule release, and the ability to transfer charge from chemical reactions. In addition, the electrical, chemical, and physical properties of CPs can be modified easily to meet the specialized demands of various applications. Together, these characteristics have enabled CPs to play a substantial role in biosensors, neural probes, tissue engineering, drug delivery, bioimaging, and bio-actuators, with both in vivo and in vitro studies supporting their potential [1, 34]. More specifically, CPs can regulate various cellular behaviors through electrical stimulation, influencing adhesion, proliferation, and the regeneration of harmed tissues. This capability further underscores their versatility and promise in biomedical applications [35–38].

There are numerous possibilities for tailoring CPs to meet the specific demands of biomedical applications. One initial challenge in using CPs in biomedical applications was their degradation issues, which could cause inflammatory responses and other adverse effects. However, recent advancements have introduced biodegradable CPs through strategies such as conducting oligomer-based architectures (e.g., linear, star-shaped, and hyperbranched designs), modified monomers, and polymeric blends using the minimum amount of conducting polymer and degradable polymer. These approaches have significantly expanded the scope of CPs for practical biomedical use. Although a significant amount of work has been done in this area, the conductivity of these CPs still needs to be enhanced [39]. In the current research endeavors, CPs are increasingly fine-tuned for controlled drug release through surface modifications and blending with biodegradable materials. This approach enhances their functionality in biomedical applications.

Building on versatile properties, CPs such as PPy and poly(3,4-ethylenedioxythiophene) (PEDOT) have gained much attention for drug delivery applications. Their ability to combine electrical conductivity with biocompatibility makes them particularly suited for regulated and responsive drug release schemes [40]. In one of the recent studies, a graphene/PPy (GN–PPy–FL) composite electrode was developed by combining a 2-D graphene film with conductive and biocompatible PPy nanoparticles. Using chemical oxidative polymerization, they loaded the nanoparticles with fluorescein sodium (FL), which is a negatively charged drug model. The composite electrode was reported to function as a drug carrier, enabling electrically controlled drug release. Interestingly, the release rate of FL could be precisely modulated by varying the applied voltage. This study highlighted the capability of such systems for multi-stimuli responsive drug delivery, offering promising applications of tuned CPs for the controlled release of drugs, peptides, and proteins [41].

Researchers have synthesized PPy nanoparticles (NPs) using a one-step micro-emulsion technique with different oxidizing agents. The resulting nanoparticles were assembled into films that exhibited significantly varied drug release profiles. Notably, while using ferric chloride, drug release could be achieved at −0.05, an order of magnitude lower than standard voltages. This advancement broadens the operational range of CPs. For instance, the chemotherapeutic methotrexate was successfully released in pulses at voltages of −0.075 V, underscoring the promise of these NPs in cancer therapy. The biocompatibility of the FePPy nanoparticles prepared using ferric chloride was confirmed through MTT assays, which also demonstrated that the bioactivity of methotrexate remained intact after stimulation. The anti-tumor effectiveness of the released dose was further validated with PC12 cell cultures, highlighting its therapeutic promise [42].

Besides an electrical stimulus, CPs are responsive to pH changes as well. This is a valuable trigger for regulated drug release. One such example is PANI, which undergoes protonation doping with acids and generates charge carriers that alter its oxidation state and facilitate the release of encapsulated drugs. They exploit the body's natural pH variations, such as the stomach (extremely acidic), small intestine (moderately acidic), colon (neutral to moderately basic), and acidic environments at spots of inflammation or tumors (slightly acidic). By targeting specific pH conditions, these tuned systems enable localized drug delivery without external stimuli, making them an excellent choice for biological applications [43]. Samanta et al. developed a versatile pH-sensitive drug delivery system using PPy nanoparticles (PPy NPs) synthesized for targeted drug release within a specific pH range. These NPs are reported to be highly monodisperse, stable for over a month in aqueous form, and exhibit excellent drug load-up capacity. Interestingly, by adjusting pH and drug charge and incorporating small amounts of charged amphiphiles, PPy-NPs can be tailored for drug release in both acidic and basic environments. They achieved the local delivery by embedding the nanoparticles in a calcium alginate hydrogel, which enabled prolonged drug release for over 21 days. This NP-hydrogel composite is promising for managing chronic conditions requiring regulated and prolonged drug delivery [43].

Recently, a conductive hybrid film was proposed for electrically controlled mRNA delivery to guide the osteogenic differentiation of human adipose-derived mesenchymal stem cells (hADMSCs). Researchers created a ppy-graphene oxide (PPy-GO) film by electropolymerizing it onto indium tin oxide substrates, optimizing its conductivity and mRNA-loading capacity. The film was reported to effectively absorb and release fluorescein Na salt, a molecule that mimics the properties of mRNA. When hADMSCs cultured on the pre-osteogenic mRNAs loaded with PPY-GO exhibited enhanced osteogenic differentiation. This system shows potential as a novel tool for stem cell therapies [44].

As mentioned earlier, CPs are considered promising electrode materials for implanted biosensors and bioelectronics. This is mainly attributed to their mechanical properties that closely resemble the properties of biological tissues. A primary challenge, however, is preventing unnecessary protein adhesion or cell binding. This vague adsorption often leads to device malfunction, posing significant challenges for long-term applications. To address these issues, studies have focused on developing antifouling CPs, offering potential solutions to mitigate nonspecific adsorption and enhance the functionality of implanted devices [45]. These advancements highlight the potential of CP-based electrodes to revolutionize long-term implantable biosensors and bioelectronics by improving their durability and reliability in biological environments. Additionally, they are known to have potential for electrotherapy applications [10].

Building on advancements in antifouling materials, Hui et al. grafted polyethylene glycol (PEG) onto PANI to create PEGylated PANI nanofibers, which demonstrated effective antifouling properties in hominid serum samples as well as in breast cancer detection. The novel PEGylated PANI nanofibers were prepared by covalently grafting PEG onto PANI nanofibers. These nanofibers retained their conductivity while exhibiting excellent antifouling capabilities, effectively preventing nonspecific adsorption from simple protein solutions and complex biological media [46]. To explore antifouling materials for biosensors, Wu et al. developed a sensitive and stable glucose electrochemical sensor poly(sulfobetaine-3,4-ethylenedioxythiophene) (PSBEDOT). It has been shown by functioning as both a three-dimensional conductive polymer matrix for glucose oxidase encapsulation and a mediator for redox reactions, PSBEDOT showcased its dual functionality in enhancing biosensor performance [47].

While CPs in drug delivery have shown potential, they faced significant limitations also due to their ability to entrap/release only tiny, charged substances that could interact efficiently with the polymeric matrix. To address these challenges, researchers have developed strategies like incorporating conducting polymer hydrogels (CPHs). These materials create systems that undergo chemical/physical transitions under applied electric fields, improving drug delivery performance. CPHs act as drug reservoirs, with their water content enhancing delivery. Combining the CPs with the hydrogels makes these systems highly versatile for biomedical applications. The hydrogel porous facilitates ion diffusion, while the CP component enables charge transport controlled by the polymer's redox state, positioning CPHs as capable candidates for expanded drug delivery systems [48]. Furthermore, CPs have shown significant potential as bioactive platforms for tissue engineering owing to their competence to transmit electrical signals, enabling the electrical stimulation of cells or tissues cultured on them. However, their inherent brittleness and limited processability present challenges for broader applications. To address these limitations, researchers have focused on tuning conductive polymers by developing composites that combine them with biocompatible and biodegradable polymers, either natural or synthetic. This approach makes them more suitable for tissue engineering. Advances in this field are achieved through methods such as electrospinning, coating, and polymerization. These tuned materials provide tailored functionalities to meet the specific demands of tissue regeneration [49]. For neuronal tissue engineering, PEDOT and PPy have been modified by incorporating bioactive molecules like nerve growth factors (NGF) and laminin peptides. These modifications significantly improve cell adhesion, differentiation, and axonal outgrowth, essential for neural regeneration. The bioactive molecules provide biochemical cues that promote neuronal growth and connectivity, making these scaffolds highly effective for treating spinal cord and peripheral nerve injuries [50].

In the advancement of hybrid CPs, researchers have combined them with substances like graphene or carbon nanotubes (CNTs). These hybrids exhibit enhanced mechanical properties, biocompatibility, and electrical conductivity. The improved features facilitate better cell proliferation and differentiation and make them apt for artificial nerves and neural implant applications. The integration of graphene or CNTs into CPs creates a collaborative result that enhances the overall operation of the material [51].

Conductive polymer hydrogels (CPHs) are another essential achievement in this field, and they can be fabricated through two main approaches, namely, polymerization and direct mixing. However, the former is the most employed method involving forming a hydrogel on a conductive substrate, drying it, and re-swelling it in a CP monomer-dopant solution. Polymerization is achieved via methods like electro- or chemical polymerization. In the latter method, the dopant ion type, concentration, and distribution in the hydrogel matrix influence electrochemical and mechanical properties. These properties can be fine-tuned by adjusting key parameters, including potential, concentrations, applied current, and the pH of the electrolyte [52]. Also, conducting polymer hydrogels, such as carbon nanotube (CNT)-incorporated gelatin methacryloyl (GelMA), have been designed for cardiac tissue engineering. These hydrogels create a conductive environment that supports myocardial tissue growth and function, enhancing synchronous beating rates and reducing the excitation thresholds of cardiac cells, which are critical factors in developing functional cardiac tissues. This approach highlights CP hydrogels' potential in creating advanced cardiac repair biomaterials. These advancements highlight the versatility of CPs as customizable materials for biomedical and drug delivery applications.

Similarly, tuning CPs for enzymatic electrochemical biosensors has enabled significant advancements in sensitivity and specificity by leveraging their unique properties in combination with metal oxide nanoparticles. One example is a biosensor for detecting amlodipine besylate (AMD), created by a biocomposite of PANI, iron oxide, and gelatin. The sensitive material was fabricated in this design, and lipase was immobilized within the biocomposite matrix using a glutaraldehyde cross-linking reagent. This setup ensured stable enzyme entrapment and enabled efficient biosensing performance, as demonstrated by cyclic voltammetry and impedometry, confirming the material's effectiveness for AMD detection [53]. Likewise, a glucose biosensor was constructed using PANI, nickel oxide nanoparticles, and graphene oxide, leveraging the complementary properties of these materials. This biosensor displayed excellent sensitivity and high selectivity, effectively detecting glucose in the presence of interfering agents. The inclusion of nickel oxide nanoparticles enhanced catalytic activity, while graphene oxide facilitated efficient electron transfer, contributing to the superior performance of the biosensor. These examples underscore the importance of tuning CPs and their composites to improve the functional capabilities of enzymatic electrochemical biosensors, broadening their applicability across biomedical and diagnostic fields [54]. Additionally, resorbable conductive materials have been developed by mixing CPs with biodegradable polymers such as polylactic acid (PLA). These scaffolds are designed to be degradable while delivering electrical stimulation to tissues, promoting tissue regeneration. Their ability to combine electrical conductivity with controlled biodegradability addresses a critical need in temporary implants, particularly in applications like wound healing, nerve repair, and cardiac tissue engineering. The resorbable nature of these materials ensures safe degradation within the body after fulfilling their function, removing the necessity for invasive removal processes and reducing potential hitches [55].

Together, these innovative approaches exemplify CPs' adaptability in addressing complex biomedical challenges and underscore their pivotal role in advancing regenerative medicine. With such advancements, tuned CP materials continue to pave the way for groundbreaking innovations in biomedical engineering.

### 2.2.3 *Tuning for Energy Applications*

CPs are reported to be valuable materials for energy-related devices owing to their ability to conduct electricity and be easily modified. They are often used in supercapacitors and batteries acting as electrodes. By tuning their properties, these electrodes can store more energy and last longer through repeated use. One way to improve their performance is by combining CPs with materials like carbon nanotubes. This creates hybrid nanocomposites that take advantage of the strength of both materials. CPs are also used in electrochromic devices, such as smart windows. These windows save energy by changing color to control the light and heat that passes through.

To improve these applications, researchers use methods like doping, where small molecules are added to CPs to improve conductivity, or combining CPs with other materials to create new composites. They also adjust the structure of CPs by changing their chemical makeup to improve flexibility and stability [3, 56]. PANI is an essential polymer of this class, but it exhibits inherent weaknesses involving instability at high temperatures and pH 7. To overcome these issues and expand their functionality, PANI-based nanocomposites have been developed by integrating nanomaterials. PANI-carbon nanotube (CNT) composites are excellent examples of such tuning, which has significantly enhanced electrical conductivity and mechanical strength. Tuning made it way more suitable for robust applications. Similarly, PANI-graphene composites have gained much attention in this field. In addition, PANI-metal oxide composites, for instance, the ones incorporating titanium dioxide ($TiO_2$) or zinc oxide (ZnO), exhibit increased activity and stability, which is ideal for sensors and coatings. PANI-clay-based nanocomposites are another excellent example of this class, offering superior mechanical strength and barrier properties. These are highly effective materials in coating and packaging applications [57–59].

Electrically conducting polymers, particularly PAc, and PPy have emerged as promising materials for energy applications as they are lightweight, high performing, and non-toxic in nature. Such properties make them suitable for compact devices and sustainable energy solutions, driving ongoing research to optimize their efficiency and durability for advanced energy storage systems [60]. Intrinsically CPs (ICPs), such as PANI, PPy, PT, and PEDOT, are also extensively used in stretchable supercapacitors due to their pseudocapacitive behavior and ability to enhance electrode conductivity. These polymers can function independently or be combined with carbon materials or metal oxides to improve performance, making them versatile for advanced energy storage applications [61, 62].

Tuned CPs offer significant advantages over conventional CPs in energy applications. Adjusting their properties is essential for improving performance. For example, the doping level in CPs directly affects their specific capacity. The exact capacity of CPs typically ranges from 0.3 to 0.5, depending on the material. For instance, PANI can reach 0.5 holes per monomer ring, PPy reaches 0.33, and PT reaches 0.25. These levels correspond to capacities of 82–147 mAh $g^{-1}$ [63–65]. CPs can be optimized for better energy storage performance by controlling doping levels. While higher doping levels can enhance the specific capacity of conducting polymers (CPs), they can also lead to irreversible polymer degradation. Another challenge is the sloping voltage profile during discharge, which arises from the voltage's dependence on doping levels [66]. However, these limitations can be addressed through molecular-level modifications that offer a promising solution for improved performance. One practical approach involves embedding redox-active groups into CPs. This can be achieved by covalently attaching these groups to the polymer backbone, forming redox CPs, or incorporating them into the polymer matrix as counterions. Redox-active molecules such as metal complexes, radicals, and organo-sulfur compounds can enhance ion insertion and extraction and complement the doping/de-doping process. This dual functionality merges the PC backbone's inherent conductivity with the redox groups' additional charge storage capacity, significantly boosting energy storage performance. However, introducing redox groups can disrupt the polymer's $\pi$-conjugation, which limits doping levels and reduces electronic conductivity. To address this, careful tuning is required to make a balance between maximizing redox activity and preserving electronic properties of the polymer for optimal performance.

CP hydrogels have emerged as promising polymeric materials for technological applications, especially for energy storage devices. This is because they combine the benefits of organic conductors and conventional polymers. This field has explored new synthetic methods, such as using acids as crosslinkers and dopants. These methods give CP hydrogels unique nanostructures with high surface area and conductivity. Their hierarchical porosity allows faster transportation. They show excellent functioning for electrochemical capacitors and high-energy lithium-ion batteries [67]. Compared to bulk conductive polymers, CPHs exhibit superior mechanical flexibility to accommodate electrochemical strain and lightweight and processable properties.

Tuning conductive polymers into CPHs optimizes their properties, making them better suited for energy conversion and storage technologies. This approach shows that CPHs can enhance existing energy systems and support the development of new and more efficient solutions.

### 2.2.4 Tuning for Smart Textiles and Wearables

CPs have revolutionized the smart textiles and wearables field, offering distinctive features such as flexibility, lightweightness, and the ability to conduct electricity. These materials are essential for developing advanced wearable technologies that monitor health, enhance performance, and provide interactive functionalities.

CPs are advancing smart textiles due to their conductive and semiconductive properties. Their mechanical responsiveness, electrical conductivity, and remarkable flexibility and stretchability have significantly stimulated recent innovations in wearable and skin-attachable devices. Unlike inorganic materials, organic electronics stand out for their lightweight nature and exceptional flexibility, making them ideal for integration into wearable sensors. Among these materials, PPy, PEDOT, and PANI are the most used, primarily due to their outstanding long-term electrical and chemical stability [68]. CPs must be tailored to ensure durability, comfort, and functionality under various conditions. This tuning process involves modifying the structure of polymers, incorporating additives, and employing different fabrication techniques to achieve the desired properties. Poly (3,4-ethylene dioxythiophene):polystyrene sulfonate (PEDOT:PSS) is one of the most emerging CPs for wearables due to its processability in aqueous solutions. Researchers have shown that conductivity can be enhanced significantly by secondary dopant treatment. This treatment is unique because the secondary dopants are intact, and their effect stays even after removal. Secondary doping is done by mixing and/or film treatment [69]. The oxidative polymerization of PANI is another direction that has been explored in the past, but the process is shown to be influenced by the pH of the medium. Strong acids have been explored for doping in this context. PANI is already used broadly in textile sensor manufacturing and corrosion protection [70].

Researchers have shown that doping counterions near the extended π-bonds can significantly enhance the CP's conductivity. This doping process allows CPs to achieve electrical conductivities from over 1 S/cm to more than 1000 S/cm, comparable to inorganic semiconductors and conductors [71]. One widespread method for such treatment is oxidative doping, which involves the insertion of anions into CPs like PPy and PEDOT, creating positive charge carriers that enable charge transport within the polymer matrix. Doped PEDOT exhibits the highest reported electrical conductivity, exceeding 8000 S/cm [72].

Additionally, tuning CPs through coating techniques has proven a highly effective approach for creating electrically conductive textiles. This method employs a specially formulated substance or fluid on cloth, including fabrics, yarns, or even individual fibers. Coating enables the integration of conductivity while preserving the inherent properties of the substrate, such as flexibility, handling, and density. This makes it apt for several fiber types, ensuring minimal alteration to the substrate's original characteristics. Typically, the process employs conductive polymer coating paste (CPC), which provides uniform conductivity and facilitates the development of high-performance smart textiles [73].

Building on these advancements, the role of CPs in smart textiles and wearables extends far beyond conductivity and flexibility. Integrating CPs into modern healthcare applications transforms how we monitor and manage health. These materials enable the development of wearable devices that can give real-time data on vitals such as heart rate, respiration, and blood oxygen levels, which are essential for personalized health monitoring and early disease detection. CP-based smart textiles also have potential therapeutic applications, including devices for chronic pain management, wound healing, and rehabilitation therapies through controlled electrical stimulation. The versatility of CPs, achieved through tuning methods like doping, coating, and structural modifications, ensures that these textiles are functional, durable, and comfortable for continuous use. As healthcare increasingly moves toward remote patient monitoring and preventive care, the need for reliable, lightweight, and adaptable wearables has become more critical. CPs address these needs by enabling seamless integration into fabrics and skin-attachable devices, ensuring that they can withstand the demands of daily wear without compromising performance.

The potential of CPs in smart textiles and wearables is not limited to healthcare alone. These materials are equally impactful in sports performance tracking, fitness monitoring, and interactive technologies, where their properties enhance user experience and functionality. With the rapid advancements in CP technology, the development of next-generation smart textiles and wearables is poised to address critical challenges in health care, presenting groundbreaking solutions that improve patient outcomes and quality of life. This growing collaboration between CPs and modern technology underscores their indispensable role in shaping the future of wearable devices.

## 2.3 Challenges and Future Directions in Conducting Polymer Tuning

The tuning of CPs has unlocked significant advancements across various applications. Yet, it also presents different challenges that must be addressed to fully realize their potential in the next generation of technologies. The ability to tune or modify CPs for enhanced flexibility, conductivity, and stability has undoubtedly opened doors in all fields, including energy, environment, healthcare, biomedical, and intelligent textiles. But the process of tuning often involves trade-offs. Factors like scalability, cost-effectiveness, environmental impact, and sustainability remain critical barriers to widespread adoption.

CPs in pure and modified forms are widely used in water decontamination. Still, challenges like low solubility, weak strength, costs, and limited active sites hinder their broader applications despite advancement in this field. Furthermore, these CPs typically degrade quickly, but it should be noted that the fast drop in thermogravimetric analysis curves limits their applicability in high temperature conditions [19], despite groundbreaking discoveries of their applications in the biomedical field. As per the ClinicalTrials.gov database, these polymers still lack enough maturity to reach the clinical stage. We must also have clearly defined boundaries by the regulatory bodies [74]. Areas such as biocompatibility need to be explored very carefully [75]. The field of nanomedicine is indeed emerging, but it still has limiting factors such as the studies of in vivo biocompatibility and biodegradability, as well as the creation of regulatory affairs and quality assurance for their use in the clinical stage [76]. These are emerging affairs; more investigation is needed to take this to the next level in real biomedical applications. Although the composition of CPs with metal oxides/sulfides boosts the performance significantly, there is still a need for further improvement in the capacitance features.

The inherent properties of CPs, such as their structural range, flexibility, and excellent strength, make them promising candidates for further research and customization to meet the demand of specific applications. In addition, the electrochromic characteristics of CPs could be an acceptable advantage for their possible use in smart supercapacitors [77]. It is seen that the material's durability is poor when CPs are made of organic components. Still, incorporating carbon materials is an appealing way to explore this direction [78]. Further research in this field could be advantageous for the new generation of electrode design.

Alongside these, challenges like large-scale production, cost-effectiveness, and recyclability also need to be addressed. Innovations in nanostructuring, doping, and blending with advanced materials like MXenes (composed of transition metals, carbon/nitrogen) and carbon nanotubes could further expand the applications of CPs in smart textiles. Collaborations across disciplines, such as electrochemistry, materials science, electronics, and fashion design, would be critical to realizing the full potential of these materials. This opens doors for collaborative research that can be more application oriented [79].

In parallel with experimental advancements, computational studies have emerged as a compelling tool for realizing and predicting material properties cost-effectively and in an environmentally friendly manner [80–87]. Researchers have successfully used computational modeling to complement experimental work, providing insights into charge distribution, band gap, charge-transfer, spin density, band gap, reactivity, thermodynamic stability, and spectroscopic features [88–106]. These properties are fundamental to the performance and functionality of CPs and computational approaches have often predicted them with remarkable accuracy, closely matching experimental results. This alignment underscores the reliability of theoretical methods in advancing the understanding of CPs while reducing the need for resource-intensive laboratory experiments. By leveraging computational modeling, researchers can adopt a guided approach to material design and synthesis and predict properties before synthesis, thus optimizing experimental efforts. This strategy saves time and resources and helps minimize environmental impacts associated with trial-and-error laboratory processes. Despite these advancements, there is still significant untapped potential in computational research for CPs [107–109]. Expanding computational studies can enable more control on material properties, leading to the model highly tailored CPS for specific applications. Integrating computational and experimental methods will be critical in driving future breakthroughs in CPs.

## 2.4 Conclusion and Future Outlook

The field of CPs has evolved significantly, tuning their properties and emerging as a cornerstone for enhancing application-based performance. This represents a pivotal step in advancing materials science across environmental remediation, biomedical devices, energy storage, and smart textiles. Through careful modifications in their chemical structure, innovative doping strategies, and integration with advanced materials, CPs have demonstrated their versatility and value across diverse applications. From environmental sustainability to healthcare, the ability to tune the CPs has unlocked opportunities for designing functional materials that meet specific demands. Despite remarkable progress, several challenges remain, particularly in scaling production, ensuring cost-efficiency, and addressing longstanding stability under diverse conditions. Additionally, the environmental impact of CP synthesis and disposal calls for more sustainable and circular approaches to material development. Integrating computational tools with experimental efforts has proven to be a powerful strategy that enables the prediction of material properties and the optimization of synthetic pathways. These approaches accelerate innovation and reduce resource dependency, underscoring the importance of interdisciplinary research.

The continuous development of CPs will rely on advancing molecular-level design, hybrid composite strategies, and fabrication techniques that balance performance with sustainability. The integration of CPs in fields like energy-efficient systems, smart textiles and wearables, and biomedical devices reflects their pivotal role in addressing critical global challenges. By pushing the boundaries of polymer science, researchers will drive the creation of next-generation materials that redefine the limits of technology and improve quality of life. Such innovations reaffirm the importance of CPs as a dynamic platform for addressing the evolving demands of science, industry, and society.

## References

1. Kenry, Liu B (2018) Recent advances in biodegradable conducting polymers and their biomedical applications. Biomacromolecules 19:1783–1803
2. Le T-H, Kim Y, Yoon H (2017) Electrical and electrochemical properties of conducting polymers. Polymers 9:150
3. Namsheer K, Rout CS (2021) Conducting polymers: a comprehensive review on recent advances in synthesis, properties and applications. RSC Adv 11:5659–5697
4. Bajpai M, Srivastava R, Dhar R, Tiwari RS (2016) Review on optical and electrical properties of conducting polymers. Indian J Mater Sci 2016:5842763
5. Chen AX, Kleinschmidt AT, Choudhary K, Lipomi DJ (2020) Beyond stretchability: strength, toughness, and elastic range in semiconducting polymers. Chem Mater 32:7582–7601
6. Ocheje MU, Selivanova M, Zhang S, Van Nguyen TH, Charron BP, Chuang C-H, Cheng Y-H, Billet B, Noori S, Chiu Y-C, Gu X, Rondeau-Gagné S (2018) Influence of amide-containing side chains on the mechanical properties of diketopyrrolopyrrole-based polymers. Polym Chem 9:5531–5542
7. Savagatrup S, Printz AD, Wu H, Rajan KM, Sawyer EJ, Zaretski AV, Bettinger CJ, Lipomi DJ (2015) Viability of stretchable poly(3-heptylthiophene) (P3HpT) for organic solar cells and field-effect transistors. Synth Met 203:208–214
8. Root SE, Savagatrup S, Printz AD, Rodriquez D, Lipomi DJ (2017) Mechanical properties of organic semiconductors for stretchable, highly flexible, and mechanically robust electronics. Chem Rev 117:6467–6499
9. Zarrintaj P, Vahabi H, Saeb MR, Mozafari M (2019) Chapter 14—Application of polyaniline and its derivatives. In: Mozafari M, Chauhan NPS (eds) Fundamentals and emerging applications of polyaniline. Elsevier, pp 259–272

10. Pan Y, Zhang J, Guo X, Li Y, Li L, Pan L (2024) Recent advances in conductive polymers-based electrochemical sensors for biomedical and environmental applications. Polymers 16:1597
11. Bhandari M, Kaur DP, Raj S, Yadav T, Abourehab MAS, Alam MS (2022) Electrically conducting smart biodegradable polymers and their applications. In: Ali GAM, Makhlouf ASH (eds) Handbook of biodegradable materials. Springer International Publishing, Cham, pp 1–24
12. Sonika SK, Verma S, Samanta AK, Srivastava S, Biswas RM, Alsharabi S (2022) Rajput, conducting polymer nanocomposite for energy storage and energy harvesting systems. Adv Mater Sci Eng 2022:2266899
13. Zhao Z, Mi Y, Lu Y, Zhu Q, Cao X, Wang N (2023) From biochemical sensor to wearable device: the key role of the conductive polymer in the triboelectric nanogenerator. Biosensors 13:604
14. Ibanez JG, Rincón ME, Gutierrez-Granados S, Chahma Mh, Jaramillo-Quintero OA, Frontana-Uribe BA (2018) Conducting polymers in the fields of energy, environmental remediation, and chemical–chiral sensors. Chem Rev 118:4731–4816
15. Janaki V, Oh B-T, Shanthi K, Lee K-J, Ramasamy AK, Kamala-Kannan S (2012) Polyaniline/chitosan composite: an eco-friendly polymer for enhanced removal of dyes from aqueous solution. Synth Met 162:974–980
16. Zaremotlagh S, Hezarkhani A (2014) Removal of textile dyes from aqueous solution by conducting polymer modified clinoptilolite. Environ Earth Sci 71:2999–3006
17. Din MI, Ata S, Mohsin IU, Rasool A, Aziz AA (2014) Evaluation of conductive polymers as an adsorbent for eradication of As (III) from aqueous solution using inductively coupled plasma optical emission spectroscopy (ICP-OES). Int J Sci Eng 6:154–162
18. Pathania D, Sharma G, Kumar A, Naushad M, Kalia S, Sharma A, Alothman ZA (2015) Combined sorptional–photocatalytic remediation of dyes by polyaniline Zr(IV) selenotungstophosphate nanocomposite. Toxicol Environ Chem 97:526–537
19. Taghizadeh A, Taghizadeh M, Jouyandeh M, Yazdi MK, Zarrintaj P, Saeb MR, Lima EC, Gupta VK (2020) Conductive polymers in water treatment: a review. J Mol Liq 312:113447
20. Jadoun S, Yáñez J, Mansilla HD, Riaz U, Chauhan NPS (2022) Conducting polymers/zinc oxide-based photocatalysts for environmental remediation: a review. Environ Chem Lett 20:2063–2083
21. Stejskal J (2020) Interaction of conducting polymers, polyaniline and polypyrrole, with organic dyes: polymer morphology control, dye adsorption and photocatalytic decomposition. Chem Pap 74:1–54
22. Stejskal J (2022) Recent advances in the removal of organic dyes from aqueous media with conducting polymers, polyaniline and polypyrrole, and their composites. Polymers 14:4243
23. Olad A, Azhar FF, Shargh M, Jharfi S (2014) Application of response surface methodology for modeling of reactive dye removal from solution using starch-montmorillonite/polyaniline nanocomposite. Polym Eng Sci 54:1595–1607
24. Xu S, Zhu Y, Jiang L, Dan Y (2010) Visible light induced photocatalytic degradation of methyl orange by polythiophene/TiO2 composite particles. Water Air Soil Pollut 213:151–159
25. Xu S, Jiang L, Yang H, Song Y, Dan Y (2011) Structure and photocatalytic activity of polythiophene/TiO2 composite particles prepared by photoinduced polymerization. Chin J Catal 32:536–545
26. Kharazi P, Rahimi R, Rabbani M (2018) Study on porphyrin/ZnFe2O4@polythiophene nanocomposite as a novel adsorbent and visible light driven photocatalyst for the removal of methylene blue and methyl orange. Mater Res Bull 103:133–141
27. Ballav N, Debnath S, Pillay K, Maity A (2015) Efficient removal of Reactive Black from aqueous solution using polyaniline coated ligno-cellulose composite as a potential adsorbent. J Mol Liq 209:387–396
28. Weidlich C, Mangold K-M, Jüttner K (2001) Conducting polymers as ion-exchangers for water purification. Electrochim Acta 47:741–745
29. Ansari R, Delavar AF (2009) Application of poly 3-methylthiophene for removal of silver ion from aqueous solutions. J Appl Polym Sci 113:2293–2300

30. Bhaumik M, Noubactep C, Gupta VK, McCrindle RI, Maity A (2015) Polyaniline/Fe0 composite nanofibers: an excellent adsorbent for the removal of arsenic from aqueous solutions. Chem Eng J 271:135–146
31. Li R, Liu L, Yang F (2013) Preparation of polyaniline/reduced graphene oxide nanocomposite and its application in adsorption of aqueous Hg(II). Chem Eng J 229:460–468
32. Eisazadeh A, Eisazadeh H, Kassim KA (2013) Removal of Pb(II) using polyaniline composites and iron oxide coated natural sand and clay from aqueous solution. Synth Met 171:56–61
33. Nurerk P, Kanatharana P, Bunkoed O (2017) Polyaniline-coated magnetite nanoparticles incorporated in alginate beads for the extraction and enrichment of polycyclic aromatic hydrocarbons in water samples. Int J Environ Anal Chem 97:145–158
34. Guimard NK, Gomez N, Schmidt CE (2007) Conducting polymers in biomedical engineering. Prog Polym Sci 32:876–921
35. Kotwal A, Schmidt CE (2001) Electrical stimulation alters protein adsorption and nerve cell interactions with electrically conducting biomaterials. Biomaterials 22:1055–1064
36. Sensharma P, Madhumathi G, Jayant RD, Jaiswal AK (2017) Biomaterials and cells for neural tissue engineering: current choices. Mater Sci Eng C 77:1302–1315
37. Mihic A, Cui Z, Wu J, Vlacic G, Miyagi Y, Li S-H, Lu S, Sung H-W, Weisel RD, Li R-K (2015) A conductive polymer hydrogel supports cell electrical signaling and improves cardiac function after implantation into myocardial infarct. Circulation 132:772–784
38. Guex AG, Puetzer JL, Armgarth A, Littmann E, Stavrinidou E, Giannelis EP, Malliaras GG, Stevens MM (2017) Highly porous scaffolds of PEDOT:PSS for bone tissue engineering. Acta Biomater 62:91–101
39. De Alvarenga G, Hryniewicz BM, Jasper I, Silva RJ, Klobukoski V, Costa FS, Cervantes TNM, Amaral CDB, Schneider JT, Bach-Toledo L, Peralta-Zamora P, Valerio TL, Soares F, Silva BJG, Vidotti M (2020) Recent trends of micro and nanostructured conducting polymers in health and environmental applications. J Electroanal Chem 879:114754
40. Naghib SM, Amiri S, Mozafari MR (2024) Stimuli-responsive chitosan-based nanocarriers for drug delivery in wound dressing applications: a review. Carbohydr Polym Technol Appl 7:100497
41. Zhu M, Hao Y, Ma X, Feng L, Zhai Y, Ding Y, Cheng G (2019) Construction of a graphene/polypyrrole composite electrode as an electrochemically controlled release system. RSC Adv 9:12667–12674
42. Samanta D, Hosseini-Nassab N, McCarty AD, Zare RN (2018) Ultra-low voltage triggered release of an anti-cancer drug from polypyrrole nanoparticles. Nanoscale 10:9773–9779
43. Samanta D, Meiser JL, Zare RN (2015) Polypyrrole nanoparticles for tunable, pH-sensitive and sustained drug release. Nanoscale 7:9497–9504
44. Kim H, Solak K, Han Y, Cho Y-W, Koo K-M, Kim C-D, Luo Z, Son H, Kim H-R, Mavi A, Kim T-H (2022) Electrically controlled mRNA delivery using a polypyrrole-graphene oxide hybrid film to promote osteogenic differentiation of human mesenchymal stem cells. Nano Res 15:9253–9263
45. Wu JG, Chen JH, Liu KT, Luo SC (2019) Engineering antifouling conducting polymers for modern biomedical applications. ACS Appl Mater Interfaces 11:21294–21307
46. Hui N, Sun X, Niu S, Luo X (2017) PEGylated polyaniline nanofibers: antifouling and conducting biomaterial for electrochemical DNA sensing. ACS Appl Mater Interfaces 9:2914–2923
47. Wu H, Lee C-J, Wang H, Hu Y, Young M, Han Y, Xu F-J, Cong H, Cheng G (2018) Highly sensitive and stable zwitterionic poly (sulfobetaine-3, 4-ethylenedioxythiophene)(PSBEDOT) glucose biosensor. Chem Sci 9:2540–2546
48. Bansal M, Dravid A, Aqrawe Z, Montgomery J, Wu Z, Svirskis D (2020) Conducting polymer hydrogels for electrically responsive drug delivery. J Control Release 328:192–209
49. Guo B, Ma PX (2018) Conducting polymers for tissue engineering. Biomacromolecules 19:1764–1782
50. Yi H, Patel R, Patel KD, Bouchard L-S, Jha A, Perriman AW, Patel M (2023) Conducting polymer-based scaffolds for neuronal tissue engineering. J Mater Chem B 11:11006–11023

51. Molino BZ, Fukuda J, Molino PJ, Wallace GG (2021) Redox polymers for tissue engineering. Front Med Technol 3
52. Horrocks MS, Zhurenkov KE, Malmström J (2024) Conducting polymer hydrogels for biomedical application: current status and outstanding challenges. APL Bioeng 8
53. Djaalab E, Samar MEH, Zougar S, Kherrat R (2018) Electrochemical biosensor for the determination of amlodipine besylate based on gelatin–polyaniline iron oxide biocomposite film. Catalysts 8:233
54. Zhuang X, Tian C, Luan F, Wu X, Chen L (2016) One-step electrochemical fabrication of a nickel oxide nanoparticle/polyaniline nanowire/graphene oxide hybrid on a glassy carbon electrode for use as a non-enzymatic glucose biosensor. RSC Adv 6:92541–92546
55. Sacchi M, Sauter-Starace F, Mailley P, Texier I (2024) Resorbable conductive materials for optimally interfacing medical devices with the living. Front Bioeng Biotechnol 12
56. Bindu M, Periyat P (2023) A review on fine-tuning of energy storage characteristics of conducting polymers. Mater Adv 4:2730–2769
57. Pyarasani RD, Jayaramudu T, John A (2019) Polyaniline-based conducting hydrogels. J Mater Sci 54:974–996
58. Kazemi F, Naghib SM, Mohammadpour Z (2020) Multifunctional micro-/nanoscaled structures based on polyaniline: an overview of modern emerging devices. Mater Today Chem 16:100249
59. G. Mandal, J. Bauri, D. Nayak, S. Kumar, S. Ansari, R.B. Choudhary, Synthesis, structural study and various applications of polyaniline and its nanocomposites, (2023)
60. Onggar T, Kruppke I, Cherif C (2020) Techniques and processes for the realization of electrically conducting textile materials from intrinsically conducting polymers and their application potential. Polymers 12:2867
61. Tang Q, Chen M, Yang C, Wang W, Bao H, Wang G (2015) Enhancing the energy density of asymmetric stretchable supercapacitor based on wrinkled CNT@ MnO2 cathode and CNT@ polypyrrole anode. ACS Appl Mater Interfaces 7:15303–15313
62. Zhao C, Wang C, Yue Z, Shu K, Wallace GG (2013) Intrinsically stretchable supercapacitors composed of polypyrrole electrodes and highly stretchable gel electrolyte. ACS Appl Mater Interfaces 5:9008–9014
63. Heinze J, Frontana-Uribe BA, Ludwigs S (2010) Electrochemistry of conducting polymers—persistent models and new concepts. Chem Rev 110:4724–4771
64. Jia X, Ge Y, Shao L, Wang C, Wallace GG (2019) Tunable conducting polymers: toward sustainable and versatile batteries. ACS Sustain Chem Eng 7:14321–14340
65. Mike JF, Lutkenhaus JL (2013) Recent advances in conjugated polymer energy storage. J Polym Sci B Polym Phys 51:468–480
66. Cui Q, Bazan GC (2018) Narrow band gap conjugated polyelectrolytes. Acc Chem Res 51:202–211
67. Shi Y, Peng L, Yu G (2015) Nanostructured conducting polymer hydrogels for energy storage applications. Nanoscale 7:12796–12806
68. Mokhtar SMA, Alvarez de Eulate E, Yamada M, Prow TW, Evans DR (2021) Conducting polymers in wearable devices. Med Dev Sens 4:e10160
69. Tseghai GB, Mengistie DA, Malengier B, Fante KA, Van Langenhove L (2020) PEDOT:PSS-based conductive textiles and their applications. Sensors 20:1881
70. Islam GMN, Ali A, Collie S (2020) Textile sensors for wearable applications: a comprehensive review. Cellulose 27:6103–6131
71. Fielding LA, Hillier JK, Burchell MJ, Armes SP (2015) Space science applications for conducting polymer particles: synthetic mimics for cosmic dust and micrometeorites. Chem Commun 51:16886–16899
72. Cho B, Park KS, Baek J, Oh HS, Koo Lee Y-E, Sung MM (2014) Single-crystal poly(3,4-ethylenedioxythiophene) nanowires with ultrahigh conductivity. Nano Lett 14:3321–3327
73. Zahid M, Anwer Rathore H, Tayyab H, Ahmad Rehan Z, Abdul Rashid I, Lodhi M, Zubair U, Shahid I (2022) Recent developments in textile based polymeric smart sensor for human health monitoring: a review. Arab J Chem 15:103480

74. Zare EN, Makvandi P, Ashtari B, Rossi F, Motahari A, Perale G (2020) Progress in conductive polyaniline-based nanocomposites for biomedical applications: a review. J Med Chem 63:1–22
75. Wei M, Gao Y, Li X, Serpe MJ (2017) Stimuli-responsive polymers and their applications. Polym Chem 8:127–143
76. da Silva AC, Córdoba de Torresi SI (2019) Advances in conducting, biodegradable and biocompatible copolymers for biomedical applications. Front Mater 6
77. Magu TO, Agobi AU, Hitler L, Dass PM (2019) A review on conducting polymers-based composites for energy storage application. J Chem Rev 1:19–34
78. Park S-J, Son Y-R, Heo Y-J (2018) Chapter 6—Prospective synthesis approaches to emerging materials for supercapacitor. In: Cheong KY, Impellizzeri G, Fraga MA (eds) Emerging materials for energy conversion and storage. Elsevier, pp 185–208
79. Virumbrales C, Hernández-Ruiz R, Trigo-López M, Vallejos S, García JM (2024) Sensory polymers: trends, challenges, and prospects ahead. Sensors 24:3852
80. Riaz U, Singh N, Banoo S (2022) Theoretical studies of conducting polymers: a mini review. New J Chem 46:4954–4973
81. Salzner U, Pickup P, Poirier R, Lagowski J (1998) Accurate method for obtaining band gaps in conducting polymers using a DFT/hybrid approach. Chem A Eur J 102:2572–2578
82. Devey A, Grau-Crespo R, De Leeuw N (2008) Combined density functional theory and interatomic potential study of the bulk and surface structures and properties of the iron sulfide mackinawite (FeS). J Phys Chem C 112:10960–10967
83. Aldaghfag SA, Yaseen M, Ambreen H, Butt M, Rehan M, Dahshan A (2021) Effect of Fe on physical characteristics of CdS: DFT investigations. Chalcogenide Lett 18:357–365
84. Reckziegel A, Kour M, Battistella B, Mebs S, Beuthert K, Berger R, Werncke CG (2021) High-spin imido cobalt complexes with imidyl radical character. Angew Chem Int Ed 60:15376–15380
85. Kour M, Taborosi A, Boyd ES, Szilagyi RK (2023) Development of molecular cluster models to probe pyrite surface reactivity. J Comput Chem 44:2486–2500
86. Liang Q, Wang S, Yao Y, Dong P, Song H (2023) Transition metal compounds family for Li–S batteries: the DFT-guide for suppressing polysulfides shuttle. Adv Funct Mater 33:2300825
87. Yamaguchi Y, Takeuchi T, Sakaebe H, Kageyama H, Senoh H, Sakai T, Tatsumi K (2010) Ab initio simulations of Li/pyrite-MS2 (M= Fe, Ni) battery cells. J Electrochem Soc 157:A630
88. Zhang T, Li D, Tao Z, Chen J (2013) Understanding electrode materials of rechargeable lithium batteries via DFT calculations. Prog Nat Sci Mater Int 23:256–272
89. Aziz EF, Vollmer A, Eisebitt S, Eberhardt W, Pingel P, Neher D, Koch N (2007) Localized charge transfer in a molecularly doped conducting polymer. Adv Mater 19:3257–3260
90. Sharma N, Kour M, Gupta R, Bansal RK (2021) A new cross-conjugated mesomeric betaine. RSC Adv 11:25296–25304
91. Alemán C, Curcó D, Casanovas J (2005) Reliability of the density functional approximation to describe the charge transfer and electrostatic complexes involved in the modeling of organic conducting polymers. Phys Rev E—Stat Nonlin Soft Matter Phys 72:026704
92. Kour M, Gupta R, Saini R, Bansal R (2018) Diastereoselectivity in the Aza-Michael reaction of chiral α-methylbenzylamines with α, β-unsaturated carbonyl compounds. SynOpen 2:0122–0127
93. Akimov AV, Neukirch AJ, Prezhdo OV (2013) Theoretical insights into photoinduced charge transfer and catalysis at oxide interfaces. Chem Rev 113:4496–4565
94. Kour M, Gupta R, Bansal R (2018) Tandem Cope-rearrangement and [2+ 2] cycloaddition of cis-1, 2-diethynylcyclopropane and its mono-hetero analogues: a DFT investigation. Comput Theor Chem 1123:142–148
95. Bansal RK, Gupta R, Kour M (2018) Synergy between experimental and theoretical results of some reactions of annelated 1, 3-azaphospholes. Molecules 23:1283
96. Xiang Y, Jiang M, Xiao H, Xing K, Peng X, Zhang S, Qi D-C (2019) A DFT study of the surface charge transfer doping of diamond by chromium trioxide. Appl Surf Sci 496:143604

97. Kour M, Gupta R, Bansal RK (2017) Experimental and theoretical investigation of the reaction of secondary amines with maleic anhydride. Aust J Chem 70:1247–1253
98. Miyasaka H (2013) Control of charge transfer in donor/acceptor metal–organic frameworks. Acc Chem Res 46:248–257
99. Kour M, Sharma N, Bansal RK (2019) Theoretical study of the substituent effect of hydroxy group on tandem cope rearrangement and [2+ 2] cycloaddition in cis-1, 2-diethynylcyclopropane and its mono-hetero analogues. Struct Chem 30:1203–1210
100. Smida N, Zaidi B, Althobaiti MG (2023) Anthracene/fluorescein based semi-conducting polymer for organic photovoltaics: synthesis, DFT, optical and electrical properties. J Mol Struct 1272:134088
101. Kour M, Gupta R, Bansal RK (2017) Experimental and theoretical investigation of the reaction of amines with methyl propiolate. ChemistrySelect 2:8465–8470
102. Xu K, Sun H, Ruoko T-P, Wang G, Kroon R, Kolhe NB, Puttisong Y, Liu X, Fazzi D, Shibata K (2020) Ground-state electron transfer in all-polymer donor–acceptor heterojunctions. Nat Mater 19:738–744
103. Kaur A, Kour M, Gupta R, Bansal RK (2017) Aromaticity of 1, 4-dehydrotropylium ion and its mono-and poly-phospha analogues. Phosphorus Sulfur Silicon Relat Elem 192:674–682
104. Gupta R, Maheshwari P, Kour M (2015) Reinvestigation of the aromaticity of mono-and polyphosphacyclopentadienide ions at the DFT level. Comput Theor Chem 1060:10–16
105. Jain A, Shin Y, Persson KA (2016) Computational predictions of energy materials using density functional theory. Nat Rev Mater 1:1–13
106. Schulz P, Kelly LL, Winget P, Li H, Kim H, Ndione PF, Sigdel AK, Berry JJ, Graham S, Brédas JL (2014) Tailoring electron-transfer barriers for zinc oxide/C60 fullerene interfaces. Adv Funct Mater 24:7381–7389
107. Johansson P (2015) Computational modelling of polymer electrolytes: what do 30 years of research efforts provide us today? Electrochim Acta 175:42–46
108. Craig B, Skylaris C-K, Schoetz T, de León CP (2020) A computational chemistry approach to modelling conducting polymers in ionic liquids for next generation batteries. Energy Rep 6:198–208
109. Bhat V, Callaway CP, Risko C (2023) Computational approaches for organic semiconductors: from chemical and physical understanding to predicting new materials. Chem Rev 123:7498–7547

# Chapter 3
# Characterization of Tuned Conducting Polymers

**Manisha Rawat, Kritika S. Sharma, Priyanshi Shukla, and Rekha Sharma**

**Abstract** Conducting polymers have been the subject of extensive research because of their tunable mechanical, optical, and electrical properties, their facile synthesis and fabrication, and their enhanced environmental stabiliity relative to conventional inorganic materials. In their pure state, conducting polymers have many drawbacks, but they are solved via hybridization with other materials. The electrical industry makes extensive use of conducting polymer composites, optoelectronic and electronic, domains because of their mutually beneficial impacts. The objective of this chapter is to present transport models that elucidate conduction mechanism along with pertinent synthesis techniques, characterization approaches, and the physical properties of conducting polymers, including their optical, electrical, and mechanical properties. Additionally described are latest developments in their application in energy storage, anticorrosion coatings, and photocatalysis, and sensing applications. The performance of the composites is significantly influenced by their structural characteristics.

**Keywords** Conducting polymers · Hybridization · Electrical properties · Optical characteristics · Synthesis techniques · Characterization · Energy storage

## 3.1 Introduction

Prior to the discovery of conducting polymers, polymers were regarded exclusively as electrical insulators. Nevertheless, these organic polymers show special optical and electrical characteristics similar to those seen in semiconductors that are

---

M. Rawat · P. Shukla · R. Sharma (✉)
Department of Chemistry, Banasthali Vidyapith, Banasthali, Rajasthan, India

K. S. Sharma
School of Chemical Sciences, Central University of Gujarat, Gandhinagar, India
e-mail: kritika.sharma@cug.ac.in



S. Jadoun, N. Kumari Jangid (eds.), *Tuning of Conducting Polymers*, Springer Series on Polymer and Composite Materials,
https://doi.org/10.1007/978-3-032-00461-1_3

inorganic [1]. The optical and electrical properties of conjugated carbon chains are governed by the delocalized, polarized, and electron-dense π bonds, which alternate between single and double bonds. Polyacetylene (PA), polyfuran (PF), polypyrrole (PPy), polyaniline (PANI), poly (para-phenylene) poly (phenylenevinylene) (PPV), and polythiophene (PTH) are the examples of common conducting polymers (shown in Fig. 3.1). Electrical conductivity varies greatly across conductive polymers, ranging from insulating to highly conducting. Additionally, various external stimuli, including pressure, heat, and chemical exposure, can affect electrical conductivity. Because of these characteristics, conductive polymers show promise as materials. These materials are used in absorbers, sensors, actuators, and microwave shields. Nevertheless, these materials' primary drawbacks are their subpar mechanical qualities. We can create materials with conducting polymers' desired electrical characteristics and the great mechanical qualities of textiles by covering them with conductive polymers. Conducting polymers' optical, electrical, and mechanical properties are provided by the associated dopant ions; however, their processability and solubility are mostly determined by the side chains that are attached [2]. Conducting polymers can be both partially amorphous and crystalline. In conducting polymers, both delocalized and localized states exist and disorder is crucial for π-bond delocalization, which is essential for the creation of charge carriers such as solitons, bipolarons, and polarons, and others that cause when an insulator turns into a metal [3].

When conjugate polymers are pure, they act as an insulator by their conductivity to a semiconductor, and it increases as the focus of the dopant. When they are not doped, they exhibit the mechanical action of common polymers as well as the electrical and optical behavior of semiconductors. They behave as a quasi 1D, anisotropic electronic structure with a modest band gap of 2–3 eV, just like a

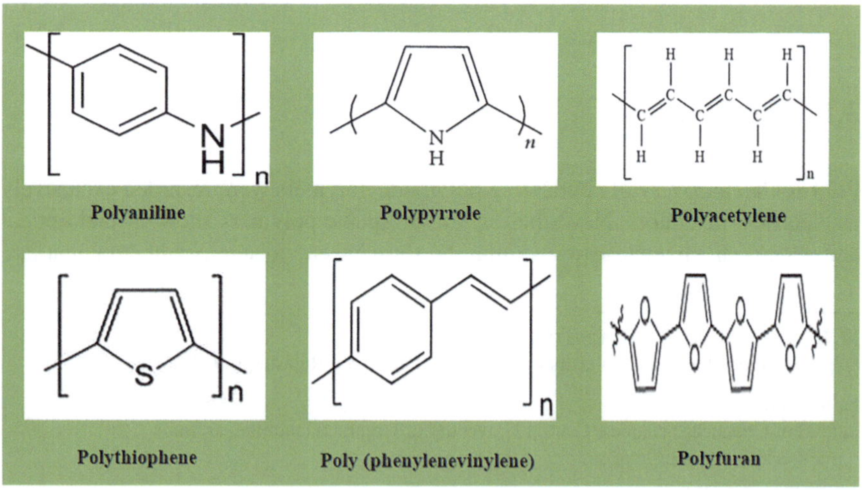


**Fig. 3.1** An example of the structure of many conducting polymers

standard semiconductor. In doped or photoexcited conjugated polymers, the π conjugated system can localize nonlinear excitation, forming bipolarons, solitons, and polarons, which allow the polymer to transition between a metallic state and a nonlinear excitation state [4–6]. We successfully cover the many synthesis techniques, traits, and applications of conducting polymers in this comprehensive overview.

## 3.2 Synthesis and Characterization of Conductive Polymer

### *3.2.1 Polyacetylene (PA)*

The recipient of the Nobel Prize was awarded for the invention and the enhancement of conductivity through doping of PA [7]. The properties of PA and its derivatives are multifunctional. Several key properties, such as liquid crystalline behavior, photoconductivity, electrical conductivity, and chiral recognition, have been extensively investigated. The components of a linear polyene chain are polyacetylene's primary chain. Because of its flexibility, it can be adorned with bunches of pendants, which allows for the formation of mono- or di-substituted PA by substituting a foreign molecule for the hydrogen molecules found in the alternate carbon [8]. Numerous techniques, such as photochemical, hydrothermal, solvo thermal, template-assisted, chemical oxidation, electrochemical polymerization, vapor phase synthesis, and electrospinning techniques, as well as plasma polymerization solid state approach, and the inclusion method, have been employed in the production of conductive polymers [9–14]. Conducting polymers frequently have poor optical and electrical properties when they are pure conductivity; but giving them dope with the right substances can improve their qualities. Polyacetylene's conductivity ranges from $10^{-5}$ S $cm^{-1}$ in its undoped state, but increasing doping level rises to $10^2$ to $10^3$ S $cm^{-1}$ [15]. Additionally, the conductivity of polyacetylene varies depending on the dopant substance, providing it with adjustable characteristics like optical mechanical or electrochemical characteristics, among other things [3, 16]. You can use polyacetylene can be produced by a number of methods, such as precursor assisted synthesis, catalytic and non-catalytic polymerization, and polymerization of other precursors. Acetylene polymerization yields PA polymer as well as by products such as vinyl acetylene oligomers and cyclo octatetraene. Among the various catalysts employed for polyacetylene synthesis, Zeigler-Natta catalysts are particularly notable for their high solubility and selectivity in organic solvents. Free-standing highly crystalline Polyacetylene films are produced on the walls of the reaction flask using a catalyst consisting of $(C_2H_5)_3A1$ and $Ti(0\text{-}n\text{-}C_4H_9)_4$ [17]. Luttinger catalysts, which consist of a combination of a hybrid reducing agent and a group [VIII] metal, like nickel chloride, are also used to make polyacetylene. High molecular weight polyacetylene is produced by these catalysts with no oligomer residues. A Luttinger catalyst, as opposed to a Ziegler–Natta catalyst, uses hydrophilic solvents, including acetonitrile or water–ethanol tetrahydrofuran (THF), for catalytic action. However, it is less catalytically active than a Zeigler–Natta catalyst, yet the physical and chemical characteristics of the product produced by these catalysts are nearly identical [18]. Acetylene electrochemical polymerization is classified as a non-catalytic polymerization process. A monomer precursor's anodic oxidation on an inert metal surface using the presence of suitable electrolytes is known as electrochemical synthesis. Numerous electrochemical techniques have been used for synthesis, including galvanostatic charge discharging, cycle-based voltammetry, galvanostatic, potentiostatic, and galvanostatic techniques. This method's primary benefit is that it permits a polymer layer to be directly deposited onto the metal, and through the modification of the electrochemical conditions, we can readily regulate the film's thickness. Korshak et al. used a metathesis catalyst, $W[OCH(CH_2Cl)_2]nCl_6{-}(C_2H_5)_2AlCl$ ($n$ = 2 or 3), to ring-open the polymerization of 1,3,5,7-cyclooctatetraene, resulting in PA films production of PA without employing an acetylene monomer [19]. Additionally, the reports of conjugated system synthesis triggered by light have surfaced, wherein acetylene gas is exposed to UV light to create PA [20].

### *3.2.2 Polyaniline*

The most encouraging and studied conducting polymer is PANI, which offers tunable conducting and optical characteristics, high stability, and excellent processability. The dopant's concentration determines the conductivity of polyaniline, which only exhibits metal-like conductivity at pH values below 3 [21]. PANI exists in a number of forms. According to their state of oxidation, they are categorized as pernigraniline, emeraldine, and leucoemeraldine; that is, leucoemeraldine is adequately reduced, while pernigraniline is entirely oxidized. Conductivity in polyaniline arises only in the partially oxidized state; in contrast, the fully oxidized from behaves as an insulator [22]. Benzoid and quinoid rings, in different proportions, constitute the backbone of the polymer. Three oxidized states result from the ratio difference: Benzoid and quinoid rings are equally distributed in the conductive emeraldine form, whereas the totally oxidized pernigraniline form is in a benzoid state and the leucoemeraldine that is completely decreased form is in a quinoid state. The dopant is present in close proximity to the chain of polymers and doesn't alter its chemical characteristics or establish a connection with the main chain. Among the most basic techniques for creating polyaniline is the chemical oxidation method, which involves mixing an oxidizing agent containing a precursor monomer of the corresponding polymer with the presence of an appropriate acid in ambient conditions to produce products. The authors' preferred oxidizing agent and doping acid are used in this procedure. The greening of the reaction media indicates the PANI synthesis. The same procedure is applied to prepare the composites. Oxidizing chemicals like ammonium peroxydisulfate, ammonium persulfate, potassium bichromate, ceric nitrate, and ceric sulfate are typically employed. The conductivity efficiently modifies the physical parameters based on an acid dopant's pH. As soon as the pH ranges from 1 to 3, the polymers and composites exhibit good conductivity [23, 24]. Interfacial polymerization represents another approach for synthesizing PANI, whereby the aniline monomer is partitioned between an oxidant solution, an organic solvent (e.g., toluene), and an aqueous phase containing a dopant acid. The process of polymerization occurs in the interphase between these two immiscible liquids when an oxidant solution is added to the monomer solution. For the production of PANI, a micro-emulsion approach is also employed, in which the polymerization occurs at the interface between two immiscible liquids; the surfactant employed differs, though [25–27]. The electro polymerization process, which also applies to polyacetylene, occurs without the influence of an oxidant [28, 29]. Electrospinning, driven by the application of a strong electric field, enables the fabrication of fibrous polymer morphologies with diameters ranging from the nanoscale to the microscale. Within this instance, the droplets of polymer are subjected to a high voltage, which causes surface tension to stretch the charge droplet. At a pivotal moment, the fluid erupts and begins to weave on the counter surface. Electrospinning and electrospraying work according to the same ideas. The only way to create a fibrous structure in bulk polymers is by electrospinning. Conducting polymers and their composites have been produced using this technique, including pure PPy, PANI, and polyaniline/carbon nanotubes and polyethylene oxide. Numerous variables affect electrospinning, encompassing the molecular weight of the viscosity, polymer, distance between the temperature, spinneret humidity, and many more [30–32].

### 3.2.3 Polypyrrole

Polypyrrole has attracted considerable economic interest owing to its high stability, improved conductivity, and versatility in forming both homopolymers and composites, originating it from many other conducting polymers. A PPy monomer was first chemically oxidized when hydrogen peroxide is present to create PPy, a black powder. When doped with halogenic electron acceptors like bromine or iodine, PPy exhibits a constant conductivity of $10^{-5}$ s $m^{-1}$, whereas in its undoped virgin condition, it acts like an insulator [33]. Although bulk PPy contains crystalline reginos in the monoclinic phase with about 15% crystallinity. It predominantly exhibits amorphous behavior and is not considered a crystalline substance [34]. Increased protonation concentration causes transformation in electrochemically produced by synthesizing PPy, which has a yellow-blackish hue and a thickness of 1 μm. It is more stable in air and has high thermal stability up to 300 °C; however, thermal degradation may occur as a result of dopant anions being lost [35]. Since the procedure and approach are comparable to those of other conducting polymers, electrochemical synthesis is a popular synthesis method for producing highly conductive polypyrrole. Due to the reduced anode size, this approach has a limited product yield. This method's primary benefit over other approaches is its ability to regulate the electrochemical parameters, which allows us to manage the thickness and morphology [36–40]. Numerous strategies for electrochemical synthesis are accessible. De-protonation in this polymerization pathway leads to the development of a free radical cation, which targets the neutral monomer unit. Following proton loss and dimeric radical reoxidation, the dimeric molecule may undergo further oxidation, leading to chain growth. Experimental evidence showed that protons were released during the oxidation of pyrrole [41]. A pyrrole monomer was first oxidized in the presence of $H_2O_2$ to produce polypyrrole, an amorphous, blackish powdered substance. Pyrrole black has limited conductivity in the $10^{-10}$ to $10^{-11}$ s $cm^{-1}$ range and is insoluble in organic solvents. It is made from peroxide and acid, and doping it with a halogen electron acceptor can boost its conductivity. As chemical oxidants, aqueous or anhydrous $FeCl_3$ or other iron (III) or copper (II) salts are frequently employed [42].

### 3.2.4 Poly (para-phenylene)s

As macromolecules consisting of aromatic benzoid nuclei linked directly by C–C bonds, Poly (para-phenylene)s are attracting significant interest owing to their exceptional air and thermal stability, superior optical properties, tunable conductivity, and facile doping behavior. Conjugation and the mechanical stiffness of the polymer backbone represent a promising advancement in nonlinear optics. When side chains that are flexible are joined to the backbone, the solubility of poly(p-phenylene)s gradually increases from its initial limit. The conductivity of poly(p-phenylene) increases 14 times when the right dopants are added, allowing for both p-type and n-type doping. The longer the dopant is exposed, the higher the conductivity [43]. Because of their superior optical characteristics and comparing blue light output to various conjugated systems, poly(p-phenylene)s are essential in the creation of organic LEDs. Temperature affects poly(p-phenylene)'s structural characteristics; at higher temperatures, it displays a planar structure; at lower temperatures, it displays a distorted planar structure along with a phase transition. It is possible to modify the optical band gap of poly(p-phenylene) and change its structure by adding side chains or suitable dopants [44]. In comparison to engineering polymers, poly(p-phenylene) has improved mechanical qualities and extremely high tensile properties. Its strength and modulus are two times greater than those of thermosetting polyimide (PI) or polyetheretherketone (PEEK). Polyparaphenylene can be used in a wide range of mechanical applications at temperatures up to 140 °C [50].

### 3.2.5 *Poly(p-phenylene vinylene) (PPV)*

Because of its outstanding optical properties, PPV was the first material employed to demonstrate electroluminescence for the fabrication of organic light-emitting diodes [45]. It was thoroughly researched to fabricate LED displays. The amorphous character of PPV was discovered by crystallographic investigations, and subsequently, and with monoclinic unit cells, the crystallites' isotropic distribution was identified [46]. In the instance of PPV, two planes were identified as the dispersion of crystallites. The electrical properties of PPV are strongly influenced by the nature and concentration of dopants, in its pristine state, however, it functions as an insulator. in its pristine. Both the reaction circumstances and the structural behavior of the pristine PPV affect its electrical characteristics. The value of its conductivity goes from $10^{-13}$ to $10^{3}$ s $cm^{-1}$ when doped. Potential uses for PPV include optoelectronic devices such as photodetectors, LEDs, and lasers. It was utilized within the LED panel industry because of its enhanced mechanical properties. Initially, the PPV panel was constructed by sandwiching it between an ITO anode and a metal cathode with a low work function [47]. In order to prepare PPV, numerous techniques are available. Numerous studies have been conducted on the Wittig coupling reaction. In this chemical process, PPV is produced by combining a salt of aromatic bis-phosphonium with a bis-aldehyde. The synthesis of PPV through the Alkyl-substituted aryl di-boron acids and di-bromo aromatic compounds are coupled by Pd was also studied using a Suzuki coupling reaction. There have also been reports of other synthesis techniques, like electropolymerization [48].

### 3.2.6 *Polythiophenes*

Polythiophene and its derivatives, in contrast to many other polymers, are conducting materials widely studied materials widely studied for their superior thermal, optical, and environmental stability. Often, polythiophenes are utilized in nonlinear optical systems, polymer LEDs, energy storage devices, anticorrosion coatings, and photochromic modules. The optical and electrical properties of polythiophenes can be tailored through chemical modifications or doping engineering. Polythiophenes with a side chains possess a tunable band gap, typically ranging from 1 to 3 eV, depending on the dopant [49]. Polysulfonates (PSS), a polyelectrolyte, are commonly added to the PEDOT matrix to improve stability. PSS uses a charge-balancing system to function as a stabilizer and a dopant. The PEDOT: PSS derivative exhibits long-term stability of heat, high conductivity, and superior mechanical flexibility. Adding a surfactant, changing the PSS content, and solvent treatment all enhance polythiophene and its derivatives' electrical properties. Another class of polythiophene compounds is poly(3-hexylthiophene), which is mostly used in the electrical and optoelectronic domains. P3HT's widespread availability, affordability, well-known morphology, and ease of processing make it a popular choice. The backbone of the semicrystalline polymer P3HT consists of linear side chains and single rings. The structural arrangement imparts flexibility, thereby enabling the sampling of conformational space. Depending on the synthesis process and sample purity, P3HT's high tensile modulus ranges from 200 MPa to 1 GPa. The glass transition temperature of P3HT is approximately12°C [50]. In the early 1980s, polythiophene was created chemically using the Yamamoto and Lin-Dudek methods. Other innovative techniques have been extensively investigated, including hydrothermal and solvothermal methods, oxidative synthesis, electropolymerization, organometallic coupling reaction [51], and direct sol-gel. The composite was created for the production of derivatives of polythiophene such as PEDOT:PSS, P3TH, and PEDOT employing a number of methods, including green synthesis electro polymerization [52], microfluid systems for synthesis, and some other novel methods [52, 53].

## 3.3 Conducting Polymer's Applications

### *3.3.1 Supercapacitors*

The global economy has a big impact on how much coal, natural gas, and other fossil fuels are used. A consequence of the degradation of fossil fuels is that number of environmental and societal concerns or issues are increasing. Conventional energy sources must be replaced due to health concerns, global warming, and climate change. As a result, researchers have worked to create high-energy, renewable, and environmental-friendly energy sources like wind, fuel cells, and supercapacitors. Super-capacitors are currently attracting a lot of commercial interest because to the potential markets for electric vehicles, wearable technology, and so on. One of the most notable distinctions between super-capacitors and normal capacitors is that super-capacitors have the capacity to store one thousand times more energy than dielectric capacitors. In addition to this, they have a cycle life that is satisfactory, a power density and high energy, and a charge-discharge cycle that is quick [54]. Super-capacitors may be classified into three basic groups, each of which is determined by the manner in which they store charge. The electrode-diode liquid chromatography (EDLC) system is a non-faradic mechanism that acts by collecting charge at the junction between the electrode and the electrolyte. This allows the system to perform its specific purpose. Capacitance is another the outcomes arisig from the formation of an electrochemical double layer of charge. Carbon-based materials with high surface areas, such as graphene, carbon nanotubes, and carbon, are often used as electrode materials for extreme-displacement liquid chromatography. In most cases, these materials are governed by this type of energy storage. It is a faradic process that is responsible for the operation of pseudo-capacitors; and a redox reaction or intercalation process is utilized in order to store charge. EDLCs have lower power and energy densities compared to pseudo-capacitors, which have higher power and energy densities [55]. They have been utilized widely as electrode materials for pseudo-capacitors due to the redox activity that they exhibit. Conducting polymers [56] and $Co_3O_4$ or $Mn_3O_4$ are the examples of these types of materials. Hybrid super-capacitors are fabricated by integrating pseudo-capacitors, and EDLCs into a single device. Pseudo-capacitors, on the other hand, have a high energy and power density, in contrast to EDLCs, which have a specific capacitance that is suitable and good cycle stability. EDLC and pseudo-capacitors are both utilized in hybrid capacitors, which allow for the advantages of both types of capacitors to be integrated [57]. As a result of their electrical properties, robust redox activity, high theoretical capacitance, excellent wave absorption [58], and outstanding electrochemical behavior, conducting polymers are utilized in a broad variety of applications [59]. The electrical, mechanical, and optical properties of conducting polymers vary depending on their morphology, which includes sheets, particles, hydrogels, rod-like structures, and more [60, 61]. The specific capacitance of a composite product is significantly influenced by the shape of the material, which results in a substantial relationship between the two. Zhang et al. conducted research on the production of porous carbon nanosheets and polyaniline nanowires, and they evaluated their performance in the context of applications for high-performance supercapacitors. Carbonaceous materials, such as activated carbon, grapheme are often composited with polyaniline in order to circumvent the problem of cycle stability that polyaniline presents. It is the tendency of polyaniline carbonaceous composite materials to agglomerate during the synthesis process that is the fundamental disadvantage of these materials. Agglomeration reduces the number of active sites and increases ion mobility, both of which have a detrimental effect on the performance of super-capacitors. Research focused on three-dimensional porous carbons as a potential answer to this problem [62]. The composite was created by the application of an in situ chemical oxidative method, which involved the utilization of ammonium persulfate. The fact that the concentration of aniline causes an increase in particular capacitance means that it also plays a substantial role in this scenario. PANI exhibits an antagonistic impact in this situation. CV and GCD were used to examine the composite's supercapacitance performance. According to the electrochemical data, the specific capacitance of PC/PANI is 512 F $g^{-1}$ when the current density is 1.0 A $g^{-1}$. This capacitance is much higher than the capacitance of PC or polyaniline, which is 250 F $g^{-1}$ and 201 F $g^{-1}$, respectively. The data that were presented revealed that when the current density is 1.0 A $g^{-1}$, the combination of PC/PANI//PC demonstrates a capacitance of 77 F $g^{-1}$, which is the appropriate value. The maximal power and energy densities are 14,000 W $kg^{-1}$ and 17.1 W h $kg^{-1}$, respectively, as compared to the numbers that were reported before [63, 64] (Fig. 3.2, Table 3.1).

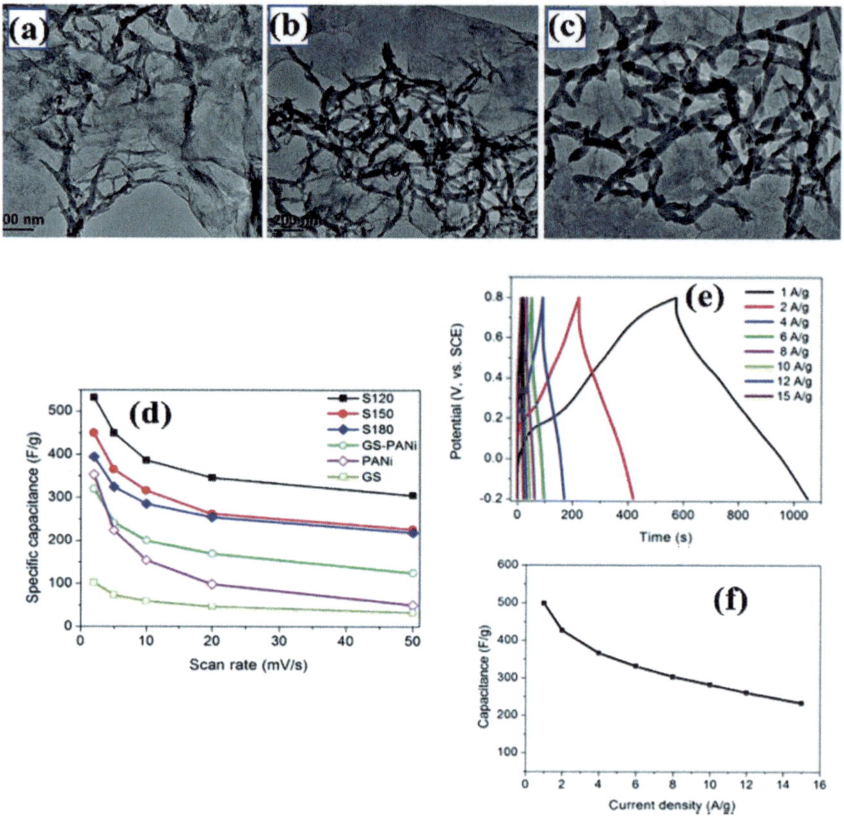


**Fig. 3.2** (**a–c**) TEM representations of GO/PANI composites at 120 °C, 150 °C, and 180 °C. (**d**) At various scan speeds, the specific capacitance of composites made of GNS/PANI was measured. (**e**) At a temperature of 120 °C, the GCD curves of GO/PANI composites were analyzed with different current densities. (**f**) Different current densities were used to determine the specific capacitance of GO/PANI composites. (Reproduced with permission [64] © 2017, scientific report License CC- BY 4.0)

## *3.3.2 Photocatalytic Applications*

$TiO_2$ is favored among photocatalysts due to its unique bandgap, straightforward synthesis, cost-effectiveness, elevated photoactivity, resistance to photocorrosion, and substantial oxidizing capability. (Brookite, rutile, and anatase). In comparison to the other three phases, anatase possesses advantageous photochemical characteristics. This can be due to the fact that its band gap is higher than the others, measuring 3.2 eV. Both hydrothermal and sol-gel techniques are very simple ways to produce titanium dioxide. Both of these procedures are able to produce titanium dioxide. PANI combined with organic semiconductors presents an intriguing avenue for exploration due to its effective electron–hole charge separation, impressive

**Table 3.1** Characterization of tuned conducting polymers

| Polymer | Electrolyte | Specific capacitance | Cycle stability | Power density energy | Ref. |
|---|---|---|---|---|---|
| rGO/PEDOT/PANI | Solid electrolyte | At 5 $mVs^{-1}$ 388.5 F $g^{-1}$ | 99% after 10,000 | 800 $Wkg^{-1}$ and 26.89 $Whkg^{-1}$ | [65] |
| $MnO_2$ /polyaniline/ hollow mesoporous Silica | PVA-KOH | At 1 $Ag^{-1}$ 248.5 $Fg^{-1}$ | >97.7% after 5000 cycles | 800 $Wkg^{-1}$ and 88.4 $Whkg^{-1}$ | [66] |
| Carbonized iron/ polyaniline/graphene Ni foam | 1 mol/L $NaNO_3$ | At 1 $Ag^{-1}$ 69.9 $Fg^{-1}$ | 91% after 10,000 cycles | 718.2 $Wkg^{-1}$ and 68 $Whkg^{-1}$ | [67] |
| Polyaniline/graphene/ $Fe_2O_3$ hydrogels | 1 mol/L $H_2SO_4$ | At 0.25 A $g^{-1}$ 1124 F $g^{-1}$ | 82% after 10,000 cycles | 58 $Wkg^{-1}$ and 14.4 $Whkg^{-1}$ | [68] |
| NiCO-MOF/ polypyrrole | 2 mol/L KOH | At 0.5 $Ag^{-1}$ 1109 $Fg^{-1}$ | 79.1% after 10,000 cycles | 375 $Wkg^{-1}$ and 41.2 $Whkg^{-1}$ | [69] |
| SDBS doped polypyrrole/HC | 6 mol.$L^{-1}$ KOH | At 5 $mVs^{-1}$ 1086 $Fg^{-1}$ | 90% after 2500 cycles | 1075 $Wkg^{-1}$ and 218.05 $Whkg^{-1}$ | [70] |
| Polyaniline/graphene | EMITFSI/ PVD -FHFP | At mA $cm^{-2}$ 87.8 $mFcm^{-2}$ | 100% after 10,000 cycles | 226.4 $\mu Wcm^{-2}$ and 12.2 $\mu Whcm^{-2}$ | [71] |
| $Ti_3C_2$ /polypyrrole | 2 mol.$L^{-1}$ $H_2SO_4$ | At 1.05 $mAcm^{-2}$ 109.4 mF $cm^{-2}$ | 96% after 10,000 cycles | 0.0845 $mWcm^{-2}$ and 3.398 $\mu Whcm^{-2}$ | [72] |
| Polythiophene-graphite graphene oxide | 1 mol.$L^{-1}$ KOH | At 1 $A^{-1}$ 971 $Fg^{-1}$ | 98% after 10,000 cycles | 7000 $Wkg^{-1}$ and 38.11 $Whkg^{-1}$ | [73] |
| CuS/C@polyaniline | 3 mol.$L^{-1}$ KCl | At 1 $Ag^{-1}$ 425.53 $Fg^{-1}$ | 89.86 after 3000 cycles | – | [74] |
| PEDOT: PSS/rGO | 1 mol/L $H_2SO_4$ | At 0.5 mA 249.5 $Fg^{-1}$ | 75% after 2000 cycles | 10.68 $Whkg^{-1}$ and 81.25 $Wkg^{-1}$ | [75] |

absorption coefficient in the visible spectrum, high charge carrier mobility, and the advantages of low-cost and straightforward synthesis. Light's contact with the photocatalyst, the creation of photo-induced holes, the accumulation of electrons on the surface of the semiconductor, the beginning of a redox reaction, and the destruction of the products that are produced are all components of the process that causes dye degradation. A hybrid system consisting of $TiO_2$ and PANI was produced by Reddy and his colleagues by the utilization of an in situ polymerization technique. The photocatalytic activity of the composite was evaluated through the degradation of Rhodamine, methylene blue, and acetic acid. Over the course of a period of 200 min, RhB underwent a deterioration rate of 80% when subjected to continuous UV irradiation for 3 h. In contrast, methylene blue and phenol demonstrated degradation rates of 67% and 51%, respectively. Following three cycles of irradiation, the photocatalyst continues to exhibit effective activity. Based on the information, it appears that the catalyst exhibits a feature that allows it to be recovered. Tests using ESR spin trapping were conducted to investigate the photocatalytic mechanism of $TiO_2$/

PANI. The results of these tests revealed that the major oxidative species that are produced by UV irradiation are holes ($h^+$), superoxide anion radicals ($O_2^{\bullet-}$), and hydroxyl radicals (•OH). Irradiation with ultraviolet light induces a transition from π to π* in PANI. Based on this information, it may be inferred that the electrons created by photochemistry are elevated to the π* orbital of PANI. Chemical bonding occurs when the energy is in phase with titania's conduction band. This helps a photogenerated electron to migrate from PANI to $TiO_2$. The photocatalyst's surface is the site of the absorbed electrons' interaction with water and oxygen, which leads to the formation of hydroxyl and superoxide anion radicals. The radicals in question make the oxidation of pollutants easier to accomplish. The valance band of PANI and $TiO_2$ both have the same energy level, which means that photogenerated pores are able to transition from the valance band of $TiO_2$ to the high-energy state of PANI. This is because the valance bands of $TiO_2$ and PANI both have the same energy levels. In addition, the holes that are produced by the photocatalyst go to the surface of the photocatalyst, where they are responsible for the oxidation of pollutants [76–81] (Fig. 3.3).

### *3.3.3 Hydrogels Made of Conducting Polymer for Use in Bioelectronics*

In order to generate certain innovative features that are not accessible for ordinary materials, conducting polymer hydrogels are a mix of conducting polymer and hydrogels. These hydrogels are utilized to develop specific applications. This combination was created in order to accomplish the desired results. These hydrogels are composed of conducting polymers, including PPy, polythiophene, and PANI, that

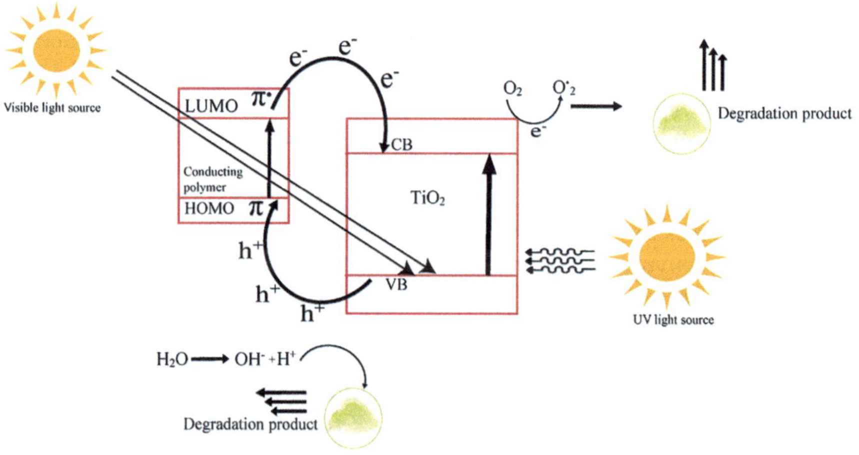


Fig. 3.3 Conductive polymer composites with a photocatalytic mechanism. (Reproduced with permission [81] © 2017, RSC advancement License CC- BY 4.0)

have been cross-linked with hydrophilic networks through either covalent or physical bonds [82, 83]. There are a variety of uses for hydrogels, including energy storage applications, sensing, bioelectronics, tissue engineering, and optically induced current (OECT). The shape of hydrogels may be readily adjusted [84–86]. Doping engineering allows hydrogels to have chemical and physical characteristics that may be altered to various degrees, which is one of the most significant benefits of these materials. When we are talking about PPy hydrogels, characteristics of these materials' morphological features may range from nanofibers to three-dimensional porous structures, depending on the kind of dopant and the amount of doping that is present. In addition to its outstanding mechanical and electrical qualities, CPH also has great self-healing and biocompatibility [87]. Bioelectronics and tissue engineering are two prominent fields that make extensive use of them. Due to the fact that conducting polymers do not have the capacity to maintain their stability over time, they are not adequate for some applications. When it comes to achieving long-term stability, a number of research organizations have suggested using conducting hydrogels rather than conducting polymers. Several techniques are employed in the process of creating CPH, including click chemistry, free radicals, condensation, and Michael-type addition polymerization. In the field of biomedicine, the chemical approaches are the subject of extensive research. As a result of the development of bioelectricity, investigators began looking into the ways in which biology and electronics communicate with one another. The electrical response of the nervous system is responsible for the capacity of people to move their muscles, remember things, and communicate with one another. Electrodes that interact with biological tissues often make up what are known as bio-electronic interfaces. Personalized self-monitoring of tissues is made possible by the collection of bio-electronic signals using these devices [88, 89]. Skin that is electronic, soft robotics, the monitoring of the activity of the brain, and implanted devices that are both flexible and wearable are just some of the applications that make extensive use of bioelectronics [90]. The years leading up to the creation of CPH were marked by the widespread use of inorganic materials in bio-electronic devices. Concerns have been raised concerning its inflammatory responses, inaccuracy in signal collecting, and cytotoxicity, and some people are pushing for its replacement. It is essential for bio-electronic devices to preserve their superior mechanical properties and conductivity in order to facilitate the movement of human body parts. This is a very difficult undertaking to complete. In the realm of bio-electronic sensing, transistors such as organic electrochemical transistors (OECTs) [91] and electrolyte-gated organic field effect transistors are often utilized. Many people in the scientific and business communities are interested in OECTs because of their biocompatibility, high trans-conductance, and processability. An organic electrochemical transistor (OECT) is made by contacting an organic film or semiconductor with two electrodes—a source and a drain and then submerging the electrode (gate) in an electrolyte. With the help of the organic film, charges and ions may be more easily transferred from the source to the drain. The channels that comprise the OECT are manufactured using PEDOT: PSS, a polypyrrole polymer that has been doped with p-type dopants [92]. For the purpose of tissue response monitoring, bioelectronic devices serve as the fundamental building block through which wearable and implantable devices are constructed. Considering that conducting polymer hydrogels have the capability to convert mechanical data into electrical signals, they are used in the manufacturing process of pressure and strain sensors. Conducting polymer hydrogels are widely used to create flexible electrodes that can monitor electrophysiological signals. These signals include electrocardiogram (ECG) [93], electromyography, and electroencephalography [94]. In the majority of instances, electrophysiological signals are not particularly strong. During the process of developing the device, the most significant obstacle that must be overcome is the enhancement of the conductivity and the signal-to-noise ratio. Additionally, a number of additional aspects, such as biocompatibility, long-term stability, and skin adhesion are taken under careful consideration. Different electrophysiological signals may be detected using PAA/PEDOT hydrogels, which have been created for this purpose. The glucose and lactate levels may be monitored in real time using hydrogels that are made from PANI [95]. Electroencephalography, electrocardiography (ECG), and electromyography are all the examples of electrophysiological signals that may be monitored with OECT. Adding conductive fillers to hydrogels increases their conductivity.

## 3.4 Future Aspects

A significant number of investigations have been carried out up to this point on the subject of the manufacturing of conducting polymers. This review article's primary objective has been to investigate the most recent developments in the study of conducting polymers, including their characteristics and applications. Because of their superior physical qualities and the metallic conductivity that they exhibit when doped, conducting polymers and their composites are perpetually in the spotlight. It was discovered via the examination of published works that conducting polymers, when in their purest state, exhibit the characteristics of either an insulator or a semiconductor. Only after being doped with an appropriate dopant or after being formed into a composite with extraneous materials do they exhibit metallic conductivity. The morphology of conducting polymers is another factor that influences the chemical and physical characteristics of these materials. There are several morphologies which allow for the production of unique features. In order to create conducting polymers for various uses, it is essential to understand their basic characteristics. Understanding the doping mechanism and the source of conductivity is vital since theoretical modeling conducting polymers and their derivatives is challenging due to their complicated design. The formation of charge carriers on the backbone of a polymer takes place when electrons are either added to or withdrawn from the delocalized $\pi$-bond. The transport qualities of a material are influenced by a wide variety of factors, including temperature, dopants, and structural features at the very least. Because of their metallic conductivity, processability, flexibility, and simplicity of manufacture, conducting polymers are utilized extensively in the production of supercapacitors as electrode materials.

## 3.5 Conclusion

In light of the public health and environment concerns associated with coatings consisting of chromate conversion, researchers have dedicated their efforts to discovering an alternative. In the recent decades, conducting polymers have found extensive application in the prevention of corrosion in metals and the alloys of those metals. They serve as pigments for organic coatings as well. Additionally, corrosion inhibitors are multilayers of conducting polymers and composites made from those polymers. During the process of the polymer being reduced, the dopants that are incorporated within the core of the polymer liberate blockers that are necessary. By manipulating the electrochemical parameters, conducting polymer films that have been electrochemically synthesized in the presence of appropriate electrolytes may be made to operate as effective corrosion inhibitors. Additionally, the thickness of the film can be altered. An improvement in the corrosion protection property of conducting polymers can be achieved by the application of organic coatings such as polyurethane, acrylic, and epoxy other substances that are analogous to these. Inhibiting corrosion is a crucial function that is played by morphology. The most significant benefit of using polymer coatings is that they allow for the regulated release of inhibitors following reduction. Additionally, in harsh settings, they produce a layer of metal oxide beneath the topcoat, and this layer of metal oxide is able to efficiently and effectively resist corrosion. When doing a study of previous studies, it is recommended that future advancements concentrate on conducting polymer-incorporated organic coatings with the purpose of enhancing the performance of corrosion inhibition. Utilizing organic coatings in the compositing process results in an improvement in the longevity of coatings. The efficiency of conducting polymer-based photocatalysts is determined by the rate at which photoelectrons are sent to the CB by particles that are embedded on the polymer backbone. Examples of such particles include metal oxide, titanium dioxide, and others. A significant amount of research has demonstrated that in order to produce a high level of photocatalytic activity, it is necessary to reduce the amount of electron–hole combination. Literature review findings indicate that structural arrangements are only one of several elements influencing the performance of conducting polymer sensitized $TiO_2$ photocatalysts.

## References

1. Nezakati T, Seifalian A, Tan A, Seifalian AM (2018) Conductive polymers: opportunities and challenges in biomedical applications. Chem Rev 118(14):6766–6843
2. Kanicki J, Skotheim TA (1986) Handbook of conducting polymers. Dekkker, New York, p 543
3. Le TH, Kim Y, Yoon H (2017) Electrical and electrochemical properties of conducting polymers. Polymers 9(4):150
4. Miyata S (1997) Organic electroluminescent materials and devices. CRC Press
5. Nalwa HS (1997) Handbook of organic conductive molecules and polymers. No Title
6. Skotheim TA (ed) (1997) Handbook of conducting polymers. CRC Press

7. Marsh G (2001) Electronic polymers revolution. Mater Today 4(1):4–6
8. Liu J, Lam JW, Zhong Tang B (2010) Synthesis and functionality of substituted polyacetylenes. In: Design and synthesis of conjugated polymers, pp 1–43
9. Martin CR (1995) Template synthesis of electronically conductive polymer nanostructures. Acc Chem Res 28(2):61–68
10. MacDiarmid AG, Jones Jr WE, Norris ID, Gao J, Johnson Jr AT, Pinto NJ et al (2001) Electrostatically-generated nanofibers of electronic polymers. Synth Met 119(1–3):27–30
11. Tang Q, Wu J, Sun X, Li Q, Lin J (2009) Shape and size control of oriented polyaniline microstructure by a self-assembly method. Langmuir 25(9):5253–5257
12. Deronzier A, Moutet JC (1996) Polypyrrole films containing metal complexes: syntheses and applications. Coord Chem Rev 147:339–371
13. Park CS, Kim DH, Shin BJ, Kim DY, Lee HK, Tae HS (2016) Conductive polymer synthesis with single-crystallinity via a novel plasma polymerization technique for gas sensor applications. Materials 9(10):812
14. Vandesteeg NA (2007) Synthesis and characterization of conducting polymer actuators. Doctoral dissertation, Massachusetts Institute of Technology
15. Bredas JL, Street GB (1985) Polarons, bipolarons, and solitons in conducting polymers. Acc Chem Res 18(10):309–315
16. Håkansson E, Lin T, Wang H, Kaynak A (2006) The effects of dye dopants on the conductivity and optical absorption properties of polypyrrole. Synth Met 156(18–20):1194–1202
17. Ito T, Shirakawa H, Ikeda S (1974) Simultaneous polymerization and formation of polyacetylene film on the surface of concentrated soluble Ziegler-type catalyst solution. J Polym Sci Polym Chem 12(1):11–20
18. Luttinger LB, Colthup EC (1962) Hydridic reducing agent—group VIII metal compound. A new catalyst system for the polymerization of acetylenes and related compounds. II. J Org Chem 27(11):3752–3756
19. Korshak YV, Korshak VV, Kanischka G, Höcker H (1985) A new route to polyacetylene. Ring-opening polymerization of 1, 3, 5, 7-cyclooctatetraene with metathesis catalysts. Makromol Chem Rapid 6(10):685–692
20. Simionescu CI, Percec V (1982) Progress in polyacetylene chemistry. Prog Polym Sci 8(1–2):133–214
21. Wang Y, Levon K (2012, August) Influence of dopant on electroactivity of polyaniline. In: Macromolecular symposia, vol 317, No. 1. WILEY-VCH Verlag, Weinheim, pp 240–247
22. Bhandari S (2018) Polyaniline: structure and properties relationship. In: Polyaniline blends, composites, and nanocomposites. Elsevier, pp 23–60
23. Boeva ZA, Sergeyev VG (2014) Polyaniline: synthesis, properties, and application. Polym Sci Ser C 56(1):144–153
24. Vivekanandan J, Ponnusamy V, Mahudeswaran A, Vijayanand PS (2011) Synthesis, characterization and conductivity study of polyaniline prepared by chemical oxidative and electrochemical methods. Arch Appl Sci Res 3(6):147–153
25. Yang L, Yang L, Wu S, Wei F, Hu Y, Xu X et al (2020) Three-dimensional conductive organic sulfonic acid co-doped bacterial cellulose/polyaniline nanocomposite films for detection of ammonia at room temperature. Sensors Actuators B Chem 323:128689
26. Österholm JE, Cao Y, Klavetter F, Smith P (1993) Emulsion polymerization of aniline. Synth Met 55(2–3):1034–1039
27. Zeng F, Qin Z, Liang B, Li T, Liu N, Zhu M (2015) Polyaniline nanostructures tuning with oxidants in interfacial polymerization system. Prog Nat Sci Mater Int 25(5):512–519
28. El-Basaty AB, Moustafa E, Fouda AN, El-Moneim AA (2020) 3D hierarchical graphene/CNT with interfacial polymerized polyaniline nano-fibers. Spectrochim Acta A Mol Biomol Spectrosc 226:117629
29. AlMashrea BA, Abla F, Chehimi MM, Workie B, Han C, Mohamed AA (2020) Polyaniline coated gold-aryl nanoparticles: electrochemical synthesis and efficiency in methylene blue dye removal. Synth Met 269:116528

30. Ashokkumar SP, Vijeth H, Yesappa L, Niranjana M, Vandana M, Devendrappa H (2020) Electrochemically synthesized polyaniline/copper oxide nano composites: to study optical band gap and electrochemical performance for energy storage devices. Inorg Chem Commun 115:107865
31. Kang TS, Lee SW, Joo J, Lee JY (2005) Electrically conducting polypyrrole fibers spun by electrospinning. Synth Met 153(1–3):61–64
32. Cardenas JR, De França MGO, De Vasconcelos EA, De Azevedo WM, Da Silva EF (2007) J Phys D Appl Phys 40:1068–1071
33. Laforgue A, Robitaille L (2008) Fabrication of poly-3-hexylthiophene/polyethylene oxide nanofibers using electrospinning. Synth Met 158(14):577–584
34. Kiebooms R, Menon R, Lee K (2001) Synthesis, electrical, and optical properties of conjugated polymers. In: Handbook of advanced electronic and photonic materials and devices. Academic Press, pp 1–102
35. Vernitskaya TV, Efimov ON (1997) Polypyrrole: a conducting polymer (synthesis, properties, and applications). Успехи химии 66(5):502–505
36. Kang ET, Neoh KG, Ong YK, Tan KL, Tan BTG (1991) Thermal stability and degradation of some chemically synthesized polypyrrole complexes. Thermochim Acta 181:57–70
37. Istakova OI, Konev DV, Glazkov AT, Medvedeva TO, Zolotukhina EV, Vorotyntsev MA (2019) Electrochemical synthesis of polypyrrole in powder form. J Solid-State Electrochem 23:251–258
38. Ozkazanc H, Zor S (2013) Electrochemical synthesis of polypyrrole (PPy) and PPy|metal composites on copper electrode and investigation of their anticorrosive properties. Prog Org Coat 76(4):720–728
39. Koinkar PM, Patil SS, Moriga T, More MA (2016) Electrochemical synthesis of conducting polypyrrole film on tin substrate: structural, chemical and field emission investigations. J Nano Res 36:44–50
40. Tang X, Raskin JP, Kryvutsa N, Hermans S, Slobodian O, Nazarov AN, Debliquy M (2020) An ammonia sensor composed of polypyrrole synthesized on reduced graphene oxide by electropolymerization. Sensors Actuators B Chem 305:127423
41. Chandler GK, Pletcher D (1986) The electrodeposition of metals onto polypyrrole films from aqueous solution. J Appl Electrochem 16(1):62–68
42. Yussuf A, Al-Saleh M, Al-Enezi S, Abraham G (2018) Synthesis and characterization of conductive polypyrrole: the influence of the oxidants and monomer on the electrical, thermal, and morphological properties. Int J Polym Sci:4191747
43. Shacklette LW, Chance RR, Ivory DM, Miller GG, Baughman RH (1980) Electrical and optical properties of highly conducting charge-transfer complexes of poly (p-phenylene). Synth Met 1(3):307–320
44. Ambrosch-Draxl C, Majewski JA, Vogl P, Leising G (1995) First-principles studies of the structural and optical properties of crystalline poly (para-phenylene). Phys Rev B 51(15):9668
45. Friedrich K, Burkhart T, Almajid AA, Haupert F (2010) Poly-para-phenylene-copolymer (PPP): a high-strength polymer with interesting mechanical and tribological properties. Int J Polym Mater 59(9):680–692
46. Ahlskog M, Reghu M, Heeger AJ (1997) The temperature dependence of the conductivity in the critical regime of the metal-insulator transition in conducting polymers. J Phys Condens Matter 9(20):4145
47. Burroughes JH, Bradley DD, Brown AR, Marks RN, Mackay K, Friend RH et al (1990) Light-emitting diodes based on conjugated polymers. Nature 347(6293):539–541
48. Peres LO, Varela H, Garcia JR, Fernandes MR, Torresi RM, Nart FC, Gruber J (2001) On the electrochemical polymerization of poly (p-phenylene vinylene) and poly (o-phenylene vinylene). Synth Met 118(1–3):65–70
49. Kaloni TP, Giesbrecht PK, Schreckenbach G, Freund MS (2017) Polythiophene: from fundamental perspectives to applications. Chem Mater 29(24):10248–10283

50. Dubal DP, Lee SH, Kim JG, Kim WB, Lokhande CD (2012) Porous polypyrrole clusters prepared by electropolymerization for a high-performance supercapacitor. J Mater Chem 22(7):3044–3052
51. Babudri F, Farinola GM, Naso F (2004) Synthesis of conjugated oligomers and polymers: the organometallic way. J Mater Chem 14(1):11–34
52. Kiari M, Berenguer R, Montilla F, Morallón E (2020) Preparation and characterization of montmorillonite/pedot-pss and diatomite/pedot-pss hybrid materials. Study of electrochemical properties in acid medium. J Compos Sci 4(2):51
53. Kim Y, Kim J, Lee H, Park C, Im S, Kim JH (2020) Synthesis of stretchable, environmentally stable, conducting polymer PEDOT using a modified acid template random copolymer. Macromol Chem Phys 221(5):1900465
54. Meng C, Gall OZ, Irazoqui PP (2013) A flexible super-capacitive solid-state power supply for miniature implantable medical devices. Biomed Microdevices 15:973–983
55. Roldán S, Barreda D, Granda M, Menéndez R, Santamaría R, Blanco C (2015) An approach to classification and capacitance expressions in electrochemical capacitors technology. Phys Chem Chem Phys 17(2):1084–1092
56. Zhao C, Jia X, Shu K, Yu C, Wallace GG, Wang C (2020) Conducting polymer composites for unconventional solid-state supercapacitors. J Mater Chem A 8(9):4677–4699
57. Grama A (2018) Laboring along: industrial workers and the making of postwar Romania, vol 4. Walter de Gruyter GmbH & Co KG
58. Huo J, Wang L, Yu H (2009) Polymeric nanocomposites for electromagnetic wave absorption. J Mater Sci 44:3917–3927
59. Heinze J, Frontana-Uribe BA, Ludwigs S (2010) Electrochemistry of conducting polymers persistent models and new concepts. Chem Rev 110(8):4724–4771
60. Armes SP, Aldissi M, Hawley M, Beery JG, Gottesfeld S (1991) Morphology and structure of conducting polymers. Langmuir 7(7):1447–1452
61. Bansal V, Sharma PK, Sharma N, Pal OP, Malviya R (2011) Applications of chitosan and chitosan derivatives in drug delivery. Adv Biol Res 5(1):28–37
62. Meng Y, Wang K, Zhang Y, Wei Z (2013) Hierarchical porous graphene/polyaniline composite film with superior rate performance for flexible supercapacitors. Adv Mater 25(48):6985–6990
63. Zhang W, Jin X, Chai H, Diao G, Piao Y (2018) 3D hybrids of interconnected porous carbon Nanosheets/vertically aligned polyaniline nanowires for high-performance supercapacitors. Adv Mater Interfaces 5(11):1800106
64. Wang R, Han M, Zhao Q, Ren Z, Guo X, Xu C et al (2017) Hydrothermal synthesis of nanostructured graphene/polyaniline composites as high-capacitance electrode materials for supercapacitors. Sci Rep 7(1):44562
65. Hür E, Varol GA, Arslan A (2013) The study of polythiophene, poly (3-methylthiophene) and poly (3, 4-ethylenedioxythiophene) on pencil graphite electrode as an electrode active material for supercapacitor applications. Synth Met 184:16–22
66. Huang Y, Bao S, Lu J (2020) Flower-like MnO2/polyaniline/hollow mesoporous silica as electrode for high-performance all-solid-state supercapacitors. J Alloys Compd 845:156192
67. Rantho MN, Madito MJ, Manyala N (2020) High-performance symmetric supercapacitor device based on carbonized iron-polyaniline/nickel graphene foam. J Alloys Compd 819:152993
68. Gupta A, Sardana S, Dalal J, Lather S, Maan AS, Tripathi R et al (2020) Nanostructured polyaniline/graphene/Fe2O3 composites hydrogel as a high-performance flexible supercapacitor electrode material. ACS Appl Energy Mater 3(7):6434–6446
69. Liu Y, Wang Y, Chen Y, Wang C, Guo L (2020) NiCo-MOF nanosheets wrapping polypyrrole nanotubes for high-performance supercapacitors. Appl Surf Sci 507:145089
70. Bhardwaj P, Singh S, Kharangarh PR, Grace AN (2020) Surfactant decorated polypyrrole-carbon materials composites electrodes for supercapacitor. Diam Relat Mater 108:107989
71. Zhang M, Wang X, Yang T, Zhang P, Wei X, Zhang L, Li H (2020) Polyaniline/graphene hybrid fibers as electrodes for flexible supercapacitors. Synth Met 268:116484

72. Zhang C, Xu S, Cai D, Cao J, Wang L, Han W (2020) Planar supercapacitor with high areal capacitance based on $Ti_3C_2$/Polypyrrole composite film. Electrochim Acta 330:135277
73. Li Y, Zhou M, Wang Y, Pan Q, Gong Q, Xia Z, Li Y (2019) Remarkably enhanced performances of novel polythiophene-grafting-graphene oxide composite via long alkoxy linkage for supercapacitor application. Carbon 147:519–531
74. Liu Q, Zhang S, Xu Y (2020) Two-step synthesis of CuS/C@ PANI nanocomposite as advanced electrode materials for supercapacitor applications. Nano 10(6):1034
75. Jia L, Guanghai C, Yiqun C, Jietao J, Xiao X, Qiang W et al (2023) Boosting the supercapacitance performance of mesostructured carbon nanocages by enlarging pore sizes via carbothermal reduction★. Acta Chim Sin 81(7):709
76. Reddy KR, Karthik KV, Prasad SB, Soni SK, Jeong HM, Raghu AV (2016) Enhanced photocatalytic activity of nanostructured titanium dioxide/polyaniline hybrid photocatalysts. Polyhedron 120:169–174
77. Deng X, Chen Y, Wen J, Xu Y, Zhu J, Bian Z (2020) Polyaniline-TiO2 composite photocatalysts for light-driven hexavalent chromium ions reduction. Sci Bull 65(2):105–112
78. Wang D, Wang Y, Li X, Luo Q, An J, Yue J (2008) Sunlight photocatalytic activity of polypyrrole–TiO2 nanocomposites prepared by 'in situ' method. Catal Commun 9(6):1162–1166
79. Mitra M, Ahamed ST, Ghosh A, Mondal A, Kargupta K, Ganguly S, Banerjee D (2019) Polyaniline/reduced graphene oxide composite-enhanced visible-light-driven photocatalytic activity for the degradation of organic dyes. ACS Omega 4(1):1623–1635
80. Gulce H, Eskizeybek V, Haspulat B, Sarı F, Gülce A, Avcı A (2013) Preparation of a new polyaniline/CdO nanocomposite and investigation of its photocatalytic activity: comparative study under UV light and natural sunlight irradiation. Ind Eng Chem Res 52(32):10924–10934
81. Namsheer K, Rout CS (2021) Conducting polymers: a comprehensive review on recent advances in synthesis, properties and applications. RSC Adv 11(10):5659–5697
82. Stejskal J, Sapurina I, Trchová M (2010) Polyaniline nanostructures and the role of aniline oligomers in their formation. Prog Polym Sci 35(12):1420–1481
83. Malti A, Edberg J, Granberg H, Khan ZU, Andreasen JW, Liu X et al (2015) An organic mixed ion–electron conductor for power electronics. Adv Sci 3(2):1500305
84. Stejskal J (2017) Conducting polymer hydrogels. Chem Pap 71:269–291
85. Liu X, Liu J, Lin S, Zhao X (2020) Hydrogel machines. Mater Today 36:102–124
86. Yuk H, Lu B, Zhao X (2019) Hydrogel bioelectronics. Chem Soc Rev 48(6):1642–1667
87. Zhao F, Bae J, Zhou X, Guo Y, Yu G (2018) Nanostructured functional hydrogels as an emerging platform for advanced energy technologies. Adv Mater 30(48):1801796
88. Fu F, Wang J, Zeng H, Yu J (2020) Functional conductive hydrogels for bioelectronics. ACS Mater Lett 2(10):1287–1301
89. Kim Y, Chortos A, Xu W, Liu Y, Oh JY, Son D et al (2018) A bioinspired flexible organic artificial afferent nerve. Science 360(6392):998–1003
90. Khodagholy D, Doublet T, Quilichini P, Gurfinkel M, Leleux P, Ghestem A et al (2013) In vivo recordings of brain activity using organic transistors. Nat Commun 4(1):1575
91. Chen S, Surendran A, Wu X, Lee SY, Stephen M, Leong WL (2020) Recent technological advances in fabrication and application of organic electrochemical transistors. Adv Mater Technol 5(12):2000523
92. Khau BV, Scholz AD, Reichmanis E (2020) Advances and opportunities in development of deformable organic electrochemical transistors. J Mater Chem C 8(43):15067–15078
93. Yang H, Ji S, Chaturvedi I, Xia H, Wang T, Chen G et al (2020) Adhesive biocomposite electrodes on sweaty skin for long-term continuous electrophysiological monitoring. ACS Mater Lett 2(5):478–484
94. Wang Q, Pan X, Lin C, Lin D, Ni Y, Chen L et al (2019) Biocompatible, self-wrinkled, antifreezing and stretchable hydrogel-based wearable sensor with PEDOT: sulfonated lignin as conductive materials. Chem Eng J 370:1039–1047
95. Li L, Pan L, Ma Z, Yan K, Cheng W, Shi Y, Yu G (2018) All inkjet-printed amperometric multiplexed biosensors based on nanostructured conductive hydrogel electrodes. Nano Lett 18(6):3322–3327

# Chapter 4
# Tuning of Conducting Polymers for Removal of Toxic Microorganisms

**Paramjit Singh, Gyan Singh, Jorge Yáñez, and Sapana Jadoun**

**Abstract** Conducting polymers are defined as a class of electrically conductive organic materials. Conducting polymer nanocomposites are the general term for the nanomaterials doped conducting polymers. These nanocomposites have served as significant substitutes for standard drug delivery in treating bacterial infections caused by harmful microorganisms. Primary causes for the rise of mortality and morbidity in the world is bacterial infection. In addition, the illness burden has increased due to the advent of multi-drug resistance in toxic microorganism. Considering advancements in antimicrobial therapy, developing novel antibiotics has not achieved significant results. Many nanocomposites based on sulfur and nitrogen-containing conducting polymers have recently been studied as antibacterial platforms against toxic microorganisms. The most frequently used metals in nanoparticles include copper, titanium oxide, zinc oxide, iron oxide, gold, silver, and palladium. Still, there is much to learn about the precise mechanism behind their antibacterial activity. Some research findings clearly show that polymer-based nanocomposites are acceptable matter, generating excellent choices for developing drugs to treat infectious diseases. The antibacterial activity of conducting polymer nanocomposites has also been used to investigate antibacterial activity. Not much research has been done on the application of polymer-based nanocomposites against bacterial infections. In this chapter, we will impart an outline of the latest explora-

P. Singh · J. Yáñez (✉)
Facultad de Ciencias Químicas, Departamento de Química Analítica e Inorgánica, Universidad de Concepción, Concepción, Chile
e-mail: jyanez@udec.cl

G. Singh
Nanomaterials and Polymer Energy Laboratory, Centre for Nanoscience and Nanotechnology, New Delhi, India

S. Jadoun (✉)
Sol-ARIS, Departamento de Química, Facultad de Ciencias, Universidad de Tarapacá, Arica, Chile
e-mail: sjadoun@academicos.uta.cl



S. Jadoun, N. Kumari Jangid (eds.), *Tuning of Conducting Polymers*, Springer Series on Polymer and Composite Materials,
https://doi.org/10.1007/978-3-032-00461-1_4

tion on the application of polymer-based nanocomposites to treat toxic microorganisms. Moreover, we will also urge an upcoming approach to overcome current constraints, and researchers exploring antimicrobials and nanomedicines will find this chapter useful.

**Keywords** Conducting polymers · Nanomaterials · Nanocomposites · Microorganisms · Toxicity

## 4.1 Introduction

A polymer is a macro-molecule made up of several replicating subunits. In electrical applications, they are used as insulators [1]. They are useful material for many everyday applications because of their high flexibility, high resistivity moldability, durability, thermal stability, and low-cost processing [2]. Conducting polymers (CPs), beforehand effectively constructed in 1970s, have attracted a lot of interest because they can conduct electricity while maintaining the practical qualities of traditional polymers, like flexibility. More than 25 conducting polymer systems exist now [3]. Giulio Natta's use of a Zeigler–Natta catalyst to polymerize acetylene gas in the 1960s marked the beginning of conducting polymer research [4]. With a conductivity of around $10^{-7}$ S/cm$^2$, the resulting polyacetylene was subsequently shown to be, at most, semi-conducting. Shirakawa, MacDiarmid, and Heeger took up this work 10 years later. They synthesized polyacetylene as a thin, glossy film and found that oxidizing it with iodine vapor may improve its conductivity by up to 103 S/cm [2, 5]. "Doping" refers to reducing or oxidizing a polymer to boost its conductivity [6].

On the other hand, such conducting materials polydispersity and non-biodegradability in a system continue to be a backbone for continuing in vivo toxicity. However, CPs such as polyaniline (PANI), polypyrrole (PPy), polythiophene (PTh), and their derivatives have numerous applications in the biomedical and clinical fields due to their attractive properties, which include tissue compatibility, ease of manipulation, and electrical regulation of chemical and physical characteristic using dopant surface functionalization [7].

Although the polymer was initially categorized as an insulator due to its predominant insulating capabilities, CPs also exhibit significant conductive capacity, comparable to those metals and semiconductors. In these polymers, the conduction of electricity is caused by the conjugation of the double bonds in the polymeric chains as well as the doping process. The conjugated $\pi$ electron and dopant ions enable electron mobility in CPs to function as charge carriers and delocalize into the conduction band, increasing the metallic behavior of the polymer.

In general, there are two well-known ways to dope CPs namely adding (n-doping) or reducing (p-doping) an electron employ a distinct kind of dopant, as shown in Fig. 4.1. CPs have a lot of promising qualities, there are still some dubious problems that come up, like process challenges, poor stability, and insolubility problems.

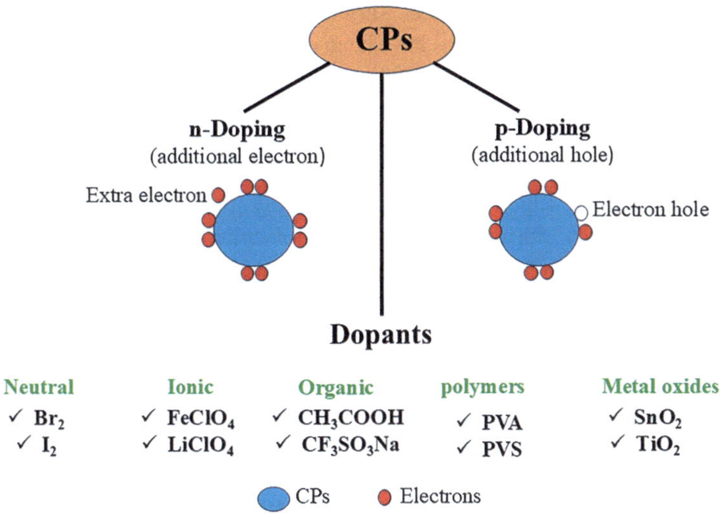


**Fig. 4.1** Various dopants used in n-doping and p-doping of CPs

However, to address those shortcomings in respect of merging, different changes can be made to the structure and functional group [8].

PTh, PPy, poly(3,4-ethylenedioxythiophene) (PEDOT), PANI, and polycarbazole (PC) are the polyheterocycles that are the focus of current research on CPs [9].

### *4.1.1 Polyaniline (PANI)*

One of the conducting polymer family members, polyaniline (PANI), stands out from the others due to its simple synthesis process and good environmental stability. The application of PANI in electronics and engineering is facilitated by the existence of conductivity validate emeraldine salt and many oxidation states [10].

For their widespread use in various fields, the chemical structure and physical morphology are composed in compliance with the requirements of the synthesis method, the kind of oxidants used, the quantity of dopants and solvents associated, and the temperature and time regimes [11, 12].

### *4.1.2 Polypyrrole (PPy)*

Among the most researched conducting polymers is PPy, which has been used in various applications such as brain probes [13], blood conduits, and tissue engineering biomaterials [14]. This polyheterocycle conducting polymer has also been used in medication delivery systems, microsurgical instruments, and corrosion prevention. Numerous beneficial qualities have been discovered [15]. PPy exhibits remarkable stability in both air and water, biocompatibility in vitro and in vivo, strong electrical properties under physiological settings, and a moderate water solubility [16]. Vast surface areas, different porosities, and vast volumes of PPy can be synthesized using a variety of solvents. Additionally, PPy makes it simple to incorporate bioactive compounds, which improves its eligibility for usage in biomedical applications [17]. However, during synthesis, PPy molecular structure renders it insoluble, brittle, complex, and non-thermoplastic, making additional polymer processing challenging. PPy has also been widely employed in manufacturing electrochemical biosensors and batteries due to these properties, notably its conductivity [18]. PPy effectiveness as an antibacterial agent has been relatively well studied, which is covered in more detail in the following sections [19].

### *4.1.3 Poly(3,4-Ethyldioxythiophene) (PEDOT)*

The bicyclic monomer 3,4-ethylenedioxythiophene is polymerized to yield PEDOT derived from polythiophene [15]. Compared to PTh, PEDOT has superior properties due to a dioxy alkylene linking group between its heterocyclic rings 3 and 4 positions [20]. Its band gap, reduction potential, and oxidation potential have all been found to decrease when this bridging group is present, improving conductivity and stability. Due to its in vivo biocompatibility, PEDOT has been used in biosensing and bioengineering applications, such as in heart muscle patches and nerve grafts [21].

Like PTh, research on PEDOT antibacterial properties has been limited compared to the field evaluations of PANI and PPy. However, PEDOT-based composites, like PEDOT: poly(styrene-sulfonate) [22], are currently being evaluated for their antibacterial efficacy.

### *4.1.4 Polythiophene (PTh)*

The two compounds share a structural similarity with the replacement of the S group in polythiophene for the NH group in PPy. Polythiophene has demonstrated good thermal and environmental stability, making it appropriate for various uses. These include the application of PTh in organic thin-film transistors and the manufacturing of electrochromic devices [23]. PEDOT is one of the several polythiophene derivatives that have been synthesized chemically or electrochemically to attain higher solubilities and capacitor behavior [24]. There is relatively little research on using PTh in its pure form for antimicrobial therapy. However, CPs have been added to nanocomposites that contain metals like copper and gold [25] to increase the antibacterial potential.

### *4.1.5 Polycarbazole (PC)*

A five-membered nitrogen-involving ring along two six-membered rings on either side makes up the electroactive conducting polymer of carbazole monomers known as PC [26]. Like PTh, polycarbazole has been used to manufacture batteries and electrochromic devices [24]. It has been demonstrated that, in comparison to the other CPs discussed, Pristine PC exhibits negligible antibacterial activity. However, its incorporation into nanocomposites, such as PC/$TiO_2$, has increased this activity in every instance, exhibiting antimicrobial activity on par with the well-known antibiotic Ciprofloxacin [27].

## 4.2 Antimicrobial Effect of Conducting Polymers

PPy and PANI are two of the many CPs being increasingly utilized for various purposes, including chemical sensors, photocatalysts, and antibacterial agents.

Another study that synthesized a co-polymer of chitosan functionalized PANI-PPy (CS/PANI/PPy) using a one pot method of oxidative polymerization evaluated the antibacterial potential of PPy when used in conjunction with PANI. To analyze the polymer different analytical spectroscopy were used like XRD, SEM, and FTIR spectroscopy, which was found to have spherical particles with a size of roughly 0.2 μm. Its antibacterial efficacy against *Enterobacter agglomerans* and *Escherichia coli* (*E. coli*) was evaluated by Kumar et al. The same zone of inhibition (ZOI) and percentage viability were obtained when both bacterial species were treated with CS/PANI/PPy instead of PANI/PPy. Nevertheless, improved antibacterial activity was made possible by the addition of PANI to PPy. In contrast, *E. coli* probability was decreased to 5% by treatment with the same PANI/PPy concentration. *E. coli* viability dropped to less than 1% when the PANI/PPy concentration was increased to 75 μg/mL [19]. According to this study, pristine PANI has more substantial antibacterial qualities against *E. coli* than pristine PPy.

The antibacterial effectiveness of PANI in its pure form has also been extensively researched. The development of both *B. cereus and E. coli* was completely inhibited by the same quantity of PANI (3500 μg/mL). The cytotoxicity of colloidal PANI to a human keratinocyte cell line (HaCat) was examined in this work. This demonstrated that PANI at lower doses (70 μg/mL) did not cause harm to human cells. Higher quantities (345 μg/mL) impacted the feasibility and anatomy of HaCat cells, even though they are typically considered biocompatible [28].

Polythiophene (PTh) has also been studied for its antibacterial qualities in addition to PANI and PPy. A suspension of *E. coli* was put on the films after spin-coating and oxidative polymerization, and the CFU that emerged were counted. A killing efficiency of 65% was reported following 20 min of irradiation. Additionally, it was shown that adding porphyrin to PTh film improved its antibacterial activity [29]. The chemical structure of a few example conducting polymers, which may offer a

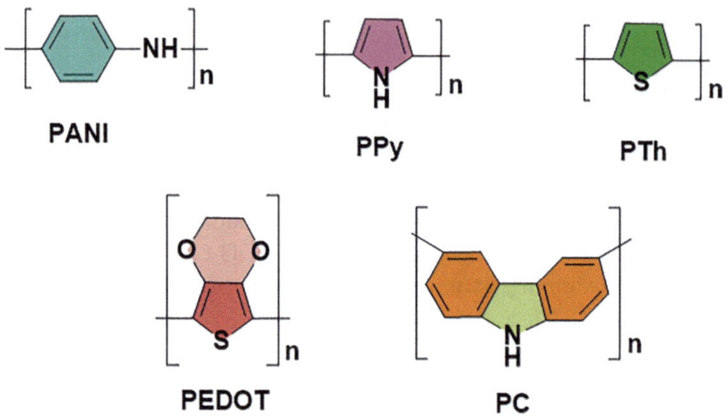


**Fig. 4.2** Some conducting polymers chemical structure

range of characteristics, including high conductivity, for diverse practical applications, is shown in Fig. 4.2.

## 4.3 Pathogenic Microorganism

A new period of various-drug resistance has evolved, making it extremely difficult to treat these infections and creating a pressing need for the creation of innovative treatments [30].

This is particularly correct for toxic microorganism, which may be one of the significant dangerous causes contributing to the emergence of infectious diseases and severe food poisoning [31]. Even though a lot of work has gone into creating different tests to report the splash levels of toxic microorganism effectively, generally can hardly identify a single species of bacteria and cannot react to many pathogenic bacteria at once [32]. In actuality, many unhealthy bacteria might coincide in a contaminated pattern, with varying and low quantities of these germs [33]. Because of this, it is challenging to effectively manage diseases by depending solely on detecting individual bacterium, which poses a risk to human health. As an alternative, an innovative sensing platform should be able to identify several pathogenic bacteria at once. This is important to regulate the pathogenic bacteria and protect human health precisely [34]. The ability to accurately identify and quantify harmful bacteria should be done while carefully considering the analytical presentation of response sensitivity, accuracy, strength, duration, value, and so on, which are two essential detection components [35]. These include *Acinetobacter baumanni (A. baumannii), P. aeruginosa, Klebsiella pneumoniae (K. pneumonia), Enterococcus faecium (E. fecium), S. aureus*, and *Enterobacter species* [36].

Immunocompetent and immunocompromised people can become infected with the virulent gram-positive bacterial species *S. aureus* [37]. While *S. aureus* is a typical element of the human body flora, it is an acquisitive microorganism and the cause of several illnesses, including bloodstream infections, infections of the skin and soft tissues, and infections of the skin that frequently need to be treated immediately [38].

Hypervirulent strains of *K. pneumonia*, resistant to many drugs, have recently become more isolated, restricting the available treatment options. This gram-negative bacterium is widely distributed in natural settings like soil and water, and in addition to pneumonia, it can also cause bloodstream and urinary tract infections [39]. Following their non-pathogenic existence on mucosal surfaces, *K. pneumonia* bacteria spread to other tissues, where they cause infections that result in death and morbidity [40]. Since *Klebsiella* species are called as a stock of antimicrobial resistance genes, they can transfer these genes to other gram-negative bacteria [41]. Furthermore, it was discovered that more than one-third of the isolated *K. pneumoniae* were resistant to at least one class of antibiotics. *K. pneumoniae* can develop β-lactamases, enzymes that break apart the β-lactam ring of β-lactam antibiotics and deactivate the molecules antibacterial effects, among other resistance mechanisms [42]. It has been observed that *A. baumannii* isolates resistant to multiple drugs can develop further antibiotic resistance quickly [43].

In general, the creation of novel antimicrobial drugs capable of circumventing these acquired resistance mechanisms is desperately needed. For this reason, novel approaches, such as the application of CPs and their nanocomposites, have been studied more and more.

## 4.4 Conducting Polymer Nanocomposites (CPNC)

Materials comprising two or more constituents, at least one nanoscale, are called "nanocomposites." Polymer nanocomposites have attracted much attention lately and are now the mainstay of contemporary nanotechnology. Progressive functional materials, or CPNCs, are created when nanoparticles (NPs) are dispersed inside polymer matrices or coated by a polymer, combining the beneficial qualities of its constituents [44]. CPs have been investigated for use in various fields, such as biomedical applications ranging from cell culture technologies to drug delivery systems. However, these materials have drawbacks, like limited processability, degradability, and biocompatibility. Combining CPs with NPs has also been shown to reduce the thermal and mechanical stability issues associated with CPs [45].

Additionally, the usage of NPs has received a lot of interest lately from various sectors, including electronics and medical. When combined with medications, they are helpful in the treatment of several illnesses, including bacterial infections. More precisely, because of their appealing catalytic, electrical, and antibacterial properties, metal and metal oxide NPs have been thoroughly studied as curing agents in the medical field. Figure 4.3 shows the different applications of CPNC. For instance, antimicrobials based on silver exhibit low volatility, high temperature stability, and

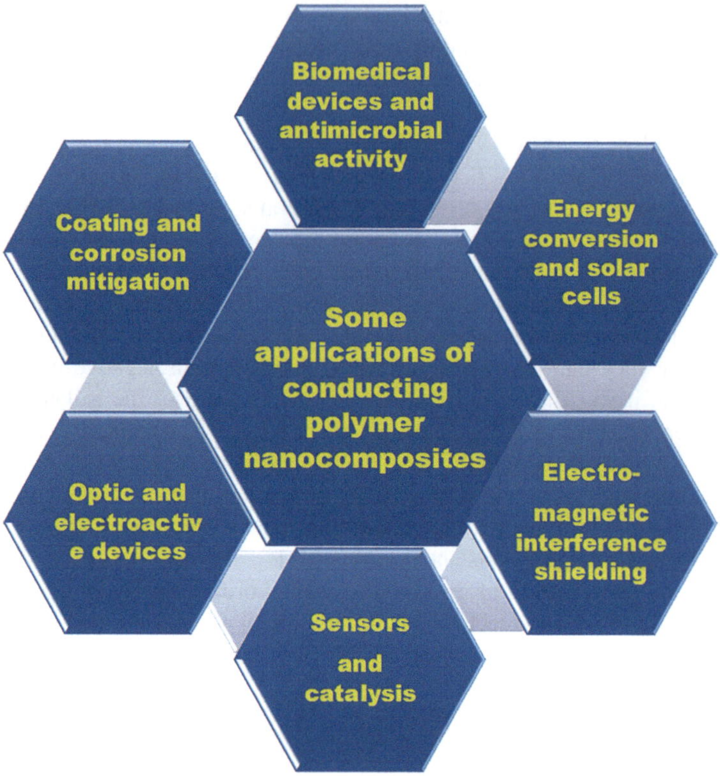


**Fig. 4.3** Various applications of conducting polymer nanocomposites

sustained biocidal activity. However, they are somewhat undesirable due to their restricted overall stability and aggregation tendency [46].

CPNC preparation has been accomplished in several ways. Techniques for synthesizing CPNC include in situ reduction of metal salts inside a polymer matrix [47] and in situ polymerization of a monomer in the sight of a metal [48]. Careful material selection and combination make optimizing the intended composite possible [49].

## 4.5 Antimicrobial Effect of Conducting Polymer Nanocomposites

### *4.5.1 Polyaniline Based Nanocomposites*

The antibacterial potential of polyaniline (PANI)-based nanocomposites has been the subject of increased research. Vishwakarma et al. prepared them [50]. Explore the antibacterial properties of the PANI-coated cobalt ferrite ($CoFe_2O_4$) [$CoFe_2O_4$/

PANI] nanocomposite against *S. aureus*, *B. subtillis* and *E. coli*, *P. aeruginosa* bacterial strains.

Images from a transmission electron microscope confirm that $CoFe_2O_4$ NPs having particle size (average) of 5 nm have been incorporated into the PANI matrix. Compared to $CoFe_2O_4$ NPs, the magnetization of $CoFe_2O_4$/PANI nanocomposites was 50% (54 emu/g to 27 emu/g). Using the Agar well diffusion method, the antimicrobial activity of PANI-coated $CoFe_2O_4$ nanocomposites was investigated. The findings demonstrated a distinct inhibitory zone, and $CoFe_2O_4$/PANI nanocomposites had strong antibacterial action against the growth of an alternative bacterial cell [50].

In a different work, Jebarshia et al. used a straightforward chemical oxidative polymerization process to successfully create co-doped polyaniline (PANI) composites with different organic acids, including DBSA, phthalic acid, oxalic acid, and Calotropis procera latex. From the remark analysis peaks of PANI in all the composites, the characterisation using XRD and FTIR analysis verified the synthesis of the emeraldine salt form of PANI. The synthesis of PANI in emeraldine form was also confirmed by the observed PL emission peaks and UV-visible absorption. For the PANI composites, the average ZOI value is determined to be between 21 and 24 nm. All PANI composites have a minimum inhibitory concentration (MIC) of 1 mg/L, while PANI (DBSA-Phthalic acid) and PANI (DBSA-Oxalic acid) composites have minimum bactericidal concentrations (MBC) of 60 mg/L and 80 mg/L, respectively. The produced co-doped PANI composites ability as an efficient antimicrobial agent is confirmed by a broader and more distinct ZOI [51]. Another example demonstrates the antibacterial activity of polyaniline-coated copper sulfide nanocomposites (CuS@PANI NCs), which efficiently produce alternative bactericides using CuS and PANI as reactants. According to the respective zone inhibition assays, the in vitro case-control studies show that the CuS@PANI NCs exhibit exceptional bactericidal performance against both Gram-positive (*S. aureus* and *S. pneumoniae*) and Gram-negative (*E. coli* and *P. aeruginosa*) bacteria.

Electron spin resonance (ESR) studies, together with radical trapping experiments, confirmed that superoxide radicals ($^{\cdot}O_2^{-}$) and photogenerated holes ($h^+$) are the main active species in the system. These primary species further drive the formation of secondary reactive oxygen species (ROS), including hydroxyl radicals ($^{\cdot}OH$) and hydrogen peroxide ($H_2O_2$), which play a crucial role in the antibacterial mechanism. The participation of $^{\cdot}OH$ and $H_2O_2$ enhances oxidative stress, leading to effective bacterial cell damage. The in vitro antibacterial performance of the synthesized polymer@inorganic composite materials was further supported by in silico molecular simulations, which revealed favorable interactions with essential bacterial proteins such as DNA gyrase (*E. coli*) and β-lactamase (*E. coli*). These combined experimental and computational insights strongly suggest that polymer@inorganic composites hold significant promise as effective antibacterial agents, capable of combating antibiotic resistance and reducing the risk of infectious diseases [52].

The ability of TiO-Ag to evade bacterial resistance pathways has also been evaluated. To obtain photocatalytic performance activated by visible light, the $TiO_2$ photocatalyst was synthesized in this work as both two- and three-component composites with Ag NPs and polyaniline (PANI) or poly(aniline-co-3-aminobenzoic acid) (fPANI). Both two- and three-component composite systems visible light photocatalytic capabilities were validated by the UV-DRS results and conductivity tests verified the presence of charge carriers (unpaired electrons) in the ESR spectra. The ESR peak intensity rises almost three times when comparing PANI/fPANI-TiO2-Ag composites to pure PANI/fPANI. $TiO_2$ may have contributed to the slight increase in conductivity of pure PANI and fPANI in three-component systems. In contrast, conductive Ag Nps caused the value to increase by about three times in the PANI/fPANI-Ag two-component system. When tested in opposite to three pathogens, *E. coli*, *S. aureus*, and a model virus strain called Phi X 174-the composites containing PANI had better antibacterial efficacy than PANI composites [53].

### 4.5.2 Polypyrrole-Based Nanocomposites

Copper nanoparticles (CuNPs) and polypyrrole (PPy) combined to create a new and potent antibacterial agent. A copper/polypyrrole (Cu/PPy) nanocomposite with spherical CuNPs, evenly distributed all over PPy matrix, was created using a simple in situ polymerization process in water. In this survey, it was partly anticipated that the different characteristics of CuNPs and PPy would work in concert to combat bacteria. While 10 ppm has a cidal value of 99.99% at the same time, 2 ppm of Cu/PPy with only 9.45 weight percent of Cu is already working nearly entirely cidal, with 99.17% bacterial growth suppression, after a 24 h contact with *S. aureus*. Cu/PPy concentrations that are both safe and environmentally acceptable suggest that it is a practical antibacterial agent for a broader range of applications [54].

The novel approach put forth by Cabezon et al. combines two well-known, simple, and affordable preparation methods: electropolymerization to coat these substrates with the composite conductive polymers of polypyrrole-silver nanoparticles (PPy-Ag NPs) to improve their corrosion resistance, biocompatibility, and enhanced antibacterial activity. The deposited PPy-based coatings have a structure like cauliflower and are firmly attached to the porous substrates. In contrast to conventional thick substrates made by powder metallurgy, the superior adherence of the conductive polymer is caused by Ti SH substrates macroporosity and rough inner pore surface. The bioactivity of the coatings was assessed through immersion testing, and hydroxyapatite was formed following a 90-day immersion in PBS. Following a 90-day immersion in PBS, a relocation of the polarization curves to more noble potentials and a diminish in the current density in both PPy and PPy-Ag NPs composite coatings demonstrated that the defensive nature of the coating was preserved. The Kirby-Bauer disk-diffusion method was used to evaluate the antibacterial activity against *S. aureus* (ATCC 25923). The possible application of porous Ti implants via SH method, which envisions replacing tiny, damaged bone structures, such as in malignancies, would be significantly aided by this property, which is linked to increased corrosion resistance and biocompatibility [55].

The enhanced Sol-gel technique created another example demonstrating the antibacterial activity of mixed metal oxide nanocomposites of $TiO_2$, ZnO, and CuO. PPy was polymerized via a chemical oxidative process. The synthesis of mixed metal/PPy nanocomposites was done in situ. Polymeric nanocomposites' electrical, magnetic, antibacterial, anticorrosive, and photocatalytic qualities were examined. The ability of polymeric nanocomposites to degrade methylene blue dye under visible light was investigated. The relationship between the dyes degradation period and changes in pH and COD was monitored. Over time, as dye photodegradation progresses, pH rises and COD decreases. The antibacterial capabilities of metal oxide and polymeric nanocomposites were examined against *B. subtilis* and *E. coli* and contrasted with six antibiotics (Hexa disk). Tauc plots examined the optical band gap of metal oxide nanocomposites, PPy, and PPy/mixed metal oxide nanocomposites. With a velocity constant of 0.0685 $min^{-1}$, the photocatalytic degradation of methylene blue exhibits pseudo first-order kinetics. The efficacy of the mixed metal/PPy nanocomposites in preventing corrosion is 80.1 ± 0.2%. When tested against *E. coli*, mixed metal nanocomposites demonstrated a maximal zone of inhibition of 12.5 ± 0.1 mm [56].

Ag NPs are added to PPy co-doped with tartrate (TA) and salicylate (SA) anions for biodegradable implant applications. This creates a multifunctional Ag NPs-PPy(TA + SA)/Zn bilayer coating on ZK60 magnesium alloy with strong antibacterial and protective qualities. The composition, structure, antibacterial, and protective qualities of Ag NPs-PPy(TA + SA)/Zn bilayer coating are examined about Ag NPs. The electrochemical growth conditions of PPy(TA + SA) are affected by the incorporation of Ag NPs, which may increase the flaws in PPy chains. It is comparable to PPy(TA + SA) regarding surface morphology and film thickness. Only the PPy surface layer (hundreds of nm) contains Ag species (Ag+/Ag0 = 0.1), with some adsorbed and agglomerated Ag NPs present. Ag NPs-PPy(TA + SA) degrades slowly in Hanks solution at 37 °C, and the Ag concentration gradually drops to zero. Compared to PPy(TA + SA) (54 ± 1.5 days), its protection period is greatly extended to 113 ± 2.6 days, indicating the critical role of Ag NPs. ZK60 exhibits typical local corrosion following the local annihilation of Ag NPs-PPy(TA + SA). Ag NPs-PPy(TA + SA) displayed outstanding cytocompatibility and antibacterial activities [57].

### *4.5.3 PEDOT-Based Nanocomposites*

Given their numerous applications, nanocomposites with specific characteristics are becoming more and more critical. The processes of pyrolysis and in situ chemical oxidative polymerization are used to create PEDOT and functionalized carbon NPs (f-CNPs). By altering the PEDOT concentration, f-CNPs-PEDOT nanocomposites were created, and their antibacterial, dielectric, and thermal stability were assessed. Synthesized f-CNPs-PEDOT nanocomposites were tested for their antibacterial properties against two distinct bacterial strains: Gram-negative *E. coli* and Gram-positive *S. aureus*. The results of this study may pave the way for the use of f-CNPs-PEDOT nanocomposites in antibacterial and physical scientific applications [58].

The development of bio-synthetic polymers with metallic biomaterials is evident in orthopedic applications. In orthopedics, zirconium is a developing implant material. Zirconia nanotubes (ZNTs) were created in this work using the electrochemical anodization process. ZNTs and bio-synthetic polymers together may enhance zirconium characteristics for load-bearing applications. Thus, a Chitosan-PEDOT (ChP) composite was applied to the produced ZNTs using the dip-coating technique. For 7 days, ChP-ZNTs underwent in vitro biomineralization in Hanks solution. Comparing ChP-ZNTs to ZNT and bare sample, the MTT assay indicates a higher viable percentage for both cell viability and proliferation rate. As a result, the produced ChP-ZNTs have improved biocompatibility and corrosion resistance for orthopedic applications.

This study used FTIR, XRD, and Raman analysis to characterize the PEDOT/FHA nanocomposite development. Surface spread plate techniques were used to evaluate the composites antibacterial properties against *B. subtilis* and *E. coli*. It was shown that uncoated TNZ had the least antibacterial action and permitted the most bacterial growth.

Human osteosarcoma cells (MG63) were used in in vitro tests to evaluate the nanocomposites' biocompatibility. The cells were planted on both coated and uncoated TNZ substrates, and the number of cells was then counted. The PEDOT coating enabled MG63 cells to multiply to a higher cell density uniformly over the coated substrate, unlike uncoated TNZ, where the proliferation rate was very low. After 5 days, it was shown that cells seeded on the PEDOT/FHA coating multiplied more quickly than cells grown on PEDOT-coated and uncoated TNZ [59].

In addition, the application of near infrared (NIR) induced photothermal treatment against harmful microorganisms has received a lot of interest lately. The innocuous and permeable nature of NIR irradiation in live tissues at wavelengths between 700 and 1100 nm is partly to blame for this [40]. Ko et al. evaluated the photothermal antibacterial activity of PEDOT: PSS/agarose (40 v/v%) using *E. coli* and *S. aureus*. Before being exposed to an NIR laser (808 nm) for t3 min, bacterial suspensions were disseminated onto PEDOT: PSS/agarose plates and pure agarose plates. Neither bacterial species' capacity to survive on pure agarose was impacted by NIR irradiation.

Remarkably, the survivability of *S. aureus* and *E. coli* on the nanocomposite hydrogel decreased to 37% and 44% viability, respectively, following 1 min of NIR irradiation. Two minutes of NIR irradiation resulted in complete bacterial mortality [22].

### 4.5.4 Polythiophene-Based Nanocomposites

Myrick et al. emphasize the production of graphene oxide (GO) modified polythiophene (PTh) as a visible light active photocatalyst and antibacterial agent. PTh was created using the interfacial polymerization technique with $FeCl_3$ as an initiator. It was then combined with graphene oxide (GO) in several weight ratios to create nanohybrids, which included 1 wt.%, 3 wt.%, and 5 wt.%. While UV-visible investigations validated the shifting of PTH electronic transitions upon alteration with GO, the polymerization of PTh and its modification with GO were confirmed by IR experiments. Using PTh and GO/PTh as catalysts, sonophotocatalysis was used to break down the dyes Congo Red (CR), Alizarin Yellow (AY), and Basic Blue (BB) for 60 min under visible light. Compared to PTh, the GO/PTh showed a greater degrading efficiency. The disk diffusion inhibitory assay was conducted in light and dark environments to assess its efficacy as an antibacterial agent against Bacillus subtilis. Regardless of light exposure, the results demonstrated the bactericidal capabilities of PTh and GO/PTh nanohybrids. The 3-GO/PTh nanohybrid had the lowest minimal inhibitory concentration (MIC), according to the results of the MIC experiment. 1-GO/PTh and 5-GO/PTh were next in line [60].

Poly-3-butyl thiophene (P3BT) was coated with $TiO_2$@ZnO NPs to create unique hybrid nanocomposites. Using the disc diffusion method, ZOI against *S. bacillus* and *E. coli* bacterial cells was used to assess the produced hybrid nanocomposites crystalline phase, microstructure, zeta potential, water contact angle, and antibacterial activity. The outcomes of the statistical analysis and the standard medication ciprofloxacin were compared with the in vitro antibacterial activity results. Zeta potential investigations and measurements of the water contact angle were used to assess the stability and surface dispersion of the NPs. As the concentration of $TiO_2$@ZnO NPs on the P3BT polymer rose, an irregular shape was seen, and phase analysis showed that the hexagonal crystal structure had a mixed composition.

$TiO_2$@ZnO concentrations on P3BT polymer increased due to a redshift in optical absorption spectra. As the concentration of $TiO_2$@ZnO NPs increased, a rise in hydrophobicity was noted, along with an increase in the Rrms value. When compared to $TiO_2$@ZnO NPs, it was discovered that P3BT:$TiO_2$@ZnO hybrid nanocomposites (10 wt.% and 15 wt.%) showed significantly better antibacterial activity against *S. bacillus* and *E. coli* bacterial cells. The P3BT:$TiO_2$@ZnO hybrid nanocomposites demonstrated a wide surface area, improved hydrophobicity, and excellent stability in basic solution. $TiO_2$@ZnO:P3BT hybrid nanocomposites are a superior option for antimicrobial therapy, according to the results of antibacterial property experiments that are compared with statistical results and show strong agreement with other results [61].

Through electrostatic interaction, Gunavathana et al. examined the synthetically integrated palladium (Pd) nanospheres embedded polythiophene (PT) nanocomposite for its antibacterial activity. In a Brunauer–Emmett–Teller (BET) analysis, the specific surface area and regular pore size measurements of the integrated PT-Pd nanocomposite were determined to be 5.870 $m^2\ g^{-1}$ and 3.5 nm, respectively, based on the $N_2$ adsorption/desorption isotherm. The inhibitory effects of the generated compounds were tested over a 24 h period using the Agar well diffusion technique. With no effective activity against fungal species, this investigation demonstrated the best response against gram-negative organisms compared to gram-positive organisms. However, to enhance the antibacterial activity, the electrochemical characteristics of the generated materials, such as conductivity and electroactivity, were examined using cyclic voltammetry and electrochemical impedance spectroscopy.

The PT-Pd based nanocomposite created in our current framework is a promising option for various biological and environmental applications, according to the observed results [62].

Using an oxidative chemical polymerization process, polythiophene (PTh) and graphene nanoplatelets (GNPs) nanocomposites with varying GNP ratios (10–50 wt.%) were created and used for pathogen control and photocatalytic breakdown of bromophenol blue (BPB). Energy dispersive X-ray analysis, UV-Vis spectroscopy, XRD, FTIR spectroscopy, and scanning electron microscopy (SEM) were used to analyzed the as-prepared nanomaterials. The PTh-GNPs(50) outperformed the pristine PTh (31.3%) by more than three times, achieving the maximum photocatalytic degradation efficiency for BPB (94.1%). The better charge-transport and longer recombination time of $e^-/h^+$ couples from PTh NPs to GNPs were credited with the increased photocatalytic activity of PTh-GNPs. Remarkably, PTh-GNPs(50) demonstrated a strong inhibitory effect against *E. coli* and *S. aureus*. These findings suggest that PTh-GNPs(50) nanocomposites could be used to remove BPB and prevent the development of bacteria [63].

### *4.5.5 Polycarbazole-Based Nanocomposites*

Bhat et al. synthesized a conducting polymer-based silver phosphate polypyrrole (Ag-P/PPy) core-shell nanocomposite using in situ chemical oxidative polymerization with antibacterial and antioxidant properties. In a 1,1-diphenyl-2-picrylhydrazyl (DPPH) radical scavenging test, the produced nanocomposite demonstrated 91.2% radical scavenging activity. Using the agar well diffusion technique, the antibacterial activity of the Ag-P/Ppy core-shell nanocomposite against gram-positive (*S. aureus*) and gram-negative (*P. aeruginosa*, *E. coli*, and *K. pneumoniae*) bacteria was evaluated. It inhibited *S. aureus*, *E. coli*, *K. pneumoniae*, and *P. aeruginosa* by 82.35%, 79.5%, 63.63%, and 76.7%, respectively, compared to the positive control. To exclude Cr (VI) ions from water, the Ag-P/PPy core–shell nanocomposite may be a powerful and adaptable adsorbent [64].

In a study by Iram et al., poly carbazole-based nanocomposites were assessed for their antimicrobial activity. Polycarbazole titanium dioxide nanocomposites (PC/$TiO_2$-8) were synthesized using in situ oxidative polymerization with ammonium persulfate (APS) as an oxidizing agent in $TiO_2$ NPs. The antibacterial ability of the nanocomposite was assessed in disk diffusion assays against the gram-positive bacteria Bacillus megaterium and Bacillus subtilis, as well as the gram-negative pathogens Proteus vulgaris and *P. aeruginosa* [25]. While PC/$TiO_2$ showed antimicrobial activity like that of Ciprofloxacin, an antibiotic used as a positive control in this work, Pristine PC provoked limited antibacterial activity on the species evaluated.

PC demonstrated a 10%–30% area of inhibition against the evaluated bacterial pathogens. By contrast, PC/$TiO_2$ produced an area of inhibition against the microorganisms of 90%–100%. Therefore, adding $TiO_2$ to the composite was thought to be responsible for the inhibitory effects. Additional research was done to examine DNA binding using biophysical methods. It was determined that, in comparison to PC, PC/$TiO_2$ had better DNA binding properties. Using molecular docking simulation, the interaction mechanism between PC/$TiO_2$ and DNA helices was identified in this study as "groove binding," a non-covalent contact mode [27].

## 4.6 Conclusion

The capacity of electrically CPs to conduct electricity while maintaining the characteristics of standard polymers and their appropriateness as matrices for the immobilization of NPs have recently drawn considerable attention in the field of nanotechnology. As a result, CPNC has been synthesized more frequently, and its antibacterial efficacy has been evaluated.

In their pure forms, CPs—primarily PANI and PPy—have been shown to have antimicrobial qualities. In contrast to their pristine counterparts, conducting polymer-based nanocomposites have generally been found to be more effective antibacterial agents. Furthermore, it has been observed that adding CPs to the CPNC enhances the antibacterial activity of the nanomaterial, presumably by stabilizing and immobilizing NPs and preventing their agglomeration.

The exact mechanism by which CPNC produces these antimicrobial properties is unknown. In these findings, the cytotoxicity and biocompatibility of CPNC have not been thoroughly examined. However, research indicates that the impacts on human cells are negligible. These characteristics combine to make CPNC a promising target for antibacterial agent research. These materials would be helpful in looking into their effects on other pathogens in treating numerous infectious diseases that currently represent severe hazards to society, provided that biocompatibility is further investigated.

**Acknowledgments** Authors are grateful for the support of the National Research and Development Agency of Chile (ANID) for FONDECYT Postdoctoral projects 3240619, 3250135, and ANID/FONDAP/1523A0006, Solar Energy Research Center (SERC).

## References

1. Patil AO (2000) Electrically conducting polymers. In: Applied polymer science: 21st century
2. Ramakrishnan S (2011) From a laboratory curiosity to the market place. Resonance 16:1254–1265
3. Yi N, Abidian MR (2016) Conducting polymers and their biomedical applications. In: Biosynthetic polymers for medical applications. Elsevier, pp 243–276. https://doi.org/10.1016/B978-1-78242-105-4.00010-9
4. Abdelnasir S, Anwar A, Anwar A (2021) Conducting polymer-based nanocomposites against pathogenic bacteria. In: Advances in hybrid conducting polymer technology, pp 271–310. https://doi.org/10.1007/978-3-030-62090-5_11

5. Singh P, Mirza AU, Mondal AH, Mukhopadhyay K, Nishat N (2022) Functionalization of PMMA/$TiO_2$ nanocomposites: synthesis, characterization and their antioxidant and antibacterial evaluation. J Appl Polym Sci 139
6. Ateh D, Navsaria H, Vadgama P (2006) Polypyrrole-based conducting polymers and interactions with biological tissues. J R Soc Interface 3:741–752
7. Lee J-W, Serna F, Nickels J, Schmidt CE (2006) Carboxylic acid-functionalized conductive polypyrrole as a bioactive platform for cell adhesion. Biomacromolecules 7:1692–1695
8. Kaur G, Adhikari R, Cass P, Bown M, Gunatillake P (2015) Electrically conductive polymers and composites for biomedical applications. RSC Adv 5:37553–37567
9. Guimard NK, Gomez N, Schmidt CE (2007) Conducting polymers in biomedical engineering. Prog Polym Sci 32:876–921
10. Aksu Z, Kutsal T (1991) A bioseparation process for removing lead(II) ions from waste water by using C. vulgaris. J Chem Technol Biotechnol 52:109–118
11. Boeva ZA, Sergeyev VG (2014) Polyaniline: synthesis, properties, and application. Polym Sci Ser C 56:144–153
12. Maráková N et al (2017) Antimicrobial activity and cytotoxicity of cotton fabric coated with conducting polymers, polyaniline or polypyrrole, and with deposited silver nanoparticles. Appl Surf Sci 396:169–176
13. Gomez N, Lee JY, Nickels JD, Schmidt CE (2007) Micropatterned polypyrrole: a combination of electrical and topographical characteristics for the stimulation of cells. Adv Funct Mater 17:1645–1653
14. Bettinger CJ, Bruggeman JP, Misra A, Borenstein JT, Langer R (2009) Biocompatibility of biodegradable semiconducting melanin films for nerve tissue engineering. Biomaterials 30:3050–3057
15. Balint R, Cassidy NJ, Cartmell SH (2014) Conductive polymers: towards a smart biomaterial for tissue engineering. Acta Biomater 10:2341–2353
16. Cetiner S, Kalaoglu F, Karakas H, Sarac AS (2010) Electrospun nanofibers of Polypyrrole-poly(acrylonitrile- co -vinyl acetate). Text Res J 80:1784–1792
17. Akkouch A, Shi G, Zhang Z, Rouabhia M (2010) Bioactivating electrically conducting polypyrrole with fibronectin and bovine serum albumin. J Biomed Mater Res Part A 92A:221–231
18. Ghasemi-Mobarakeh L et al (2011) Application of conductive polymers, scaffolds and electrical stimulation for nerve tissue engineering. J Tissue Eng Regen Med 5:e17–e35
19. Kumar R, Oves M, Almeelbi T, Al-Makishah NH, Barakat MA (2017) Hybrid chitosan/polyaniline-polypyrrole biomaterial for enhanced adsorption and antimicrobial activity. J Colloid Interface Sci 490:488–496
20. Lota K, Khomenko V, Frackowiak E (2004) Capacitance properties of poly(3,4-ethylenedioxythiophene)/carbon nanotubes composites. J Phys Chem Solids 65:295–301
21. Das TK, Prusty S (2012) Review on conducting polymers and their applications. Polym Plast Technol Eng 51:1487–1500
22. Ko Y et al (2019) Antibacterial poly (3,4-ethylenedioxythiophene):poly(styrene-sulfonate)/agarose nanocomposite hydrogels with thermo-processability and self-healing. Carbohydr Polym 203:26–34
23. Zou Y, Sang G, Wu W, Liu Y, Li Y (2009) A polythiophene derivative with octyloxyl triphenylamine-vinylene conjugated side chain: synthesis and its applications in field-effect transistor and polymer solar cell. Synth Met 159:182–187
24. Ates M, Karazehir T, Sezai Sarac A (2012) Conducting polymers and their applications. Curr Phys Chem 2:224–240
25. Ma G et al (2014) Cu-doped zinc oxide and its polythiophene composites: preparation and antibacterial properties. Chemosphere 100:146–151
26. Sangwan W, Paradee N, Sirivat A (2016) Polycarbazole by chemical oxidative interfacial polymerization: morphology and electrical conductivity based on synthesis conditions. Polym Int 65:1232–1237

27. Iram N e et al (2015) Interaction mode of polycarbazole–titanium dioxide nanocomposite with DNA: molecular docking simulation and in-vitro antimicrobial study. J Photochem Photobiol B Biol 153:20–32
28. Kucekova Z et al (2014) Colloidal polyaniline dispersions: antibacterial activity, cytotoxicity and neutrophil oxidative burst. Colloids Surf B Biointerfaces 116:411–417
29. Liu L, Chen J, Wang S (2013) Flexible antibacterial film deposited with polythiophene–porphyrin composite. Adv Healthc Mater 2:1582–1585
30. Mulani MS, Kamble EE, Kumkar SN, Tawre MS, Pardesi KR (2019) Emerging strategies to combat ESKAPE pathogens in the era of antimicrobial resistance: a review. Front Microbiol 10
31. Hameed S, Xie L, Ying Y (2018) Conventional and emerging detection techniques for pathogenic bacteria in food science: a review. Trends Food Sci Technol 81:61–73
32. Cui F, Ye Y, Ping J, Sun X (2020) Carbon dots: current advances in pathogenic bacteria monitoring and prospect applications. Biosens Bioelectron 156:112085
33. Song C et al (2016) Development of a lateral flow colloidal gold immunoassay strip for the simultaneous detection of Shigella boydii and Escherichia coli O157:H7 in bread, milk and jelly samples. Food Control 59:345–351
34. Lv W et al (2020) Recent advances in electrochemiluminescence-based simultaneous detection of multiple targets. TrAC Trends Anal Chem 123:115767
35. Cesewski E, Johnson BN (2020) Electrochemical biosensors for pathogen detection. Biosens Bioelectron 159:112214
36. Rice LB (2008) Federal funding for the study of antimicrobial resistance in nosocomial pathogens: no ESKAPE. J Infect Dis 197:1079–1081
37. Jansen KU, Girgenti DQ, Scully IL, Anderson AS (2013) Vaccine review: "Staphyloccocus aureus vaccines: problems and prospects". Vaccine 31:2723–2730
38. Tong SYC, Davis JS, Eichenberger E, Holland TL, Fowler VG (2015) Staphylococcus aureus infections: epidemiology, pathophysiology, clinical manifestations, and management. Clin Microbiol Rev 28:603–661
39. Bengoechea JA, Sa Pessoa J (2019) Klebsiella pneumoniae infection biology: living to counteract host defences. FEMS Microbiol Rev 43:123–144
40. Paczosa MK, Mecsas J (2016) Klebsiella pneumoniae: going on the offense with a strong defense. Microbiol Mol Biol Rev 80:629–661
41. Lam MMC et al (2018) Genetic diversity, mobilisation and spread of the yersiniabactin-encoding mobile element ICEKp in Klebsiella pneumoniae populations. Microb Genom 4
42. Holt KE et al (2015) Genomic analysis of diversity, population structure, virulence, and antimicrobial resistance in Klebsiella pneumoniae, an urgent threat to public health. Proc Natl Acad Sci 112
43. Lee C-R et al (2017) Biology of Acinetobacter baumannii: pathogenesis, antibiotic resistance mechanisms, and prospective treatment options. Front Cell Infect Microbiol 7
44. Naseri M, Fotouhi L, Ehsani A (2018) Recent progress in the development of conducting polymer-based nanocomposites for electrochemical biosensors applications: a mini-review. Chem Rec 18:599–618
45. Bhat NV, Gadre AP, Bambole VA (2001) Structural, mechanical, and electrical properties of electropolymerized polypyrrole composite films. J Appl Polym Sci 80:2511–2517
46. Tamayo LA et al (2014) Release of silver and copper nanoparticles from polyethylene nanocomposites and their penetration into Listeria monocytogenes. Mater Sci Eng C 40:24–31
47. Upadhyay J, Kumar A, Gogoi B, Buragohain AK (2015) Antibacterial and hemolysis activity of polypyrrole nanotubes decorated with silver nanoparticles by an in-situ reduction process. Mater Sci Eng C 54:8–13
48. Khan AAP, Khan A, Rahman MM, Asiri AM, Oves M (2017) Sensor development of 1,2 Dichlorobenzene based on polypyrole/Cu-doped ZnO (PPY/CZO) nanocomposite embedded silver electrode and their antimicrobial studies. Int J Biol Macromol 98:256–267

49. Dallas P, Sharma VK, Zboril R (2011) Silver polymeric nanocomposites as advanced antimicrobial agents: classification, synthetic paths, applications, and perspectives. Adv Colloid Interf Sci 166:119–135
50. Vishwakarma AK, Sen Yadav B, Singh J, Sharma S, Kumar N (2022) Antibacterial activity of PANI coated CoFe2O4 nanocomposite for gram-positive and gram-negative bacterial strains. Mater Today Commun 31:103229
51. Jesbin Jebarshia J, Preethi T, Ashokan S, Geetha N, Senthil K (2024) Co-doped polyaniline composites for biological application: morphological, functional, optical and antibacterial activities. Inorg Chem Commun 161:111971
52. Shah BA et al (2023) Photoresponsive CuS@polyaniline nanocomposites: an excellent synthetic bactericide against several multidrug-resistant pathogenic strains. Inorg Chem Front 10:6339–6356
53. Jose A, Yadav P, Svirskis D, Swift S, Gizdavic-Nikolaidis MR (2024) Antimicrobial photocatalytic PANI based-composites for biomedical applications. Synth Met 303:117562
54. Mašojević D et al (2023) Evaluation of the antibacterial effectiveness of novel copper/polypyrrole nanocomposite. Mater Lett 338:134051
55. García-Cabezón C, Godinho V, Pérez-González C, Torres Y, Martín-Pedrosa F (2023) Electropolymerized polypyrrole silver nanocomposite coatings on porous Ti substrates with enhanced corrosion and antibacterial behavior for biomedical applications. Mater Today Chem 29:101433
56. Yadav A et al (2023) Mixed metal oxide decorated polypyrrole nanocomposites for multifunctional applications. Inorg Chem Commun 158:111701
57. Luo W, Qi K, Qiu Y, Guo X (2024) Silver nanoparticles composited co-dopped polypyrrole/zinc coating for enhancing anti-corrosion and antibacterial properties of ZK60 magnesium alloy in Hank's solution. Appl Surf Sci 669:160521
58. Taj M et al (2024) Functional carbon nanoparticles modified poly(3,4-ethylenedioxythiophene) nanocomposites with enhanced dielectric and antibacterial properties. Hybrid Adv 6:100204
59. Madhan Kumar A et al (2019) PEDOT/FHA nanocomposite coatings on newly developed Ti-Nb-Zr implants: biocompatibility and surface protection against corrosion and bacterial infections. Mater Sci Eng C 98:482–495
60. Myrick S et al (2024) Graphene oxide-modified polythiophene nanohybrids: antibacterial properties and sonophotocatalytic degradation of dyes under visible-light irradiation. Int J Polym Mater Polym Biomater 73:1451–1460
61. Adimule V, Kendrekar P, Yallur BC, Sawant AD (2023) Synthesis, characterization, and antibacterial activity of novel Poly-3-butyl thiophene embedded TiO2@ZnO hybrid nanocomposites. BioNanoScience 13:2049–2065
62. Dheepthi GunaVathana S et al (2022) Palladium nanospheres incorporated polythiophene nanocomposite: investigation of potency promising antimicrobial efficacy. Inorg Chem Commun 144:109867
63. Noreen H et al (2021) Synthesis of graphene nanoplatelets/polythiophene nanocomposites with enhanced photocatalytic degradation of bromophenol blue and antibacterial properties. Mater Res Bull 142:111435
64. Bhat AH, Chisti H-T-N (2023) Fabrication of versatile Ag-P/Ppy core-shell nanocomposite: adsorption efficacy for chromium (VI) removal and assessment of antioxidant and antibacterial activities. J Environ Chem Eng 11:110664

# Chapter 5
# Tailored Conducting Polymers for Photocatalytic Treatment of Water

**Suman Chirra, Radhamanohar Aepuru, and Manda Sathish**

**Abstract** Industrial waste, agricultural residues, and urban waste have become significant issues concerning water pollution and its permissible remediation solutions. Organosynthetically-accessible conducting polymers (CPs) like polyaniline (PANI), polypyrrole (PPy), and poly(3,4-ethylenedioxythiophene) (PEDOT) have shown immense potential in photocatalytic water treatment due to their optical activity, electrical properties, and physicochemical characteristics. These polymers are solution-processable, biocompatible, and can be effectively incorporated into hybrid semiconductor nanocomposites with inorganic materials, including titanium dioxide ($TiO_2$), to improve visible light absorption and reduce the recombination of photogenerated charge carriers. This chapter discusses recent developments in polymer-based photocatalysts, synthesis strategies, optoelectronic properties, and pollutant degradation mechanisms. These materials' applications, challenges, and future directions in global environmental remediation are also critically assessed.

**Keywords** Conducting polymers · Photocatalysts · Polyaniline (PANI) · Heterojunction · Green chemistry

---

S. Chirra (✉)
Department of Physical Chemistry, Faculty of Chemical Technology, University of Pardubice, Pardubice, Czech Republic

R. Aepuru
Departamento de Ingeniería Mecánica, Facultad de Ciencias Físicas y Matemáticas, Universidad de Chile, Santiago, Chile
e-mail: venkata.aepuru@uchile.cl

M. Sathish (✉)
Centro de Investigación de Estudios Avanzados del Maule (CIEAM), Vicerrectoría de Investigación y Postgrado, Universidad Católica del Maule, Talca, Chile
e-mail: smanda@ucm.cl



S. Jadoun, N. Kumari Jangid (eds.), *Tuning of Conducting Polymers*, Springer Series on Polymer and Composite Materials,
https://doi.org/10.1007/978-3-032-00461-1_5

## 5.1 Introduction

Water contamination remains one of the most critical challenges facing humanity today. Industrialization, urban sprawl, and the expansion of agricultural activities have led to the widespread discharge of pollutants into water bodies, adversely impacting aquatic ecosystems and public health. Pollutants such as organic dyes, heavy metals, pharmaceutical residues, and other emerging contaminants have proven resistant to traditional water treatment technologies, necessitating the development of advanced and sustainable remediation strategies [1].

The significance of clean water cannot be overstated. It is fundamental to human health, agriculture, industry, and environmental sustainability. However, with increasing industrial effluents, urbanization, and agricultural runoff, water resources worldwide are under unprecedented stress. Contaminants such as pharmaceuticals, persistent organic pollutants (POPs), and heavy metals often exceed safe levels in water bodies, posing severe health and environmental risks. Conventional treatment technologies, including filtration, coagulation, and adsorption, are usually inadequate for addressing complex pollution challenges. Moreover, these technologies generate secondary waste, highlighting the urgent need for innovative, green, and efficient solutions [2].

Photocatalysis has emerged as a promising approach to address these issues. It leverages light energy, often from solar sources, to activate catalytic processes that degrade pollutants into benign substances like carbon dioxide and water. This process aligns seamlessly with global goals for environmental sustainability by being both energy-efficient and eco-friendly. However, achieving high photocatalytic efficiency requires materials with specific properties, such as broad-spectrum light absorption, efficient charge separation, and robust chemical stability under operational conditions [3].

Conducting polymers (CPs) such as polyaniline (PANI), polypyrrole (PPy), and poly(3,4-ethylenedioxythiophene) (PEDOT) have recently gained attention as versatile photocatalytic materials. Unlike conventional photocatalysts, CPs offer unique optoelectronic properties, including intrinsic conductivity, tunable bandgaps, and strong absorption in the visible light spectrum. These features, combined with their chemical stability and environmental compatibility, make CPs ideal for photocatalytic applications [4–13].

### *5.1.1 The Need for Advanced Water Treatment*

Traditional water treatment methods have limitations in addressing complex pollutants. For example, coagulation and filtration processes effectively remove particulate but fail to degrade dissolved organic contaminants. Though efficient for specific pollutants, adsorption techniques suffer from issues of regeneration and disposal of spent adsorbents. Biological treatments, while eco-friendly, are often inadequate for non-biodegradable substances such as pharmaceuticals and pesticides. These limitations necessitate exploring advanced oxidation processes (AOPs), with photocatalysis being a leading contender [14–28].

### 5.1.2 Principles of Photocatalysis

Photocatalysis involves activating materials using light energy, typically in the ultraviolet (UV) or visible spectrum. This process relies on generating electron–hole pairs upon light absorption, which initiate redox reactions that degrade pollutants. Factors influencing photocatalytic efficiency include light absorption, charge separation, and chemical stability. Conventional photocatalysts, such as titanium dioxide ($TiO_2$), are effective under UV light but are limited under visible light due to their wide bandgap. Consequently, alternative materials, such as conducting polymers, have emerged as promising candidates due to their narrower bandgaps and enhanced visible light absorption [29].

### 5.1.3 Unique Properties of Conducting Polymers

Conducting polymers exhibit conjugated π-electron systems, enabling electron delocalization and efficient charge transport. These properties enhance solar energy utilization, improve charge carrier dynamics, and allow for chemical versatility through doping and functionalization, optimizing their photocatalytic performance for specific applications [30].

### 5.1.4 Integration of Conducting Polymers with Inorganic Semiconductors

Hybrid systems combining conducting polymers with inorganic semiconductors demonstrate improved photocatalytic efficiency. Examples include:

- Polyaniline (PANI) and $TiO_2$ hybrids for extended light absorption and dye degradation.
- PEDOT and ZnO composites for enhanced charge carrier mobility and reduced photodegradation.
- g-$C_3N_4$ and conducting polymer hybrids leverage visible light activity and efficient charge transfer.

## 5.2 Fundamentals of Conducting Polymers and Their Role in Photocatalysis

Conducting polymers (CPs) represent a groundbreaking class of materials that have reshaped the landscape of photocatalysis. Their remarkable combination of optoelectronic properties, adaptability in structure, and eco-friendly characteristics has established them as pivotal components in advanced material science. These polymers leverage their conjugated π-electron systems for charge mobility, making them exceptionally suited for energy and environmental applications. By integrating electrical conductivity with strong light absorption in the visible spectrum, conducting polymers have become indispensable in driving innovation in photocatalytic processes.

Conducting polymers are characterized by their conjugated backbone, where alternating single and double bonds allow the delocalization of π-electrons. This structure supports intrinsic electrical conductivity and facilitates interactions with light and charged species. Polymers such as polyaniline (PANI), polypyrrole (PPy), and poly(3,4-ethylenedioxythiophene) (PEDOT) exhibit diverse functionalities. These materials are chemically versatile, allowing the incorporation of functional groups to tailor their properties for specific applications, such as solar energy harvesting and environmental remediation.

The unique properties of conducting polymers make them highly effective in photocatalysis. Their electrical conductivity can be modulated through doping with electron donors or acceptors, ranging from semiconducting to metallic behavior. Their strong absorption in the visible light spectrum aligns with the energy distribution of solar radiation, enabling efficient utilization of solar energy. Furthermore, conducting polymers exhibit chemical stability, resisting oxidative and reductive degradation under operational conditions. They also offer structural flexibility, allowing the formation of films, coatings, and nanostructures through solution-based processing techniques.

Doping plays a critical role in enhancing the functionality of conducting polymers. The polymers can achieve either p-type or n-type behavior by introducing electron donors or acceptors. P-type doping, achieved by introducing electron acceptors, increases hole carrier density, while n-type doping, facilitated by electron donors, enhances electron carrier mobility. These modifications significantly improve charge transport, making conducting polymers more efficient for photocatalytic applications.

Conducting polymers are integral to a broad spectrum of applications, ranging from organic electronics to renewable energy systems. Their potential in photocatalysis is particularly significant, addressing challenges in pollutant degradation, energy conversion, and environmental sustainability. By leveraging their intrinsic properties and exploring innovative synthesis techniques, conducting polymers holds the potential to revolutionize photocatalytic technologies, contributing to a more sustainable and energy-efficient future [31] (Table 5.1).

**Table 5.1** Comparison of physicochemical properties of common conducting polymers used in photocatalysis [32]

| Conducting polymer | Bandgap (eV) | Conductivity (S/cm) | Chemical structure |
|---|---|---|---|
| PANI | 3.2 | $30–2 \times 10^2$ | 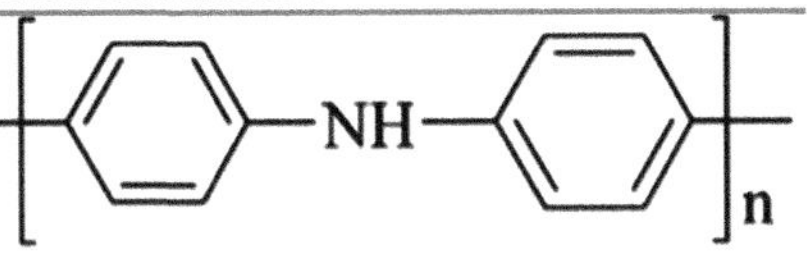 Polyaniline |
| PPy | 3.1 | $10^2–7.5 \times 10^3$ | 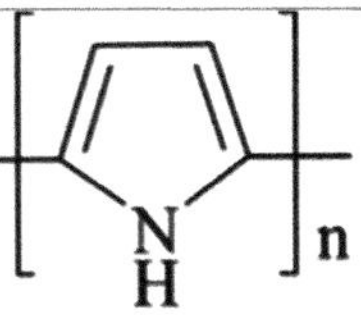 Polypyrrol |
| PTh | 2.1 | $10^2–1.7 \times 10^5$ | 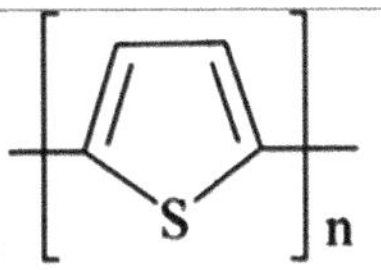 Polythiophene |
| PPP | 2.8 | $10^{-1}–10^3$ | 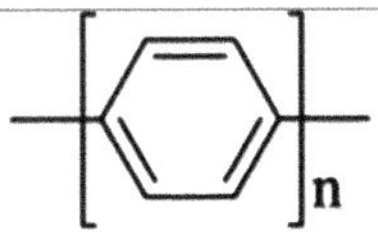 poly(p-phenylene) |
| PA | 1.5 | Trans: $4.4 \times 10^5$,<br>Cis: $1.7 \times 10^{-9}$ | 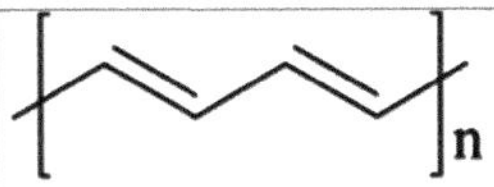 Polyacetylyne |
| PPV | 2.5 | $10^{-13}$ | 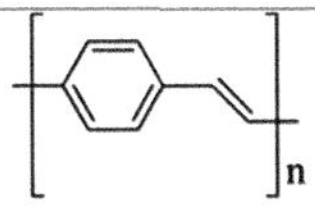 poly(p-phynylene vinylene) |

## 5.2.1 *Mechanisms of Photocatalysis in Conducting Polymers*

The photocatalytic efficiency of conducting polymers is rooted in their ability to absorb light, generate and separate charge carriers, and drive redox reactions for pollutant degradation. These mechanisms directly result from their unique electronic structure and interaction with light. Conducting polymers harness visible light effectively, promoting electrons from the valence band to the conduction band, where they participate in photocatalytic processes (Fig. 5.1). The π-conjugated electron system inherent to these polymers ensures efficient charge separation by minimizing recombination losses, a standard limitation in conventional photocatalysts [33].

Photogenerated electron–hole pairs in conducting polymers interact with water and oxygen molecules in the environment, producing reactive oxygen species (ROS) such as hydroxyl radicals (·OH) and superoxide anions ($O_2^-$·). These ROS are potent oxidizing agents, capable of breaking down complex organic pollutants into less harmful or inert compounds. This makes conducting polymers particularly suitable for applications in environmental remediation, such as water purification and air quality improvement.

The conjugated structure of conducting polymers supports rapid charge transport, a critical factor for maintaining photocatalytic efficiency. By facilitating the movement of electrons and holes through the material, this structure ensures sustained activity under light exposure. The alignment of energy bands in conducting polymers with the redox potential of the target pollutants is another crucial aspect of their photocatalytic performance. Proper energy band alignment allows for the efficient transfer of photogenerated charges to the pollutant molecules, enhancing the overall reaction efficiency.

Conducting polymers stand out in their ability to integrate light absorption, charge dynamics, and redox chemistry within a single material. These mechanisms not only define their effectiveness in photocatalysis but also pave the way for their use in advanced applications, including solar energy conversion, hydrogen production, and sustainable environmental management. By optimizing these

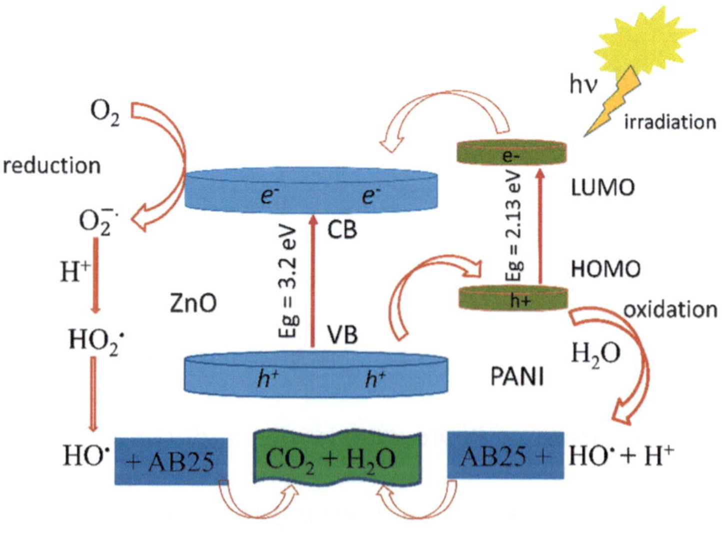


**Fig. 5.1** Schematic of conducting polymer-semiconductor hybrid photocatalyst for water treatment under visible light [34]

mechanisms through material design and doping strategies, researchers can further enhance the potential of conducting polymers as next generation photocatalysts [35].

### 5.2.2 Applications of Conducting Polymers in Photocatalysis

Conducting polymers have demonstrated significant potential in a wide range of photocatalytic applications, serving as both standalone photocatalysts and integral components of hybrid systems. Their strong light absorption, redox capabilities, and adaptability make them indispensable in addressing critical environmental and energy challenges.

As standalone photocatalysts, conducting polymers such as polyaniline (PANI) and polypyrrole (PPy) have been effectively utilized in degrading organic pollutants like dyes and pharmaceuticals. Their ability to harness visible light and generate reactive oxygen species enables efficient breakdown of complex organic compounds. Additionally, conducting polymers play a pivotal role in heavy metal reduction, converting highly toxic forms such as Cr(VI) into less harmful forms like Cr(III). For instance, PANI has demonstrated exceptional performance in degrading methylene blue, achieving a removal efficiency of 92% under visible light irradiation, highlighting its practical utility in organic pollutant remediation.

In hybrid systems, conducting polymers are combined with inorganic semiconductors like titanium dioxide ($TiO_2$), zinc oxide (ZnO), and graphitic carbon nitride (g-$C_3N_4$) to overcome the limitations of conventional photocatalysts. These hybrid systems extend the light absorption range into the visible spectrum, maximizing solar energy utilization. They also enhance charge dynamics by minimizing electron–hole recombination, thereby improving the overall photocatalytic efficiency. Furthermore, conducting polymers act as protective layers for inorganic semiconductors, preventing photodegradation and increasing the durability of the system under prolonged exposure.

By leveraging the unique properties of conducting polymers and their synergistic effects in hybrid configurations, these materials offer transformative solutions in photocatalysis. Their applications span from water purification and air quality improvement to advanced energy conversion processes, demonstrating their pivotal role in addressing global sustainability challenges. Continued research and development in this domain promise to unlock even greater potential for conducting polymers in photocatalytic applications [36].

### 5.2.3 Advantages of Conducting Polymers in Photocatalysis

Conducting polymers offer a range of benefits that distinguish them from traditional photocatalytic materials, making them an attractive choice for sustainable and innovative applications. These advantages are rooted in their eco-friendly nature, chemical versatility, and adaptability to various functional and environmental demands.

Conducting polymers are inherently eco-friendly as they are derived from organic monomers and often synthesized through green chemistry approaches. This reduces the environmental footprint associated with their production and use, aligning them with sustainability goals. Their customizable properties allow for application-specific tuning through functionalization, enabling the design of polymers optimized for particular photocatalytic tasks, such as pollutant degradation or energy conversion.

These materials exhibit versatile integration capabilities, making them compatible with a wide range of substrates and hybrid systems. This adaptability facilitates their incorporation into advanced composite materials that combine the strengths of conducting polymers with the stability and efficiency of inorganic semiconductors.

From an environmental perspective, conducting polymers significantly reduce reliance on rare earth materials, which are typically required in conventional photocatalysts. This not only alleviates pressure on finite natural resources but also reduces the environmental impact of mining and extraction processes. Additionally, the processing of conducting polymers requires lower energy inputs, further enhancing their environmental benefits.

The combination of these advantages positions conducting polymers as a leading class of materials in photocatalysis. Their ability to address environmental and energy challenges while minimizing ecological impact underscores their role in the development of sustainable technologies for a cleaner and greener future [37].

## 5.3 Synthesis and Structural Modifications of Conducting Polymers

The synthesis and structural modification of conducting polymers (CPs) play pivotal roles in determining their properties and suitability for photocatalytic applications. This section explores various synthesis techniques, structural tailoring methods, and the impact of these approaches on their functionality [38].

### *5.3.1 Synthetic Techniques and Methods of Conducting Polymers*

Conducting polymers are synthesized using diverse methodologies tailored to achieve specific morphologies, electronic properties, and application requirements. These techniques include chemical, electrochemical, interfacial, and template-assisted polymerizations (Fig. 5.2).

The synthesis of conducting polymers is a critical aspect of their development, influencing their structural, optical, and electronic properties. Several established methods enable the production of these materials, each with unique advantages and

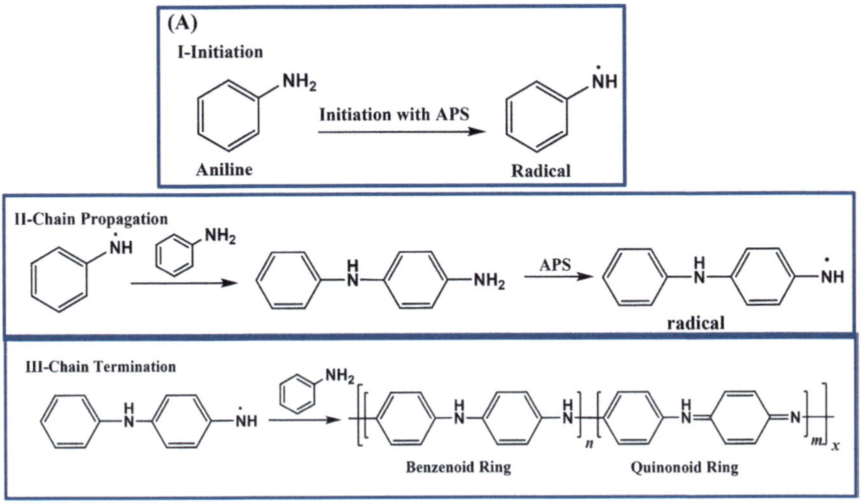


**Fig. 5.2** Three chain reaction steps involved during the chemical oxidative synthesis of CPs PANI [39]

limitations. These methods include chemical oxidative polymerization, electrochemical polymerization, interfacial polymerization, and template-assisted polymerization.

#### 5.3.1.1 Chemical Oxidative Polymerization

Chemical oxidative polymerization is widely used for its simplicity, scalability, and capability to produce high-yield, reproducible polymers. Monomers such as aniline, pyrrole, or thiophene undergo polymerization in the presence of oxidizing agents like ammonium persulfate or ferric chloride. This method allows for the production of bulk polymers with flexible reaction conditions for tailoring properties. However, post-synthesis purification is necessary, and there is limited control over molecular weight and morphology. The mechanism involves the oxidation of the monomer to form reactive intermediates, their coupling into polymer chains, and stabilization into the final polymeric product [40].

#### 5.3.1.2 Electrochemical Polymerization

Electrochemical polymerization provides precise control over polymer film thickness, morphology, and doping levels. Monomers are polymerized on an electrode surface under an applied potential, making this technique suitable for fabricating thin films for electronic and sensor applications. This method offers high-purity and uniform polymer films with real-time monitoring of growth. However, its application is typically limited to small-scale processes and requires conductive substrates [41].

#### 5.3.1.3 Interfacial Polymerization

Interfacial polymerization is employed to produce ultrathin films by reacting to monomers at the interface of two immiscible phases, usually an aqueous phase containing the monomer and an organic phase containing the oxidizing agent. This simple and scalable process is particularly effective for fabricating hybrid systems and membranes for water treatment. It offers high uniformity and is suitable for specific applications where thin, consistent films are required [42].

#### 5.3.1.4 Template-Assisted Polymerization

Template-assisted polymerization utilizes nanostructured membranes or surfactants as templates to direct polymer growth, resulting in nanostructures like nanowires, nanotubes, or nanorods. This method enhances the surface area, making it ideal for photocatalytic applications, and allows for precise control over morphology. However, the complexity of template removal and the high costs associated with production present challenges for widespread adoption.

The choice of synthesis method depends on the desired properties, scalability requirements, and specific application of the conducting polymer. Advancements in these techniques continue to expand the potential of conducting polymers in various fields, from energy conversion to environmental remediation [43].

### *5.3.2 Structural Modifications of Conducting Polymers*

Structural modifications play a vital role in enhancing the intrinsic properties of conducting polymers, enabling their optimization for photocatalytic and other advanced applications. Key strategies such as doping, functionalization, nanostructuring, and hybridization with inorganic semiconductors significantly expand the functionality and efficiency of these materials.

### *5.3.3 Doping*

Doping involves the introduction of electron donors or acceptors into the polymer matrix, substantially improving the electrical conductivity and optical properties of conducting polymers.

- **P-Type Doping** introduces electron acceptors, creating holes as charge carriers, while **N-Type Doping** incorporates electron donors, increasing electron density.
- These modifications enhance charge carrier mobility and light absorption, particularly in the visible spectrum.
- For example, doping polyaniline with sulfonic acid groups enhances its solubility and conductivity, making it highly effective in hybrid photocatalytic systems for pollutant degradation and energy applications [44].

### *5.3.4 Functionalization*

Functionalization involves attaching specific chemical groups (e.g., -OH, -COOH, $-NH_2$) to the polymer backbone. This enhances the interaction of conducting polymers with pollutants and their compatibility with inorganic materials for hybrid systems.

- Functionalization improves pollutant adsorption and chemical stability, making these materials suitable for targeted applications.
- Functionalized conducting polymers are widely employed in water treatment, effectively addressing contaminants like heavy metals and organic pollutants [45].

### *5.3.5 Nanostructuring*

Nanostructuring increases the surface area and introduces more active sites, boosting the photocatalytic performance of conducting polymers.

- Techniques such as electrospinning, template-guided growth, and self-assembly create structures like nanofibers, nanotubes, and nanosheets.
- These nanostructures improve light absorption efficiency and reduce recombination losses, significantly enhancing photocatalytic activity [46].

### *5.3.6 Hybridization with Inorganic Semiconductors*

Hybridization combines conducting polymers with inorganic materials such as $TiO_2$, ZnO, or $g\text{-}C_3N_4$ to improve their overall photocatalytic efficiency.

- This strategy extends light absorption into the visible spectrum, enhances charge separation and transport, and improves stability under operational conditions.
- These hybrids are particularly effective in pollutant degradation, solar energy harvesting, and hydrogen generation applications [47].

## 5.4 Hybrid Nanocomposites for Photocatalysis

Hybrid nanocomposites, combining conducting polymers (CPs) with inorganic semiconductors, represent a transformative class of materials in photocatalysis. By synergizing the unique optical and electronic properties of CPs with the stability and catalytic efficiency of inorganic materials, these composites achieve enhanced photocatalytic performance. This integration enables extended light absorption, improved charge separation, and reduced recombination losses, positioning hybrid nanocomposites as pivotal in advancing photocatalytic technologies [48].

### *5.4.1 Fundamentals of Hybrid Nanocomposites*

Hybrid nanocomposites are engineered materials that leverage the complementary properties of conducting polymers and inorganic semiconductors. Conducting polymers act as photosensitizers, extending light absorption into the visible spectrum and facilitating efficient charge transport due to their conjugated π-electron systems. Inorganic semiconductors, such as $TiO_2$, ZnO, and g-$C_3N_4$, contribute stability, crystallinity, and intrinsic catalytic activity, creating a robust platform for photocatalysis. This hybridization enhances solar spectrum utilization, chemical and photostability, and functionalization for targeted applications, broadening their utility across environmental and energy domains.

### *5.4.2 Synthesis Strategies for Hybrid Nanocomposites*

The synthesis of hybrid nanocomposites is designed to achieve optimal interaction between CPs and inorganic semiconductors, ensuring uniform interfaces and maximizing synergistic effects. Key synthesis methods include:

#### 5.4.2.1 In Situ Polymerization

This approach synthesizes conducting polymers directly on the surface of inorganic semiconductors, forming strong interfacial bonds and ensuring a uniform distribution of CPs. It is particularly effective in applications such as the photodegradation of organic pollutants, where close integration between the components is essential for efficient charge transfer [49].

#### 5.4.2.2 Physical Blending

Physical blending mechanically mixes CPs and inorganic semiconductors, followed by deposition onto a substrate. While this method is simple and scalable, it often results in weaker interfacial interactions, which may limit charge transfer efficiency. However, the preservation of individual material properties makes it suitable for applications requiring distinct functionality from each component [50].

#### 5.4.2.3 Solvothermal and Hydrothermal Methods

These methods employ high-temperature and high-pressure environments to facilitate strong hybridization between CPs and semiconductors. They enable the production of highly crystalline and pure hybrid nanostructures, such as nanowires and nanosheets, tailored for specific applications. Although energy-intensive, these techniques are invaluable for creating hybrids with precise structural control [51].

#### 5.4.2.4 Electrodeposition

Electrodeposition involves coating inorganic semiconductors with thin films of conducting polymers, offering precise control over the thickness and morphology of the CP layer. This method is particularly suitable for electronic and optoelectronic applications, including electrochemical water splitting, where uniform films are critical for consistent performance.

Hybrid nanocomposites provide a versatile platform for addressing critical challenges in photocatalysis, from environmental remediation to renewable energy conversion. By optimizing their synthesis and exploiting their synergistic properties, these materials hold the potential to revolutionize sustainable technologies. Continued innovation in this field promises to unlock new applications and further enhance the efficiency and practicality of photocatalytic systems [52].

### *5.4.3 Mechanisms of Enhanced Photocatalysis*

The integration of conducting polymers (CPs) with inorganic semiconductors significantly enhances photocatalytic performance through several synergistic mechanisms. These mechanisms exploit the unique properties of both components to extend light absorption, improve charge carrier dynamics, and amplify the generation of reactive species essential for catalytic reactions [53].

#### 5.4.3.1 Extended Light Absorption

Conducting polymers serves as sensitizers, broadening the light absorption range of semiconductors into the visible spectrum. This extension increases the utilization of solar energy, enabling more efficient photocatalysis under natural light conditions. For instance, PANI/$TiO_2$ composites demonstrate absorption up to 700 nm, in contrast to pristine $TiO_2$, which is restricted to UV light absorption. This enhanced light harvesting allows hybrid nanocomposites to target a wider array of photocatalytic applications, including pollutant degradation and solar-driven chemical transformations [54].

#### 5.4.3.2 Improved Charge Separation

The heterojunction formed at the interface of CPs and semiconductors facilitates efficient charge carrier separation, reducing recombination losses that often limit photocatalytic efficiency.

**Mechanism:** Photogenerated electrons in the conducting polymer are transferred to the conduction band of the semiconductor, while the holes remain in the polymer. This spatial separation of charge carriers ensures that electrons and holes are available for subsequent reduction and oxidation reactions, respectively.

This charge separation mechanism is critical for sustaining long-term photocatalytic activity and increasing reaction rates in environmental and energy applications.

#### 5.4.3.3 Generation of Reactive Oxygen Species (ROS)

Hybrid nanocomposites enhance the production of reactive oxygen species (ROS), which are key agents in the degradation of organic pollutants and other photocatalytic reactions (Fig. 5.3).

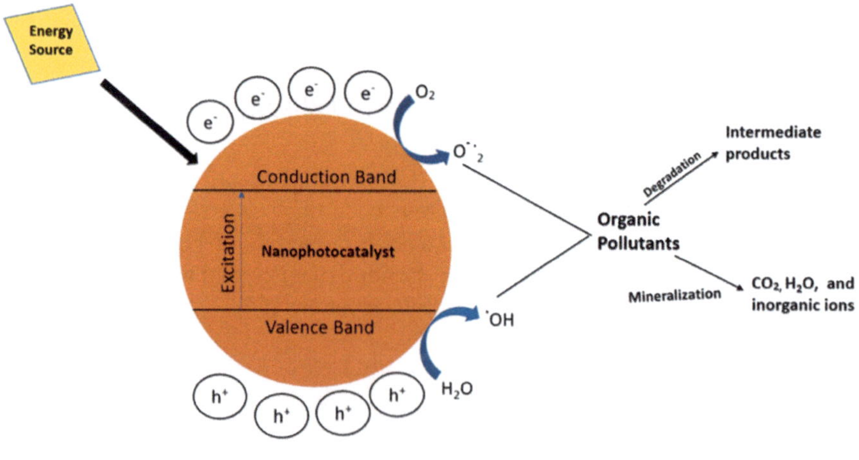


**Fig. 5.3** Electron excitation and recombination [55]

- **Electron Transfer Pathway:** Electrons donated by the CP interact with oxygen molecules, forming superoxide anions ($O_2^{-}\cdot$).
- **Hole Reaction Pathway:** Holes in the CP react with water molecules, generating hydroxyl radicals (·OH).

These ROS are highly reactive and capable of breaking down complex organic compounds into smaller, less harmful molecules, making hybrid systems highly effective in applications such as water purification and air quality improvement.

### *5.4.4 Applications of Hybrid Nanocomposites*

Hybrid nanocomposites, combining conducting polymers with inorganic semiconductors, have emerged as versatile materials with applications spanning environmental remediation, renewable energy, and advanced functional systems. Their enhanced photocatalytic properties, derived from synergistic interactions between the organic and inorganic components, make them highly effective in tackling modern challenges.

#### 5.4.4.1 Environmental Remediation

Hybrid nanocomposites play a pivotal role in water and air purification:

- **Water Treatment:** They efficiently degrade organic dyes, pharmaceuticals, and other pollutants while removing heavy metals through reduction or adsorption processes. For example, a PANI/$TiO_2$ composite demonstrated remarkable performance, achieving 95% degradation of methylene blue under visible light in just 2 h, significantly outperforming pure $TiO_2$.
- **Air Purification:** These materials break down volatile organic compounds (VOCs) and convert harmful nitrogen oxides (NOx) into harmless byproducts, making them invaluable for improving air quality in urban and industrial environments.

#### 5.4.4.2 Renewable Energy

Hybrid nanocomposites contribute significantly to renewable energy initiatives:

- **Hydrogen Production:** Composites such as CP/g-$C_3N_4$ hybrids facilitate solar water splitting, enabling efficient hydrogen evolution, a clean energy source for the future [56].
- **Carbon Dioxide Reduction:** Polypyrrole (PPy)-based hybrids efficiently reduce $CO_2$ to valuable products like methanol, addressing both greenhouse gas mitigation and sustainable energy generation [57].

#### 5.4.4.3 Advanced Functional Materials

The unique properties of hybrid nanocomposites enable their application in cutting-edge material systems:

- **Sensors:** Their enhanced conductivity and sensitivity make them ideal for use in gas sensors and biosensors, providing rapid and reliable detection capabilities for various applications [58].
- **Energy Storage:** Hybrid nanocomposites are integrated into supercapacitors, offering improved charge storage capacity, stability, and energy density, advancing energy storage technologies [59].

The broad applicability of hybrid nanocomposites underscores their potential to address critical global challenges in environmental sustainability, renewable energy, and advanced materials (Table 5.2). Through continued research and development, these materials promise to revolutionize technology across multiple sectors, driving innovation and fostering a cleaner, more sustainable future.

**Table 5.2** Summarized properties and applications of conducting polymer/metal oxide composites as photocatalysts

| Composite | Bandgap energy (eV) | Photocatalytic properties | Applications | Reference |
|---|---|---|---|---|
| Mesoporous PANI/$TiO_2$ | – | Enhanced water oxidation efficiency under sunlight irradiation, reaching about two-fold higher photocurrent densities than pure $TiO_2$ nanoparticles | Water splitting | [60] |
| PANI/$TiO_2$ nanorods | 3.1 | Quantum yield: $9.86 \times 10^{-5}$ molecules/photon and $2.82 \times 10^{-5}$molecules/photon for PANI/$TiO_2$ and $TiO_2$, respectively | Degradation of organic pollutants (Bisphenol A) | [61] |
| PANI nanobelt/ $TiO_2$ | 2.77 | The photocatalytic degradation rate of rhodamine B was 99% | Degradation of rhodamine B | [62] |
| PANI/$TiO_2$ nanotubes | – | The photocatalytic activity can easily be tuned using a particular type and concentration of the acid dopant in the redoping process | Degradation of rhodamine B | [63] |
| PANI/ZnO | – | The composite exhibits a dramatic photocatalytic activity both under ultraviolet and visible light irradiation. The photo-corrosion of ZnO was successfully inhibited | Degradation of methylene blue (MB) | [64] |
| PANI/ZnO | 2.13–2.22 | The composite photocatalysts' activity was broadened into the Vis region | Degradation of acid blue, degradation of MB | [65] |
| PANI/$Fe_3O_4$ | – | The adsorption process prevails in relation to photodegradation | Degradation of MB | [66] |

**Table 5.2** (continued)

| Composite | Bandgap energy (eV) | Photocatalytic properties | Applications | Reference |
|---|---|---|---|---|
| PANI/$SnO_2$ | 2.7 | Increased photocatalytic activity for visible light is due to its electrical conductivity and efficient charge separation | Degradation of direct blue 15 | [67] |
| PANI/$Sn_3O_4$ | 2.06 | Photocatalytic activity for visible light is 2.27 times higher than that of $Sn_3O_4$ alone | Degradation of rhodamine B | [68] |
| PANI/ $CoFe_2O_4$ | – | Photocatalytic activity under visible light irradiation with $CoFe_2O_4$/PANI was 80 times greater than for $CoFe_2O_4$ | Degradation of methyl orange (MO) | [69] |
| PEDOT/ $TiO_2$ | 3.01–3.05 | PEDOT infused $TiO_2$ nanofiber exhibits the highest degradation enhancement (125%) | Degradation of phenazopyridine | [70] |
| PPy/$TiO_2$ | – | The photocatalytic activity of the nanocomposite arises from the electron transfer from excited PPy to $TiO_2$ nanoparticles | Degradation of MB | [70] |
| PPy/$TiO_2$ | 3.08–3.11 | The photocatalytic activity of nanocomposites increased by 84% compared with pure $TiO_2$ | Degradation of RhB and $CO_2$ | [71] |
| ZnO-microrods/ PPy | 1.7 | Composite films achieve a much higher photocatalytic efficiency in comparison with pure ZnO-microrod arrays | Degradation of MB | [72] |
| $TiO_2$-$CoFe_2O_4$-PANI | – | The ternary $TiO_2$-$CoFe_2O_4$-PANI composite shows highly enhanced photocatalytic activity in visible light | Degradation of methyl orange | [73] |
| $ZnFe_2O_4$-$TiO_2$-PANI | – | The decontaminating efficiency of composites on MO and RhB reached up to 98% | Degradation and adsorption of MO and RhB | [74] |
| rGO-$ZnFe_2O_4$-PANI | – | The photocatalyst shows enhanced photocatalytic performance in the photodegradation of MO (almost 100%) | Degradation of RhB | [75] |
| ZnO/rGO/ PANI | – | The ternary composite with Z-scheme heterojunction properties displayed outstanding adsorption properties and photocatalytic activities | Degradation of methyl orange | [76] |
| $Cu_2O$/ ZnO-PANI | 2.68 | The photocatalytic activity still stays above 90% after five recycles | Degradation of Congo red | [77] |
| PANI/$TiO_2$/ graphene | 2.1 | High photocatalytic activity is partly due to the sensitizing effect of PANI | Degradation of MB | [78] |
| PANI-rGO-$MnO_2$ | 1.92 | The ternary composite exhibited significantly enhanced catalytic and photocatalytic activity under visible light irradiation | Degradation of MB | [79] |
| g-$C_3N_4$/ $TiO_2$/PANI | 2.58 | Greatly enhanced photocatalytic degradation and high reusability | Degradation of Congo red | [80] |

## 5.5 Environmental Applications of Conducting Polymer-Based Hybrid Photocatalysts

Environmental pollution caused by industrialization, urbanization, and agricultural expansion has prompted the search for sustainable and innovative remediation strategies. Conducting polymer-based hybrid photocatalysts have proven to be a transformative class of materials capable of addressing various environmental issues, including water purification, air pollution control, and solid waste management, while also contributing to renewable energy solutions. These materials leverage the unique properties of conducting polymers, such as tunable optoelectronic characteristics, visible light activation, and enhanced charge separation, to provide efficient and eco-friendly solutions to pressing environmental challenges.

One of the most critical applications of hybrid photocatalysts lies in the domain of water treatment. Water pollution, caused by the discharge of industrial effluents, agricultural runoff, and domestic waste, has led to the accumulation of organic pollutants, heavy metals, and microbial contaminants in water bodies. Traditional treatment methods, including chemical coagulation, filtration, and adsorption, often fall short in addressing these complex contaminants. Hybrid photocatalysts, however, offer a promising alternative by harnessing light energy to drive photocatalytic reactions that degrade pollutants into benign substances [81].

### 5.5.1 *Degradation of Organic Pollutants*

The degradation of organic pollutants, such as dyes, pesticides, and pharmaceuticals, is a primary focus in water treatment applications. Polyaniline (PANI)/titanium dioxide ($TiO_2$) composites have shown exceptional performance in degrading methylene blue (MB), achieving over 95% efficiency under visible light irradiation. This high efficiency can be attributed to the synergistic effects of PANI and $TiO_2$. PANI extends the light absorption range into the visible spectrum, while $TiO_2$ provides a robust platform for photocatalytic activity. Upon visible light activation, the hybrid system generates electron–hole pairs, which subsequently interact with water molecules and dissolved oxygen to produce reactive oxygen species (ROS), such as hydroxyl radicals and superoxide ions. These ROS attack the chemical bonds of the organic pollutants, breaking them down into non-toxic byproducts like carbon dioxide and water.

A similar mechanism is observed in the degradation of rhodamine B using poly(3,4-ethylenedioxythiophene) (PEDOT)/graphitic carbon nitride (g-$C_3N_4$) hybrids. The integration of PEDOT enhances the charge transport properties of g-$C_3N_4$, reducing recombination losses and increasing photocatalytic efficiency. Studies have reported that such hybrids can achieve over 89% degradation efficiency under simulated solar light, highlighting their potential for real-world applications [82] (Fig. 5.4).

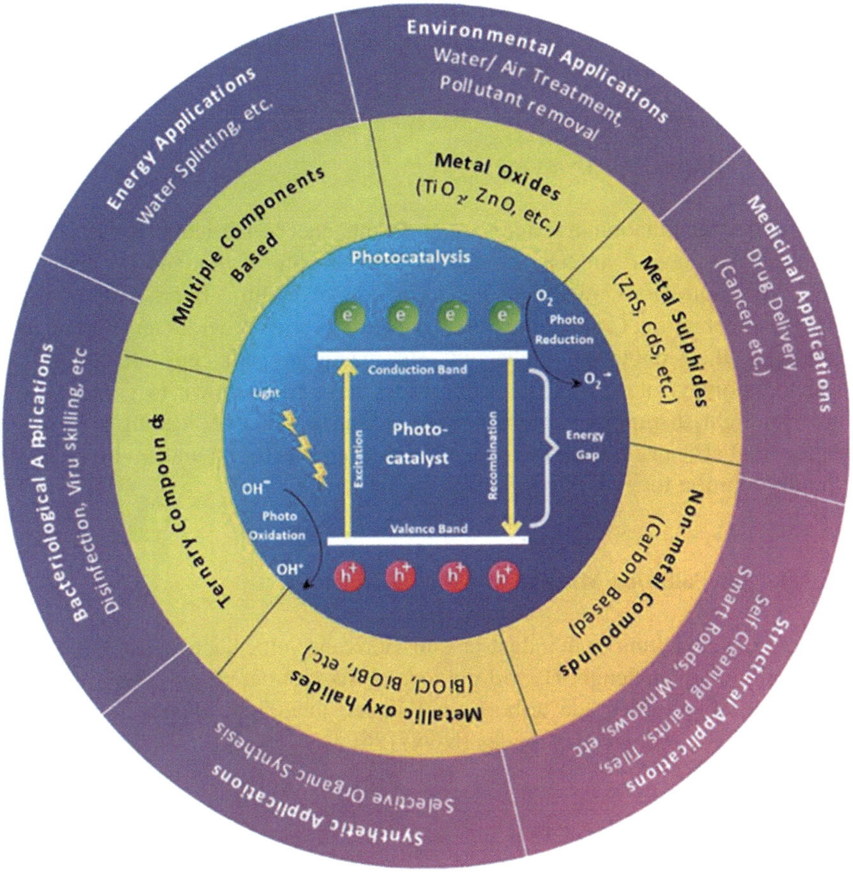


**Fig. 5.4** This topical review attempts to highlight the roles of semiconducting polymer (polyaniline and polypyrrole)-titanium dioxide interfaces in visible light photocatalysis. Progress and prospects of a visible light-activated inorganic-organic semiconductor (conducting polymer)-based photocatalysts are comprehensively reviewed to highlight their importance and scopes [83]

#### 5.5.1.1 Heavy Metal Reduction

Heavy metals, such as chromium (Cr), lead (Pb), and cadmium (Cd), are persistent environmental contaminants with severe health implications. Conducting polymer-based hybrids offers an effective solution for their removal. For example, ZnO/PANI composites have demonstrated remarkable efficiency in reducing Cr(VI) to Cr(III) under solar irradiation. This process involves the transfer of photogenerated electrons from the conducting polymer to the Cr(VI) ions, facilitating their reduction to the less toxic Cr(III) state. Such systems not only detoxify heavy metals but also minimize secondary pollution due to the stability of the hybrid catalysts.

In addition to chromium, PEDOT-based hybrids have shown high affinity for adsorbing and reducing $Pb^{2+}$ and $Cd^{2+}$ ions. The functional groups on the PEDOT backbone enhance the interaction with heavy metal ions, ensuring efficient removal from water systems. These systems represent a scalable and sustainable approach to addressing heavy metal contamination [84].

#### 5.5.1.2 Pathogen Inactivation

Microbial contamination in water is a significant concern, particularly in developing regions. Traditional disinfection methods, such as chlorination and UV treatment, often have limitations, including the formation of harmful byproducts and high energy consumption. Conducting polymer-based hybrids provides an alternative by leveraging their photocatalytic activity to inactivate pathogens effectively. For instance, polypyrrole (PPy)/$TiO_2$ composites have been shown to generate ROS under sunlight, disrupting microbial cell membranes and DNA, leading to over 99% bactericidal efficiency. Such systems offer an energy-efficient and environmentally friendly solution for ensuring safe drinking water [85].

#### 5.5.1.3 Air Pollution Mitigation

Air pollution, resulting from industrial emissions and vehicular exhaust, has become a critical global challenge. Hybrid photocatalysts are increasingly being explored for their ability to degrade airborne pollutants, including volatile organic compounds (VOCs) and nitrogen oxides (NOx) [86].

#### 5.5.1.4 Decomposition of VOCs

VOCs, such as formaldehyde and toluene, contribute to indoor air pollution and pose serious health risks. PEDOT/$TiO_2$ hybrids have been shown to decompose formaldehyde into carbon dioxide and water under sunlight, achieving an 80% reduction within 3 h. The integration of PEDOT enhances the light absorption capabilities of $TiO_2$, extending its activity into the visible spectrum and enabling efficient VOC degradation.

#### 5.5.1.5 Reduction of NOx

Nitrogen oxides are precursors to smog and acid rain, necessitating their removal from the atmosphere. Hybrid photocatalysts, such as PANI/$TiO_2$, convert NOx into harmless nitrates through photocatalytic reactions. These systems not only improve air quality but also contribute to reducing the environmental impact of industrial emissions.

#### 5.5.1.6 Solid Waste Management

The management of solid waste, particularly microplastics and persistent organic pollutants (POPs), is a growing concern. Conducting polymer-based hybrids offer innovative solutions for degrading these pollutants into less harmful components.

#### 5.5.1.7 Degradation of Microplastics

PANI/$TiO_2$ hybrids have shown potential in breaking down polyethylene microplastics under visible light. The photocatalytic process generates ROS that attack the polymer chains of microplastics, reducing them into smaller, biodegradable molecules. This approach provides a sustainable method for addressing the global microplastic crisis.

#### 5.5.1.8 Degradation of POPs

Persistent organic pollutants, such as pesticides and industrial chemicals, are resistant to conventional treatment methods. PEDOT/ZnO hybrids have demonstrated high efficiency in degrading these pollutants, highlighting their applicability in solid waste management.

#### 5.5.1.9 Renewable Energy Applications

In addition to environmental remediation, conducting polymer-based hybrids are being explored for renewable energy applications. Solar water splitting for hydrogen production is one of the most promising areas. Hybrid systems like g-$C_3N_4$/PANI have shown significant enhancements in hydrogen evolution rates, leveraging the synergistic effects of the components to maximize efficiency.

Carbon dioxide reduction is another critical application, where hybrid photocatalysts convert $CO_2$ into valuable fuels such as methane and methanol. These processes not only address greenhouse gas emissions but also provide sustainable energy solutions.

## 5.6 Challenges and Limitations of Conducting Polymer-Based Hybrid Photocatalysts

While conducting polymer-based hybrid photocatalysts hold immense promise in addressing environmental challenges and advancing renewable energy technologies, several critical issues hinder their widespread adoption and practical implementation. These challenges can be broadly categorized into material-related limitations, scalability issues, environmental and health impacts, and economic feasibility [87].

## 5.6.1 *Material Stability and Durability*

One of the most significant challenges faced by conducting polymer-based hybrid photocatalysts is their stability and durability under operational conditions. Prolonged exposure to light, heat, and reactive species can degrade the materials, reducing their photocatalytic efficiency over time [88].

### 5.6.1.1 Photostability Issues

Conducting polymers, such as polyaniline (PANI) and polypyrrole (PPy), are prone to photodegradation under prolonged light exposure. This degradation often results in discoloration, loss of conductivity, and structural breakdowns. Reactive oxygen species (ROS), which are crucial for pollutant degradation, can also attack the polymer backbone, further accelerating degradation [89].

### 5.6.1.2 Strategies to Enhance Stability

- **Surface Coatings:** Protective coatings, such as silica or graphene oxide, can shield conducting polymers from direct exposure to ROS.
- **Cross-Linking:** Introducing cross-linked structures within the polymer matrix can enhance mechanical and chemical stability.
- **Functionalization:** Chemical modification of polymers to introduce stabilizing groups can reduce their susceptibility to degradation [90].

## 5.6.2 *Scalability of Production*

The synthesis of conducting polymer-based hybrids often involves complex processes that are resource-intensive and challenging to scale up for industrial applications.

### 5.6.2.1 Resource-Intensive Synthesis Techniques

Techniques like hydrothermal, solvothermal, and template-assisted synthesis require precise control over reaction conditions, including temperature, pressure, and pH. These processes are energy-intensive and may involve hazardous chemicals, making them less feasible for large-scale production.

#### 5.6.2.2 Proposed Solutions

- **Green Synthesis:** Developing environmental-friendly methods using renewable precursors and non-toxic solvents.
- **Automated Production:** Leveraging automation and high-throughput techniques to standardize and scale up production processes.
- **Cost Optimization:** Reducing reliance on expensive precursors and equipment by exploring alternative synthesis pathways.

### *5.6.3 Environmental and Health Concerns*

#### 5.6.3.1 Nanoparticle Release

During operation or disposal, hybrid photocatalysts may release nanoparticles into the environment, raising concerns about secondary pollution and long-term ecological impacts. The lack of established recycling or disposal methods for spent photocatalysts exacerbates this issue.

#### 5.6.3.2 Toxicity of Components

Certain dopants or additives used to enhance photocatalytic performance may be toxic to aquatic and terrestrial ecosystems. For example, heavy metal-based dopants can bioaccumulate, posing risks to the food chain.

#### 5.6.3.3 Mitigation Strategies

- **Lifecycle Analysis:** Comprehensive studies to assess the environmental impact of photocatalyst production, use, and disposal.
- **Recyclable Materials:** Designing hybrid photocatalysts with recyclable or biodegradable components.
- **Safe Handling Guidelines:** Developing standardized protocols for the safe use and disposal of nanomaterials.

### *5.6.4 Energy Efficiency*

#### 5.6.4.1 Low Solar-to-Chemical Conversion Efficiency

Despite advancements, many hybrid photocatalysts exhibit suboptimal conversion efficiencies when utilizing solar energy. This inefficiency limits their practical applicability for large-scale operations.

#### 5.6.4.2 Proposed Improvements

- **Bandgap Engineering:** Designing materials with tailored bandgaps to maximize light absorption and energy conversion.
- **Coupling with Photovoltaics:** Integrating photocatalysts with solar panels to harness and utilize excess solar energy.
- **Spectral Tuning:** Employing spectral conversion techniques to optimize the utilization of available light.

### *5.6.5 Economic Viability*

#### 5.6.5.1 High Production Costs

The synthesis of conducting polymer-based hybrids often involves expensive raw materials, reagents, and specialized equipment. These factors contribute to the overall cost, making the technology less accessible for widespread use.

#### 5.6.5.2 Cost-Effective Alternatives

- **Abundant Precursors:** Exploring the use of naturally abundant or renewable feedstocks for polymer synthesis.
- **Simplified Fabrication:** Streamlining production processes to reduce energy and labor costs.
- **Public-Private Partnerships:** Collaborating with industries and governments to fund large-scale pilot projects and reduce financial barriers.

### *5.6.6 Limited Selectivity and Specificity*

#### 5.6.6.1 Broad-Spectrum Activity

While hybrid photocatalysts are effective in degrading a wide range of pollutants, they often lack specificity for certain contaminants. This broad-spectrum activity can lead to incomplete degradation or the formation of secondary pollutants.

#### 5.6.6.2 Enhancing Selectivity

- **Molecular Imprinting:** Creating polymers with specific binding sites tailored to target pollutants.
- **Functionalized Hybrids:** Introducing functional groups that selectively interact with particular contaminants.

- **Process Optimization:** Adjusting reaction conditions to favor the desired degradation pathways.

## *5.6.7 Integration with Existing Infrastructure*

### 5.6.7.1 Compatibility Issues

Incorporating hybrid photocatalysts into existing water treatment, air purification, or renewable energy systems poses challenges in terms of compatibility and efficiency.

### 5.6.7.2 Proposed Solutions

- **Modular Designs:** Developing plug-and-play systems that can seamlessly integrate with current technologies.
- **Retrofitting:** Adapting existing infrastructure to accommodate hybrid photocatalyst systems.
- **Performance Monitoring:** Employing smart sensors to monitor and optimize the performance of integrated systems.

# 5.7 Future Perspectives and Recommendations for Conducting Polymer-Based Hybrid Photocatalysts

Conducting polymer-based hybrid photocatalysts hold great promise for addressing global challenges in environmental remediation and renewable energy. Future advancements will depend on overcoming current limitations and strategically exploring innovative directions to maximize their potential. A detailed overview of future perspectives and recommendations is provided below [91].

## *5.7.1 Advancements in Material Design*

The development of multifunctional hybrids capable of performing simultaneous tasks, such as pollutant degradation and energy generation, will reduce system complexity and operational costs. Hierarchical nanostructures, including core–shell configurations, hollow spheres, and layered assemblies, will optimize the interaction between conducting polymers and inorganic semiconductors, improving light absorption and charge separation. Self-healing materials with dynamic covalent bonds or embedded repair agents will address photodegradation, extending the operational lifespan of hybrid photocatalysts.

### 5.7.2 Enhanced Stability and Longevity

Photostability and resistance to degradation under prolonged irradiation are critical for real-world applications. Strategies such as encapsulating conducting polymers with inert materials like silica or graphene and introducing cross-linked polymer structures can enhance durability. Improving chemical resistance to harsh environmental conditions, such as extreme pH or salinity, will make these materials suitable for diverse applications [92].

### 5.7.3 Scalable and Green Synthesis Techniques

Scalable production requires automation and high-throughput synthesis methods to ensure consistent quality and reduce costs. Employing green chemistry approaches, including renewable feedstocks, non-toxic solvents, and energy-efficient processes, will align material production with environmental sustainability goals. Simplified and automated processes will make hybrid photocatalysts more accessible for large-scale deployment [93].

### 5.7.4 Integration with Renewable Energy Systems

Improving solar energy utilization through bandgap engineering, spectral tuning, and doping strategies will increase the efficiency of solar-driven photocatalysis. Integrating photocatalysts with energy storage systems, such as batteries and supercapacitors, will enable dual functionality by storing excess energy generated during photocatalytic processes [94].

### 5.7.5 Improved Selectivity and Efficiency

Tailoring the surface chemistry of hybrid photocatalysts to target-specific pollutants, such as pharmaceuticals or heavy metals, will enhance their efficiency. Advanced molecular imprinting techniques and the introduction of functional groups with high pollutant affinity can achieve target-specific performance. Engineering systems to produce specific reactive oxygen species (ROS) will optimize degradation pathways for various contaminants [95].

### 5.7.6 Real-World Applications and Demonstrations

Field testing and large-scale pilot projects will validate laboratory findings and demonstrate practical feasibility. Deploying hybrid photocatalysts in industrial wastewater treatment plants, urban air purification systems, and renewable energy applications will provide valuable insights into scalability and operational performance. Modular systems designed for easy integration into existing infrastructure, such as water treatment facilities and solar farms, will encourage widespread adoption [96].

### 5.7.7 Addressing Environmental and Safety Concerns

Comprehensive lifecycle analysis is necessary to assess the environmental and economic impacts of hybrid photocatalysts from production to disposal. Developing recyclable or biodegradable materials will prevent secondary pollution and enhance environmental compatibility. Establishing standardized recycling protocols for spent materials will minimize waste and promote sustainability [97].

### 5.7.8 Interdisciplinary Collaboration

Collaboration between material scientists, environmental engineers, chemists, and industry stakeholders will accelerate the development of innovative solutions. Public–private partnerships will provide the funding and resources needed for large-scale research and development, technology transfer, and commercialization. Interdisciplinary approaches will address complex challenges and expand the range of hybrid photocatalyst applications [98].

### 5.7.9 Policy and Regulation

Establishing international standards and regulatory frameworks will ensure the safe production, use, and disposal of hybrid photocatalysts. Policies should encourage the adoption of sustainable technologies through incentives such as tax benefits, grants, and subsidies. Governments and regulatory bodies must prioritize green technology development to align with global environmental goals [99].

## 5.8 Environmental Applications of Conducting Polymer-Based Hybrid Photocatalysts

Conducting polymer-based hybrid photocatalysts have become pivotal in addressing global environmental challenges due to their ability to utilize solar energy, degrade pollutants, and integrate seamlessly with existing systems. These materials play a transformative role in water treatment, air purification, renewable energy production, solid waste management, and integrated environmental systems, offering sustainable solutions to pressing ecological problems [100].

### 5.8.1 *Water Treatment*

Water pollution from industrial, agricultural, and urban activities poses serious environmental and health threats. Hybrid photocatalysts effectively tackle various forms of water contamination. Organic pollutants, such as dyes, pesticides, and pharmaceuticals, are efficiently degraded by systems like PANI/$TiO_2$ and PPy/ZnO. Laboratory studies show over 90% efficiency in breaking down pollutants like methylene blue and tetracycline under visible light. These materials also address heavy metal contamination, with PANI/g-$C_3N_4$ hybrids converting toxic Cr(VI) to safer Cr(III) and functionalized PEDOT composites adsorbing $Pb^{2+}$ and $Cd^{2+}$ ions from wastewater. For pathogen inactivation, hybrids like PPy/$TiO_2$ generate reactive oxygen species (ROS) that achieve over 99% bactericidal efficiency, making them effective for water disinfection [101].

### 5.8.2 *Air Purification*

Industrial emissions and urbanization contribute significantly to air pollution, necessitating innovative purification technologies. Conducting polymer-based hybrids provide efficient solutions by degrading volatile organic compounds (VOCs) and reducing nitrogen oxides (NOx). Hybrids like PEDOT/$TiO_2$ break down VOCs such as formaldehyde and benzene into harmless products under visible light, while PANI-modified $TiO_2$ converts NOx into nitrates, mitigating smog and acid rain [102].

### 5.8.3 *Renewable Energy Applications*

The integration of conducting polymers into photocatalytic systems extends their functionality to renewable energy production. Hybrid materials like $TiO_2$/PANI enhance solar water splitting, enabling efficient hydrogen generation under visible light, a clean energy alternative aligned with global sustainability goals. Additionally, composites like PPy/g-$C_3N_4$ catalyze the reduction of $CO_2$ into valuable fuels like methane and methanol, contributing to carbon neutrality and energy sustainability [103].

### *5.8.4 Solid Waste Management*

Hybrid photocatalysts address solid waste challenges, including plastic degradation and the treatment of persistent organic pollutants (POPs). These systems facilitate the photodegradation of polyethylene microplastics into biodegradable molecules, reducing the ecological burden of plastic waste. Advanced oxidation processes (AOPs) driven by hybrid systems enhance the breakdown of POPs, ensuring minimal residual toxicity and environmental impact [104].

### *5.8.5 Integrated Environmental Systems*

Hybrid photocatalysts can be embedded into advanced systems for real-time environmental monitoring and remediation. Smart filters utilizing these materials enable the simultaneous removal of organic pollutants, heavy metals, and pathogens from water. Photocatalytic coatings applied to urban infrastructure degrade airborne and surface-bound pollutants under sunlight, offering a practical solution for improving urban environmental quality [105].

## 5.9 Conclusion

Conducting polymer-based hybrid photocatalysts stand as a transformative innovation in addressing the pressing environmental and energy challenges of the twenty-first century. By combining the optoelectronic properties of conducting polymers with the robustness of inorganic semiconductors, these materials have demonstrated remarkable versatility and efficiency in environmental remediation and renewable energy applications. From the degradation of persistent organic pollutants and heavy metal removal in water treatment to air purification and advanced solid waste management, their contributions to sustainable technologies are profound and impactful.

The ability of these hybrids to harness visible light, generate reactive oxygen species, and achieve efficient charge separation underpins their success in photocatalytic processes. Their applications in renewable energy, including solar-driven hydrogen production and carbon dioxide reduction, further highlight their potential to address the interconnected challenges of energy scarcity and climate change.

These advancements align closely with global goals for sustainability and carbon neutrality.

However, the widespread adoption of conducting polymer-based hybrid photocatalysts is not without challenges. Issues such as material stability under prolonged operational conditions, scalability of synthesis methods, environmental safety concerning nanomaterial release, and specificity in pollutant degradation remain critical hurdles. Addressing these challenges will require interdisciplinary collaboration, innovative material design, and the development of cost-effective, green synthesis methods. Additionally, large-scale field trials and integration with existing environmental and energy systems will be essential to validate their practical feasibility and impact.

## References

1. Wang F et al (2024) Emerging contaminants: a One Health perspective. Innovation 5(4):100612
2. O'Shea KE, Dionysiou DD (2012) Advanced oxidation processes for water treatment. J Phys Chem Lett 3(15):2112–2113
3. Nagireddi S et al (2023) Carbon dioxide capture, utilization, and sequestration: current status, challenges, and future prospects for global decarbonization. ACS Eng Au 4(1):22–48
4. Gopalan S et al (2022) Conducting polymer-based visible light photocatalytic composites for pollutant removal: progress and prospects. Environ Technol Innov 28:102698
5. Khan MI et al (2021) Conductive polymers and their nanocomposites as adsorbents in environmental applications. Polymers 13(3810)
6. Fang Y et al (2024) Advancing electrically conductive metal–organic frameworks for photocatalytic energy conversion. Adv Funct Mater
7. Wang L et al (2021) Role of conducting polymers in enhancing $TiO_2$-based photocatalytic dye degradation: a short review. J Environ Chem Eng 9(6):106709
8. Saianand G et al (2023) Hybrid nanocomposites and their applications for photocatalytic wastewater treatment. Environ Sci Technol 57:5005–5020
9. Munusamy T et al (2022) Efficient photocatalytic degradation of methylene blue using g-$C_3N_4$–PANI–PPy nanohybrids. Chem Eng J 435:134987
10. Hiragond CB et al (2023) Conducting polymers and hybrid Systems for dye degradation: comparative studies. Mater Adv 4:3456–3472
11. Nyga D et al (2024) Electrochemical properties of polyselenophenes for photocatalytic applications. Electrochem Commun 141:107388
12. Gogoi P et al (2022) Palladium oxide/palladium nanoparticles anchored on poly(3,4-ethylenedioxythiophene) and their photocatalytic investigation. Mater Chem Phys 273:125174
13. Jilani M et al (2022) Sulfonated polyaniline encapsulated graphene@g-$C_3N_4$ nanocomposites for phenol degradation. Appl Surf Sci 578:151896
14. Liu Z et al (2021) p–n Heterostructural g-$C_3N_4$/PANI composites for enhanced photocatalytic activity. ACS Appl Mater Interfaces 13(1):894–907
15. Wu D et al (2021) Design of conducting polymer-based composites for wastewater treatment applications. Water Res 189:116630
16. Chopan T, Chishti S (2023) Exploring conductive polymers for photodegradation of tetracycline hydrochloride. Mater Sci Eng B 285:115965
17. Zhang Y et al (2023) Light-activated conducting polymers for water purification: advances and challenges. Nat Commun 14:567

18. El-Bery HM et al (2023) Hybrid conducting polymers and their $TiO_2$ composites for hydrogen evolution. J Energy Chem 65:429–443
19. Dinh NT et al (2021) Enhanced photocatalytic activity of $CuWO_4$ by PPy sensitization. J Photochem Photobiol A Chem 404:112896
20. Munir M et al (2021) Development of PEDOT:photosensitization of rutile $TiO_2$ LEDs for water/water oxidation catalysis: 'green' solar fuels from PSS/g-$C_3N_4$ photocatalysts. Catal Today 376:44–52
21. Wang X et al (2023) Role of hybrid conductive polymers in visible light-driven $CO_2$ reduction. Chem Rev 123:1874–1905
22. Iqbal A et al (2022) Applications of polyaniline nanostructures in advanced photocatalysis. J Clean Prod 380:134856
23. Roy VAL et al (2023) Studies on synthesis of PANI-coated $TiO_2$ nanostructures for degradation of dyes. Mater Sci Semicond Process 149:106995
24. Saeed K et al (2023) Synergistic effects of conductive polymers and metal oxides for environmental applications. Environ Chem Lett 21:409–428
25. Rehan M et al (2022) Nanocomposites of conductive polymers for solar fuel production. J Power Sources 528:231409
26. Yin S et al (2023) Green syntheses and applications of conducting polymers for environmental remediation. Sustain Mater Technol 32:e00675
27. Mohammad F et al (2023) Functionalized polyaniline for photocatalytic applications. Polym Chem 14:2476–2489
28. Ghosh D et al (2024) Hybrid conducting polymers in advanced photocatalysis: a comprehensive review. Prog Mater Sci 142:101070
29. Chakravorty A, Roy S (2024) A review of photocatalysis, basic principles, processes, and materials. Sustain Chem Environ 8:100155
30. Hasan MB, Parvez MM, Abir AY, Ahmad MF (2025) A review on conducting organic polymers: concepts, applications, and potential environmental benefits. Heliyon 11(3):e42375
31. Saianand G, Gopalan A-I, Wang L, Venkatramanan K, Roy VAL, Sonar P, Lee D-E, Naidu R (2022) Conducting polymer-based visible light photocatalytic composites for pollutant removal: Progress and prospects. Environ Technol Innov 28:102698
32. Namsheer K et al (2021) Conducting polymers: a comprehensive review on recent advances in synthesis, properties, and applications. RSC Adv 11(10):5659–5697
33. Namsheer K, Rout CS (2021) Conducting polymers: a comprehensive review on recent advances in synthesis, properties, and applications. RSC Adv 11(10):5659–5697
34. Gilja V et al (2018) Preparation of a PANI/ZnO composite for efficient photocatalytic degradation of acid blue. Polymers 10(9):940
35. Ramesh N, Lai CW, Johan MRB, Mousavi SM, Badruddin IA, Kumar A, Sharma G, Gapsari F (2024) Progress in photocatalytic degradation of industrial organic dye by utilizing silver-doped titanium dioxide nanocomposite. Heliyon 10(24):e40998
36. Armaković S, Armaković S, Savanović M (2024) Photocatalytic application of polymers in removing pharmaceuticals from water: a comprehensive review. Catalysts 14(7):447
37. Kazim H, Sabri M, Al-Othman A, Tawalbeh M (2024) Functionalized conducting polymers in photocatalysis and opportunities for artificial intelligence applications. Nano-Struct Nano-Obj 40:101371
38. Sood Y, Singh K, Mudila H, Lokhande PE, Singh L, Kumar D, Kumar A, Mubarak NM, Dehghani MH (2024) Insights into properties, synthesis, and emerging applications of polypyrrole-based composites and future perspectives: a review. Heliyon 10(13):e33643
39. Abu-Sari et al (2022) A review of polymerization fundamentals, modification methods, and challenges of using PPy-based photocatalyst for perspective applications. J Environ Chem Eng 10(6):108725
40. Abu-Thabit NY (2016) Chemical oxidative polymerization of polyaniline: a practical approach for preparation of smart conductive textiles. J Chem Educ 93(9):1606–1611
41. Fomo G, Waryo T, Feleni U, Baker P, Iwuoha E (2019) Electrochemical polymerization. In: Polymers and polymeric composites: a reference series, pp 105–131

42. Song Y, Fan J-B, Wang S (2017) Recent progress in interfacial polymerization. Mater Chem Front 1(6):1028–1040
43. Gangwar N, Gangwar C, Sarkar J (2024) A review on template-assisted approaches & self-assembly of nanomaterials at liquid/liquid interfaces. Heliyon 10(17):e36810
44. Abel SB, Frontera E, Acevedo D, Barbero CA (2022) Functionalization of conductive polymers through covalent postmodification. Polymers 15(1):205
45. Nguyen D, Yoon H (2016) Recent advances in nanostructured conducting polymers: from synthesis to practical applications. Polymers 8(4):118
46. Tran VV, Nu TTV, Jung H-R, Chang M (2021) Advanced photocatalysts based on conducting polymer/metal oxide composites for environmental applications. Polymers 13(18):3031
47. Lee JM, Choi BG, Lee WH, Kang TH, Jung YM (2022) Conducting polymer-based composites for environmental applications. J Ind Eng Chem 103:300–317
48. Bai Y, Zhang S, Yang Y, Wang S, Zhang H (2024) Conducting polymer-based photocatalysts: synthesis, mechanisms, and environmental applications. Chem Eng J 456:141251
49. Kim HJ, Park JW, Kim JH, Yoon SH (2023) Conducting polymer-based photocatalysts: recent progress and future perspectives. Mater Today Chem 27:101133
50. Kaur H, Sharma RK (2023) Recent developments in conducting polymer-based nanocomposites for photocatalytic applications. J Mol Liq 383:122051
51. Sahoo S, Patra BC, Ghosh S, Das P (2023) Conducting polymer nanocomposites for enhanced photocatalytic applications: a review. Surf Interfaces 41:101002
52. Zhang J, Wang Y, Li L, Sun X (2023) Advances in conducting polymer-based nanomaterials for photocatalytic degradation of organic pollutants. Adv Funct Mater 33(8):2301238
53. Ghosh A, Dutta K (2024) Recent trends in metal-doped conducting polymers for advanced photocatalysis. Mater Sci Eng R Rep 152:100786
54. Yadav A, Prasad S, Tiwari A (2023) Conducting polymer-based metal oxide nanocomposites for environmental remediation. Environ Sci Technol 57(18):10245–10265
55. Xie X, Ma H (2024) Hybrid nanomaterials based on conducting polymers for enhanced photocatalysis. Nano Energy 112:107456
56. Park SY, Kim TH, Lee JW (2024) Conducting polymer-based hybrid materials for energy-efficient photocatalysis. Energy Environ Sci 17(5):2351–2378
57. Liu H, Wang Q, Zhao X (2023) Recent developments in conducting polymer-based hybrid materials for photocatalytic water treatment. Chem Soc Rev 52(11):4567–4592
58. Singh R, Kumar V (2024) Advances in conducting polymer-based nanocomposites for photocatalytic hydrogen evolution. J Energy Chem 85:123456
59. Zhao W, Sun Y, Lin C (2023) Conducting polymer-metal oxide nanocomposites for efficient degradation of organic pollutants. Environ Pollut 321:118765
60. Wang L, Li H, Xu Y (2023) Conducting polymer nanocomposites for photocatalytic CO2 reduction: current progress and challenges. Appl Catal B Environ 316:123421
61. Patel D, Shah R, Mehta B (2024) Recent advances in conducting polymer-carbon nanocomposites for wastewater treatment. J Hazard Mater 454:131567
62. Kim JW, Cho H (2023) Conducting polymer-based photocatalysts for sustainable energy applications. ACS Sustain Chem Eng 11(9):23045–23062
63. Lin Y, Zhang X, Wang Z (2024) Hybrid conducting polymer-metal nanocomposites for photocatalytic applications. Mater Adv 5(3):1124–1145
64. Sharma V, Kumar P (2023) Conducting polymer-based heterostructured materials for photocatalytic hydrogen generation. Renew Energy 212:123789
65. Xue H, Ren J (2024) Conducting polymer-derived nanomaterials for water purification. Nano Today 50:107632
66. Huang C, Zhang J (2023) Conducting polymer-based photocatalysts for CO2 conversion. Cat Sci Technol 14(2):678–701
67. Wang H, Zhang P (2023) Conducting polymer-metal sulfide nanocomposites for photocatalysis. Chem Eng J 468:142786
68. Jiao Y, Xu X (2024) Recent developments in conducting polymer-based photocatalysts for environmental remediation. Environ Sci Nano 11(4):1745–1763

69. Gupta S, Mishra R (2023) Conducting polymer-carbon nanocomposites for photocatalytic dye degradation. Colloids Surf A Physicochem Eng Asp 657:130125
70. Zhao Y, Li C (2024) Metal-organic frameworks coupled with conducting polymers for enhanced photocatalysis. Adv Funct Mater 34(7):2400123
71. Kim T, Park J (2023) Conducting polymer-graphene hybrid materials for solar-driven photocatalysis. Mater Sci Eng B 311:116045
72. Singh N, Verma A (2024) Recent trends in conducting polymer-based nanocomposites for CO2 reduction. ChemElectroChem 11(3):e202400567
73. Huang X, Wu Y (2023) Self-assembled conducting polymer nanocomposites for photocatalytic degradation of organic pollutants. J Mater Chem A 11(12):2456–2478
74. Zhao P, Lin J (2024) Conducting polymer-derived carbon materials for sustainable energy applications. Nano Energy 113:107928
75. Patel S, Desai R (2023) Advances in conducting polymer-metal nanohybrids for photocatalytic water splitting. ACS Appl Energy Mater 6(5):2145–2163
76. Wang J, Zhang L (2024) Conducting polymer-inorganic hybrid materials for visible-light-driven photocatalysis. Chem Soc Rev 53(8):5672–5691
77. Kumar A, Sharma R (2023) Role of conducting polymers in enhancing charge separation in photocatalysis. Langmuir 40(2):1023–1045
78. Liu Z, Chen M (2024) Hybrid conducting polymer-based nanomaterials for pollutant removal and energy conversion. Appl Surf Sci 658:154832
79. Gupta P, Singh V (2023) Conducting polymer-decorated metal oxide photocatalysts for hydrogen evolution. Int J Hydrog Energy 48(14):8021–8038
80. Xie Y, Luo Q (2024) Recent developments in conducting polymer-based visible-light photocatalysts. J Catal 423:123891
81. Yang C, Sun H (2023) Synergistic effects of conducting polymers in heterojunction photocatalysts. Phys Chem Chem Phys 25(9):11234–11250
82. He Z, Tang X (2024) Conducting polymer-supported photocatalysts for CO2 photoreduction. Energy Environ Sci 17(3):1456–1478
83. Zhou W, Ma J (2023) Conducting polymer-based nanocomposites for solar fuel generation. J Power Sources 635:231784
84. Singh A, Patel B (2024) Advanced hybrid conducting polymer photocatalysts for sustainable applications. Sol Energy Mater Sol Cells 256:111258
85. Lin X, Wang T (2023) Conducting polymer-derived carbon nanomaterials for water splitting applications. Carbon 214:112345
86. Zhang P, Zhao H (2024) Conducting polymer-oxide heterostructures for photocatalytic applications. J Colloid Interface Sci 631:127854
87. Sun Y, Feng L (2023) Recent advances in conducting polymer-based catalysts for electrochemical CO2 reduction. Adv Energy Mater 14(1):2304876
88. Wu J, Liang X (2024) Conducting polymer-based photocatalysts for enhanced charge transport and stability. Chem Eng J 482:123987
89. Kim H, Lee S (2023) Photocatalytic applications of conducting polymer-metal oxide nanocomposites. J Photochem Photobiol A Chem 443:114795
90. Li R, Zhou Y (2024) Hybrid conducting polymer photocatalysts for green hydrogen production. Int J Hydrog Energy 49(6):14532–14549
91. Zhao C, Xu Q (2023) Conducting polymer nanocomposites for visible-light-driven photocatalysis. Mater Today Chem 30:100904
92. Wang L, Huang B (2024) Metal-organic framework/conducting polymer hybrids for efficient solar energy conversion. ACS Appl Mater Interfaces 16(4):3176–3190
93. Patel D, Verma K (2023) Role of polyaniline-based nanocomposites in photocatalysis. J Environ Chem Eng 11(2):115430
94. Sun J, Luo C (2024) Conducting polymer-derived nanocarbon materials for solar-driven catalysis. Carbon 220:113501
95. Xie Y, Tang W (2023) Hybrid organic-inorganic photocatalysts incorporating conducting polymers. J Catal 427:124115

96. Singh V, Kumar R (2024) Conducting polymer-assisted charge transfer in semiconductor photocatalysts. Adv Mater Interfaces 11(8):2301746
97. Zhang Q, Lin M (2023) Hybrid photocatalytic systems integrating conducting polymers and metal oxides. Phys Chem Chem Phys 25(11):16423
98. He P, Ma Y (2024) Nanostructured conducting polymer composites for CO2 photoreduction. J $CO_2$ Util 75:103486
99. Feng L, Zhao H (2023) Conducting polymer heterojunctions for enhanced solar-to-fuel conversion. Sol Energy 265:112756
100. Zhang Y, Wang J (2024) Conducting polymer-based nanohybrids for photocatalytic environmental remediation. Environ Sci Technol 58(2):925–940
101. Kumar A, Sharma N (2023) Carbon-doped conducting polymers for water purification. Chem Eng J Adv 18:100579
102. Lee S, Park D (2024) Emerging trends in conducting polymer-based photocatalysts for wastewater treatment. Environ Res 242:117915
103. Chen X, Li F (2023) Conducting polymer-modified TiO2 for enhanced photocatalytic efficiency. Appl Catal B Environ 335:122589
104. Wu M, Zhang R (2024) Electrospun conducting polymer fibers as photocatalysts for organic dye degradation. J Appl Polym Sci 141(5):e53367
105. Zhao T, Liu C (2023) Conducting polymer-based quantum dots for photocatalytic water splitting. ACS Nano 18(2):1463–1478

# Chapter 6
# Tailored Conducting Polymers for Adsorption of Dyes and Heavy Metals

**Gyan Singh, Paramjit Singh, Samina Husain, and Sapana Jadoun**

**Abstract** To provide a cutting-edge investigation, this extensive chapter explores the field of conducting polymers for the adsorption of dyes and heavy metals, aiming to educate the larger materials community. These conducting polymers, well-known for their versatility, have attracted much attention because of their adaptability to various applications, such as environmental remediation. This chapter presents a thorough overview encompassing the conducting polymers, types of conducting polymers, and properties and applications. Introducing new and advanced water treatment technologies ensures the suitability of water for drinking purposes and, simultaneously, eliminates pollutants. As a result, this chapter also aims to give thorough information regarding pollutants such as heavy metals, dyes, and other pollutants, as well as how to remove heavy metals and dyes from wastewater using adsorption techniques. The procedure is thought to be the most effective way to remove dyes and heavy metals, influencing adsorption. Conducting polymers' superior qualities have enabled them to be employed in environmental remediation applications. This chapter offers essential insights for continuing improvements and

G. Singh
Nanomaterials and Polymer Energy Laboratory, Centre for Nanoscience and Nanotechnology, New Delhi, India

Facultad de Ciencias Químicas, Departamento de Química Analíticae Inorgánica, Universidad de Concepción, Concepción, Chile

P. Singh
Facultad de Ciencias Químicas, Departamento de Química Analíticae Inorgánica, Universidad de Concepción, Concepción, Chile

S. Husain (✉)
Nanomaterials and Polymer Energy Laboratory, Centre for Nanoscience and Nanotechnology, New Delhi, India
e-mail: shusain3@jmi.ac.in

S. Jadoun (✉)
Sol-ARIS, Departamento de Química, Facultad de Ciencias, Universidad de Tarapacá, Arica, Chile
e-mail: sjadoun@academicos.uta.cl



S. Jadoun, N. Kumari Jangid (eds.), *Tuning of Conducting Polymers*, Springer Series on Polymer and Composite Materials,
https://doi.org/10.1007/978-3-032-00461-1_6

research in this dynamic subject. It also focuses on present issues, recommendations, and future opportunities.

**Keywords** Conducting polymers · Adsorption · Dyes · Heavy metals

## 6.1 Introduction

Conductive polymers (CPs) are widely recognized for their exceptional properties. Conjugated π-bonds comprises of CPs with unique optical, electrical, and physical characteristics. Recent years have seen a massive increase in the usage of conductive polymers and related nanocomposites in various innovative and research domains. It is generally expected and regularly stated that they would be used as adsorbents for the adsorption and dyeing of diverse environmental contaminants as well as biosensors, gas sensors, and corrosion inhibitors in various environments and biological applications [1].

This inclusive chapter is devoted to the use and applications of conductive polymers in environmental restoration. Although it includes a section on environmental cleanup, this chapter is the only one that offers a deeper understanding of the practical uses of conductive polymers. In addition to this chapter, another author has emphasized using conductive polymers in water treatment [2]. Other studies exist, although they generally focused on conductive polymers and their composites. This chapter aims to compile recent advancements in environmental remediation using various conductive polymers (such as polypyrrole, polyaniline, and polythiophene) and their nanocomposites into a single, thorough this chapter. This chapter also sheds light on the various reported aspects of CPs as adsorbents for heavy metal removal, maximum adsorption capacities, adsorption mechanisms, adsorption, and kinetic models [3].

For the past few decades, dye and heavy metal contamination of water has been a serious public health concern, and numerous research studies have been done to reduce it. Ion exchange, chemical precipitation, membrane filtration, and other contemporary mitigation strategies are occasionally insufficient and/or expensive. To remove dyes and heavy metal ions from aqueous solution more practically and economically in the future, an attempt has been undertaken in this study to create a new conducting polymer adsorbent [4].

The most valuable element on Earth is water. The population is growing, driving demand for drinkable water, but contemporary industrialization is also producing more wastewater. Aquatics and human life are increasingly at risk from water pollution caused by different colors and heavy metal ions. Even at minimal levels, exposure to these dyes and heavy metal ions can have immediate and long-term effects on the body [5].

The discharge of colorants from the various applications of dyes in the food, paper and pulp, plastics, leather, textiles, cosmetics, and pharmaceutical industries poses a risk to aquatic life because it blocks sunlight from entering the stream.

Complex aromatic chemicals, which are more persistent and challenging to biodegrade, make up the majority of artificial dyes. Additionally, some dyes have been linked to human allergies, dermatitis, skin irritation, and even cancer. Heavy metals cannot be broken down or biodegraded, just like dyes. As a result, environmental scientists and engineers continue to face difficulties in completely and thoroughly eliminating these unwanted colors and heavy metals from water bodies [6].

Historically, dyes and heavy metal ions have been removed from wastewater or aqueous solutions using various techniques, including chemical precipitation, ion exchange, membrane filtration, adsorption, electrochemical technologies, and so on. Adsorption is frequently employed because of its many benefits for controlling water pollution, including low initial costs, straightforward construction and operation, and non-toxicity [7].

Due to their potential uses in various dyes and heavy metal ions, ease of preparation and synthesis, regeneration process, and ease of operation, conducting polymer-based adsorbents such as polyaniline, polyethelenamine, polypyrrole, and their composites have drawn a lot of attention in recent decades [8]. Because of their simplicity of synthesis, biocompatibility, and redox characteristics, onducting polymer-based adsorbents are essential for the removal of different dyes and heavy metal ions. Insufficient advancements in cutting-edge technology suggest a serious environmental catastrophe in the modern era. Out of all the current environmental problems, wastewater production in aquatic bodies is a worldwide concern [9]. Wastewater from various sources, both point and non-point, enters the freshwater body and mixes with it. A fundamental requirement of human society is the availability of clean water for every individual, but this is currently unattainable due to a significant rise in the discharge of harmful chemicals into water bodies [10], the quantity of unwanted materials or toxins in water bodies has exceeded the allowable level recommended by regulatory bodies, raising environmental concerns for the general public. Heavy metals and the high concentration of dissolved solids are the two primary contaminants of wastewater. Therefore, it is impossible to stop the microbial pathogens from growing and causing dangerous water-borne illnesses [11]. Developing low-cost, environmentally friendly, and highly effective treatment solutions is a good way to address these global issues. The industrialized process with appropriate resource management at the nanoscale level is covered by nanotechnology. Many scientific and technological fields, including energy, various industrial processes, sensors, and environmental applications, have extensively used nanotechnology in recent years [12].

According to published research, nanotechnology-based methodologies can provide effective systems and solutions for many challenging issues that traditional approaches have never been able to resolve. Polymer-based nanoadsorbents have many uses, including conducting materials, sensors, electronics, photonics, medicine, and pollution reduction [13]. Literature confirms that various adsorbents may be used to clean wastewater or aqueous solutions that contain dyes or metal ions. Of the various kinds of accessible polymers, polyaniline (PANI) and its derivatives are highly effective conducting polymers for eliminating a variety of heavy metal ions and dyes that are causing wastewater or aqueous solutions to become contaminated [14].

## 6.2 Conducting Polymers

Before their discovery, conducting polymers were employed as insulators in the electronics industry [15]. In the late 1970s, Alan Heegar, Dr. Hideki Shirakawa, and Alan MacDiarmid were awarded with 2000 Nobel Prize in Chemistry. for their ground breaking discovery of conducting polymers. Organic polymers that can conduct electricity are called conducting polymers (polyacetylene, for instance) [16–25]. The amount and mobility of electrons in each material determine the conductivity rate. According to a conventional assumption, the outer electron is free to carry and flow charge because of the electron bumping factor. These kinds of substances are typically semiconductors or metallic conductors [15]. Beyond the many benefits that conducting polymers exhibit, one of the main benefits is their "process ability," primarily due to dispersion. These polymers are organic and may effectively provide better electrical conductivity, but the polymers are generally thermoplastic (nonthermo-formable). Their electrical characteristics are fine-tuned using the sophisticated dispersion and organic synthesis approach [26].

### *6.2.1 Types of Conducting Polymers*

According to the literature, polymers that contain carbon as a backbone chain are called organic polymers, and those that do not are called inorganic polymers. However, conducting polymers are categorized according to their structure and how sulfur and nitrogen are attached. The types of conducting polymers are shown in Fig. 6.1.

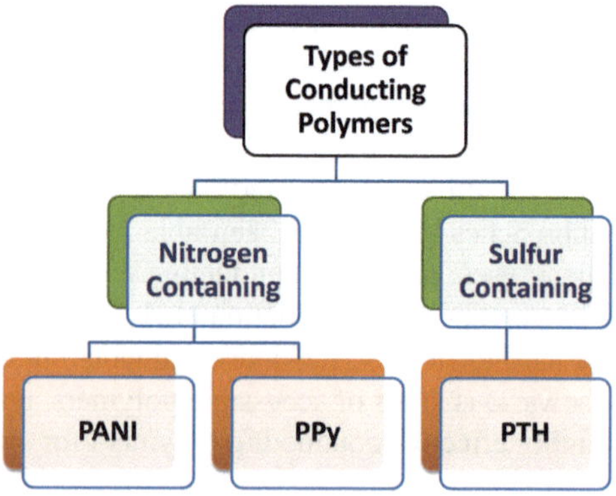


**Fig. 6.1** Types of conducting polymers

#### 6.2.1.1 Polyaniline (PANI)

The conducting polymer family includes polyaniline (PANI), which stands out due to its simple production process and strong environmental stability. PANI's application in electronics and engineering is facilitated by conductivity endorsing emeraldine salt and many oxidation states; however, its use is not often acknowledged due to its instability at neutral pH and higher temperatures. A temperature of more than 150 °C and a pH of 7 cause the polymer chain to de-dope, which reduces the polymer's conductivity. PANI has numerous distinct features. For their widespread application in various fields, the chemical structure, physical morphology, and oxidation–reduction states are generated in compliance with the requirements of the synthesis technique, the types of oxidants utilized, the type and quantity of dopants and solvents involved, and the temperature and time regimes. PANI is created by chemically or electrochemically polymerizing aniline with certain dopants present [27, 28].

#### 6.2.1.2 Polypyrrole (PPy)

Among the many conducting polymers, PPy has been investigated the most due to its facile synthesis, stability in oxidized form, high electrical conductivity, and favorable redox characteristics. This polymer is positively charged and conducts in its oxidized state, but overoxidation causes it to lose both properties. It is insulating and mechanically weak when in its neutral condition. A promising option for several applications, including supercapacitors, batteries, biosensors, antistatic coatings, textiles and fabrics, EMI shielding, and drug delivery applications, PPy has demonstrated exceptional inherent qualities. The electropolymerization conditions have a significant impact on the inherent characteristics of PPy. Because of its excellent stimulus responsiveness, PPy is an extremely intelligent biomaterial that can be dynamically controlled by applying an electric field. PPy exhibits good biocompatibility both in vitro and in vivo [29].

#### 6.2.1.3 Polythiophene (PTH)

Polythiophene exhibits excellent potential for environmental applications due to its conductivity exceeding 100 S/cm and its chemical and thermal stability [30]. The polo polythiophene can be readily assembled by direct polymerization of monomers, and they can then be used to immobilize enzymes and aptamers into the polythiophene layer [31]. It can be produced chemically and electrochemically. In recent years, polythiophene polymers have gained more attention in the synthesis and industrial sectors due to their lower band gap energy, increased environmental immovability, and superior thermal steadiness. In addition, their exciting properties, including exceptional electronic, optical, and semiconducting activities, enhanced mechanical strength, rigidity, and ease of use, have drawn much attention to polythiophene composites. Due to their potential efficiency in improving contact formation with metal electrodes and their stability in ambient settings, they are widely utilized in solar cells. Additionally, because of its transparency, it exhibits exceptional optical qualities. Because it has conjugated double bonds in its backbone, it is also regarded as a very important conducting polymer. The formation of polarons and bipolarons in the backbone of polythiophene conducting polymers through their redox states might contaminate them to improve their conducting properties [32].

### *6.2.2 Properties of Conducting Polymers*

Depending on the need in a particular industry, conductive polymers have limited large-scale applications because of their poor processability. The conducting polymer has several drawbacks, including high manufacturing costs, toxicity, low solvent solubility, inconsistent material quality, and an inability to melt directly. It shows promise for antistatic materials but may also be used in commercial displays. Recent research shows they are also advantageous instruments for organic solar cells and light-emitting diodes [33]. The quick application of conducting polymers to processable objects with adequate physical and electrical properties at a reduced cost draws attention to current research. Compared to their non-nanostructured counterparts, the capacitance values of the nanostructured conducting polymers are significantly higher. Several research studies have demonstrated that nanostructured conducting polymers can be used as efficient heavy metal ion or dye adsorbers due to their greater surface area and improved adsorption capabilities [32]. The general properties of conducting polymers are given in Table 6.1.

### *6.2.3 Application of Conducting Polymers*

#### 6.2.3.1 Pollutants in Water

Every type of life on Earth depends on water, a valuable natural resource. Utilized for forestry, industrial activity, cattle production, hydropower generation, and agriculture, it is undoubtedly a unique gift from nature. Water bodies that aren't used for cooking, drinking, bathing, or other purposes have many risks. Pollution is the term used to describe the contamination of the environment. Human activities impact the

**Table 6.1** The general properties of conducting polymers

| | |
|---|---|
| Chemical properties | Ion transportation, catalytic properties, oxidation-reduction behavior, electrochemical effect, photoactivity |
| Mechanical properties | Flexible and lightweight |

environment in many parts of the world because of widespread urbanization, industrialization, and agricultural methods [34]. Point and non-point sources are the two leading causes of pollution in bodies of water. Water bodies' primary sources of pollution are commercial, industrial, and agricultural operations [35]. Sewage leaks, population growth, industrial waste dumped into water bodies, drilling operations, floods during rainfall, heavy metals, mining operations, animal waste, deforestation, and eroded sediments are the primary causes of declining water quality (Fig. 6.2). Herbicides and pesticides are washed away with rainwater when excessive rainfall significantly threatens living things. An excellent strategy for enhancing a society without endangering the environment is sustainable development. In terms of environmental management and conservation, it gives people a better chance [36]. Protecting water quality is on society's agenda Because of the severe problems with water contamination, ranging from surface to groundwater, lead to worldwide water scarcity. The common pollutants in water bodies are shown in Fig. 6.2.

#### 6.2.3.2 Heavy Metals

The presence of heavy metals in water due to industrial influents and effluents is the leading cause for concern. It is on the list of inorganic pollutants that can be fatal to aquatic life, plants, and people [37]. Heavy metals typically enter the environment through various sources, including surface finishing, mining, animal waste, and agricultural waste. Furthermore, burning coal, using pesticides and fungicides, mining operations, the paper industry, and other businesses have produced a variety of heavy metals, including arsenic, lead, mercury, nickel, cadmium, chromium, and copper [38]. Pesticides, mining, and smelting are the sources of arsenic in the environment. Mining, sewage waste, and pesticides bring lead, copper, cadmium, and mercury [39]. Cell membrane damage, photosynthesis, and plant metabolic rate are all impacted by rising heavy metal ions [40]. Via the food chain and food web, metal toxicity impacts both human and animal life. Its effects include poor blood pressure,

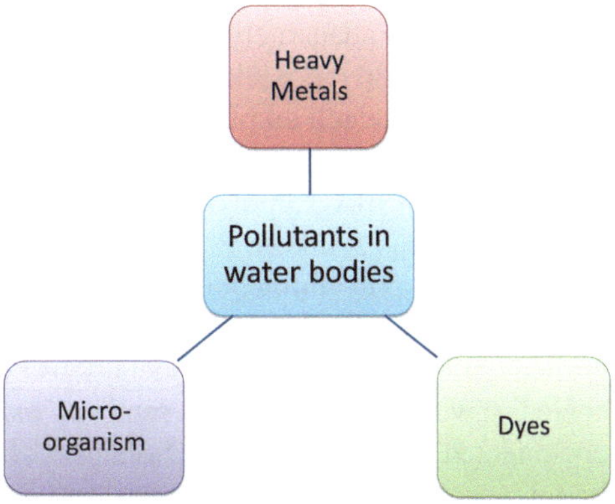


**Fig. 6.2** Common pollutants in water bodies

headaches, diarrhea, skin irritation, and vomiting [41]. Cadmium damages the kidney, lungs, and bones when it enters the body through water, and arsenic causes cancer in the lungs, liver, and gallbladder. When it comes to kids, lead causes memory loss, attention problems, and most of all, aggression. High levels of lead exposure harm men's testicles while causing miscarriages in pregnant women [42, 43]. Being a long-distance traveler and a worldwide pollutant, mercury is unique. It is ingested through the bloodstream and manifests as methyl and dimethyl mercury [44].

This chapter discusses the different polyaniline-based adsorbents separated into OPs and heavy metal ions (HMIs). A recent review [14] demonstrates how polyaniline and its derivatives can remove different HMIs and chemical colors. Under ambient settings, polyaniline-modified multiwalled carbon nanotubes [45] remove Pb (II) ions, one of the HMIs eliminated by polyaniline-based nanocomposites. It was determined that the PANI modification significantly increased the adsorption capacity because of the high affinity of the amine and imine functional groups of PANI toward Pb (II) ions. To remove Cr (VI), polyaniline was produced on the surfaces of jute fibers and reported [46]. A maximal monolayer adsorption capacity of 62.9 mg/g was reported at the ideal experimental parameters: pH 3 and 20 °C. The overall chromium adsorption decreased as the temperature rose, indicating that the chromium adsorption process is exothermic. Furthermore, Ren et al. [47] reported using polyaniline-coated polyacrylonitrile fiber mats to remove Cr (VI) and found that the PANI/PAN composite had better removal capacities for Cr (VI). It was found that the maximal adsorption capacity rose as the temperature rose, indicating that the adsorption process was endothermic. In a different study, cellulose–polyaniline composites and carbon fiber cloth coated with polyaniline were used to remove Cr (VI) [48]. They concluded that adding PANI to their substrates increased their adsorption capabilities and rates. They stated that the adsorption process is physical adsorption in nature since the pseudo-second-order kinetic model accurately describes the experimental results. A comprehensive report on removing Cr (VI) using PANI-based adsorbents was published [49]. In their research on the application of sodium alginate–polyaniline nanofibers for the adsorption of Cr (VI) ions [50], noted that the adsorption process required electrostatic interactions between the sodium alginate–polyaniline nanofibers and Cr (VI). The pseudo-second-order model adequately described the experimental data, and the Langmuir isotherm plots at 30 °C yielded a maximum adsorption capacity of 73.34 mg/g. Additionally, research employing a polyaniline nanocomposite coated on rice husk ash to remove mercury (II) from aqueous media. Ghorbani et al. [51], Li et al. [52] reported using SBA-15/polyaniline mesoporous composites to remove resorcinol from an aqueous solution. Pb (II) and Cu (II) ions are extracted from an aqueous medium using polyaniline nanofibers encapsulated on alginate microspheres [53]. Pb (II) and Cd (II) ions were removed from aqueous solution using polyaniline-grafted chitosan, and cadmium metal ions Cd (II) were removed using polyaniline-coated sawdust [54]. Cu (II) removal using PANI–clay hybrid materials was documented [55]. Furthermore, Soltani et al. [56] reported using magnetic polyaniline-doped strontium titanium nanocomposites to remove arsenic (III) ions as an adsorbent. The Langmuir isotherm model identified the maximal adsorption capacity as 67.11 mg/g, and thermodynamic investigations inferred physisorption and an exothermic adsorption mechanism for arsenic (III) ions [57].

#### 6.2.3.3 Dyes

Chemicals called dyes are extremely carcinogenic and can be found in water bodies. Because they contain highly toxic compounds such as degradable organics, hazardous dyes, detergents, inorganic salts, heavy metals, stabilizers, and stabilizers, they are regarded as one of the leading causes of wastewater formation. Approximately, 30 million tons of textiles are manufactured annually worldwide, requiring 700,000 dyes [58, 59]. Fabric sprinting, leather coloring, farming research, photo electrochemical cells, and hair complexion are only a few of its many applications. A wide range of contaminants, including metals, temperature, salts, suspended solids (SS), pH, chemical oxygen demand (COD), biochemical oxygen demand (BOD), and dense dark color, are present in the effluents produced by the dye business [60].

When so many dyes are used, their effluents are released into the environment, causing serious concerns. Most pollutants are typically produced during the dye production process's pretreatment, dyeing, printing, and finishing stages. Significant focus is being paid to finding a solution because of its hazardous chemical makeup and the challenges associated with its removal operations. The current challenge, however, is to develop cutting-edge technology that can be readily used in shorter processing times and with low-cost management plans in a sustainable, environmentally friendly manner. Only polyaniline (PANI) and its derivatives have been documented in the literature despite having notable detoxifying properties. As reported, polyaniline nanoparticles were used as an adsorbent material to remove Methylene blue (MB) dye [61]. According to their findings, the synthesized material can be employed as an effective adsorbent to remove MB from water. According to reports, the pseudo-second-order kinetic model fit the experimental data well. Additionally, they stated that both traditional PANI and PANI nanoparticles were used. The researchers observed that the pseudo-second-order kinetic model best explained the experimental results and concluded that PANI nanoparticles are more effective adsorbents for MB than traditional PANI powdered adsorbents [62, 63].

## 6.3 Removal of Heavy Metal Ions and Pollutants from Wastewater

### *6.3.1 Adsorption*

Adsorption is a widely utilized physical technique in water purification and remediation that provides a flexible and effective way to get rid of a wide variety of contaminants. The efficiency of adsorption is supported by its basic principles, which also direct the creation of sophisticated adsorbent materials for a range of uses. Adsorption is the process by which molecules, atoms, or ions from a gas, liquid, or dissolved solid stick to a surface. The whole volume of another material absorbing a substance is different from this. Adsorption in water remediation is the process by which pollutants are drawn to and build up on the surface of a solid adsorbent material. The adsorbent material provides a lot of surface area for adsorption and can be a porous solid such as polymeric resins, zeolites, or activated carbon. The pollutants are maintained on the surface of the adsorbent and are selectively eliminated as the water passes through the adsorbent [64–66].

Adsorption is a process that is used to protect the environment from water hazards as well as to produce color in chemicals. A gas or liquid solute builds up on the surface of liquid or solid adsorbents, such as zeolite, Na-betonies, and kaolinite, among others, during the adsorption process. This eventually results in the formation of an atomic or molecular film called an adsorbate [43]. It is widely utilized in commercial applications such as water purification, synthetic resin, and activated charcoal and it can function in the majority of natural physical, chemical, and biological systems. Adsorption is currently regarded as one of these techniques that works best for treating wastewater. The generalized adsorption mechanism is shown in Fig. 6.2. Compared to traditional methods, it is a fairly well-known, cost-effective, and efficient technology for removing heavy metals and dyes from a variety of industrial discharges. Adsorbents come in a variety of forms and are categorized as biological, natural, industrial, and agricultural waste [67]. For the removal of chromium (VI), the following adsorbents are frequently used: fly ash, activated carbon, and biological materials. It has been discovered that this approach is better than others in terms of design flexibility and simplicity, starting cost, operational ease, and insensitivity to harmful pollutants. Furthermore, it doesn't lead to the creation of dangerous substances. The most widely utilized adsorbents are natural zeolites, clay minerals, and activated carbon [68]. Certain adsorbents, which are inexpensive and readily accessible in large quantities, can also be made from industrial or agricultural waste. The adsorption mechanism of conducting polymers as shown in Fig. 6.3.

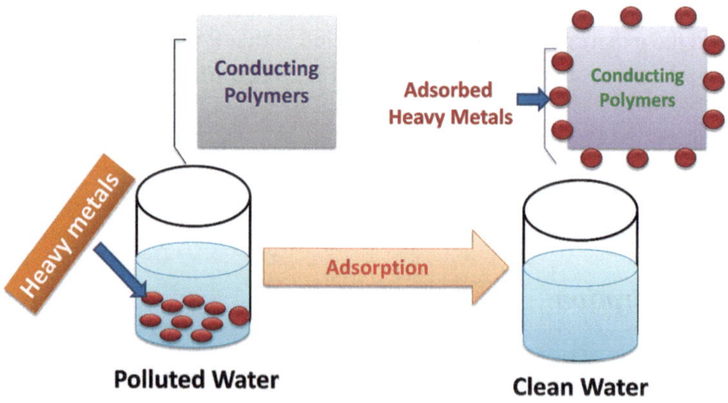


**Fig. 6.3** Adsorption mechanism of conducting polymers

### 6.3.2 *Factors Affecting Adsorption*

The adsorption process for the removal of heavy metals is affected by the following factors: time, pH, temperature, and concentration. Several important aspects impact the adsorption process' efficacy and efficiency: Surface area: a large surface area increases the adsorbent's capacity by providing more adsorption sites. Distribution of pore sizes: varied sized contaminant molecules may be accommodated by adsorbents with a range of pore sizes. Surface chemistry: the adsorbent's affinity for particular contaminants is influenced by the chemical makeup of its surface, such as the presence of hydrophilic groups. Molecular structure and size: certain hydrophilic groups and smaller molecules are more easily adsorbed. Solubility: pollutants that are less soluble tend to be absorbed more readily. Charge: Adsorption may be improved by electrostatic interactions between the contaminant and the adsorbent. pH: The adsorption process may be impacted by the pH of the solution, which may also have an impact on the charge and speciation of the contaminant and the adsorbent. Higher temperatures may reduce the adsorption capability because adsorption is often an exothermic process. Ionic strength: higher ionic strength may have an impact on the adsorbent's surface charge as well as the solubility and speciation of pollutants. Contact duration: greater contact durations give the adsorption equilibrium more time to be achieved. Mixing: effective mixing increases mass transfer and enhance the adsorbent-contaminant interaction. The factors affecting of adsorption are shown in Fig. 6.4.

#### 6.3.2.1 Effect of Adsorbent Dosages

In the adsorption process, the amount or dose of adsorbent is important because a higher adsorbent indicates a bigger number of accessible adsorption sites, which allows for the adsorption of more ions. There would be a large number of unabsorbed adsorption sites as the dosage increased, as would the ratio of adsorption sites to heavy metal ions. However, the adsorption rate can be decreased by

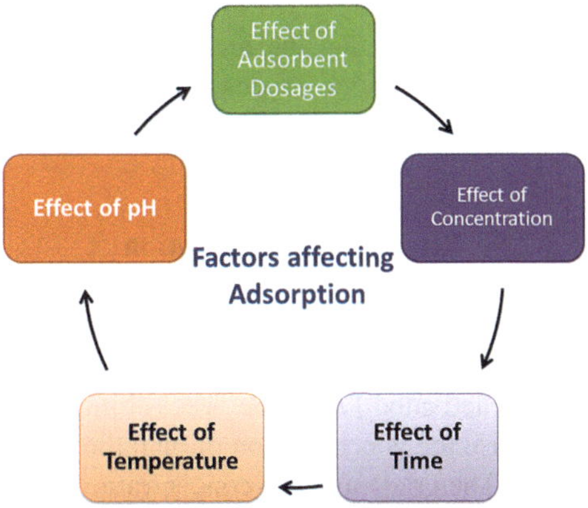


**Fig. 6.4** Factors affecting of adsorption

combining two or more adsorbents. The adsorbent's total surface area may decrease as a result of particle aggregation, which could be the cause [69].

#### 6.3.2.2 Effects of pH

One of the key factors influencing the adsorption process is the pH of the aqueous medium, which can directly impact the adsorbent's ability to absorb metal ions or dye by altering the degree of metal ionization and the surface properties of the material on which the adsorption process occurs [70]. The adsorption rate is influenced by the pH of the solution, which regulates the strength of the electrostatic charges held by the dye or ionized metal molecules. Low pH solutions generally improve the removal of anionic dyes and decrease the removal of cationic dyes in dye adsorption. On the other hand, at high pH values, the proportion of dye removal will rise for cationic dyes and fall for anionic dyes [71].

#### 6.3.2.3 Effects of Temperature

As the temperature rises, the metal ions propensity to rise causes the boundary layer's thickness to decrease, which lowers adsorption. Additionally, when the temperature increases from 40 to 80 °C, the amount of desorption decreases. This is because when the temperature rises, the force of attraction between the biomass surface and the metal ion weakens. The adsorption rate rises and the viscosity factor decreases with increasing temperature [72].

#### 6.3.2.4 Effects of Time

There are several contact times in the adsorption process. Efficiency, economical swell, and kinetics adsorption are among the adsorption processes that are impacted by contact time. Researchers used varying contact times and ranges to examine the adsorption capabilities. The adsorbate may have more time to interact with the adsorbent surface due to the extended contact period. However, other researchers have confirmed that the method's efficiency does not show further enhancement when the adsorption process' contact time reaches 4 h, which suggests the saturation limit [73].

#### 6.3.2.5 Effects of Concentration

When the mass transfer of ions from solution to the carbon surface increases due to an increase in the concentration of metal ions in a solution, the action of concentration is activated [74]. The accessibility of overloading biomass and binding sites will raise the concentration of metallic ions, which is why the removal percentages occur. The gradient force concentration caused the metallic ions to grow along with the concentration. Higher metal ion adsorption will come from the process of liquid to solid phase transfer, which will raise the likelihood of striking between the metallic ions and biomass [75]. One of the most popular techniques for removing dyes and heavy metals from wastewater is adsorption. The efficiency of dye and heavy metal removal will be improved by keeping the range within the ideal range and taking into account the variables that affect the adsorption process. Additionally, the use of nanotechnology can improve the efficiency of the adsorption process. Because of their great efficiency and removal capability, nanotechnology techniques can remove pollutants from wastewater. In contrast to their conventional counterparts, new and enhanced nanoadsorbents can be used for this purpose because of their high adsorption capacity and active adsorption sites due to their nanoscale size [76].

## 6.4 Conclusion and Future Prospects

### *6.4.1 Conclusion*

The demand for wastewater treatment is urgent due to strict waste water disposal regulations and a decrease in the supply of drinkable water. As a result, there is a pressing need for innovative and cutting-edge water purification technologies that guarantee that the water is safe to drink and get rid of any impurities or pollutants. Of the various methods used to treat water, adsorption is thought to be the most efficient. The adsorption process is the most efficient method of treating water because of its simplicity and flexibility, convenience of use, insensitivity to harmful contaminants, and cheap cost of installation, operation, and maintenance. As previously stated, a variety of parameters, such as adsorbent doses, temperature, solution pH, contact time, and the adsorbents chemical makeup, influences the adsorption process.

Furthermore, as nanotechnology has become a powerful instrument for effectively eliminating harmful substances from water or wastewater, its application improves the adsorption efficacy. The application of nanomaterials as adsorbers is receiving a lot of attention. Because of their effectiveness, affordability, and environmental friendliness, nano-engineered materials-also referred to as nanoadsorbents-such as polymer-, carbon-, inorganic-, and organic-based nanoadsorbents-offer the possibility of purifying water and wastewater more effectively than traditional techniques.

To be more precise, conducting polymer-based nanoadsorbents are possible candidates for the adsorption and exclusion of metals and dyes that are hazardous to aquatic life and the environment. They also have high electrical conductance and are environmentally stable. Because of the large number of amine and imine functional groups in the polymeric chain, PANI and its derivatives are among the conducting polymers that are utilized as efficient nano adsorbers of heavy metal ions and dyes. It is well known that conductive polymers and their composites are effective adsorbents for a variety of impurities and pollutants. Their intriguing redox properties and the inclusion of N, S, P, and O atoms in their chemical structure are mostly to blame for this. π–π interactions, hydrogen bonding, hydrophobic contacts, acid–base interactions, and electrostatic interactions are the most commonly described processes for the removal of organic dyes. The most popular methods for removing heavy metal ions include reduction, chelation, ion exchange, and electrostatic attraction. Methylene blue is said to be the most extensively researched pollution among organic dyes, while elimination of Cr (VI) is the most extensively researched contamination among heavy metals. Overall, it can be said that altering different kinds of prospective adsorbent materials with conductive polymers significantly increases the maximum adsorption capacities and adsorption rates of the unaltered adsorbents. Comparing conductive polymers to polythiophenes, polyaniline and polypyrrole have been thoroughly investigated as possible adsorption enhancers that could result in noticeably higher maximum adsorption capacities. Therefore, polythiophenes and their derivatives might offer chances for more investigation and study.

### *6.4.2 Future Prospects*

To remove a lot of heavy metals and colors from wastewater, conducting polymer-based nanoadsorbents have seen effectiveness. The extensive usage of these nano-adsorbents results in very little environmental pollution, making them environmentally beneficial. These nanoadsorbents' commercialization and processing are yet not adequately handled. Additionally, there hasn't been a thorough examination into the enhanced capabilities of conducting polymer-based nanoadsorbents. Therefore, in order to achieve improved kinetic sorption and enable the commercial operation of these effective nanoadsorbents materials, the choice of functional groups for material activation in the nanoadsorbents must be carefully considered.

One crucial factor that must be considered is the adsorbent's durability and ease of regeneration with the necessary level of recovery. Since the interactions between the metal ions would be crucial in determining the removal selectivity of the desired component, a thorough research of the behavior of multiple component systems is necessary. More thorough research is specifically needed to ensure the effectiveness of these nanoadsorbents and to bring them to market. This chapter aims to give a general review of conducting polymer-based materials for heavy metal ion adsorption in order to combat water pollution. The greatest option for wastewater treatment and for industry as well, the adsorption method of pollutant removal from wastewater is a substantial substitute for expensive conventional approaches. According to this theory, materials based on conducting polymers offer a variety of alternatives for the water-useful adsorption of heavy metals.

**Acknowledgement:** The authors are grateful for the support of the National Research and Development Agency of Chile (ANID) and the projects, FONDECYT Postdoctoral 3250134, 3240619.

## References


1. Khan MI et al (2021) Conductive polymers and their nanocomposites as adsorbents in environmental applications. Polymers (Basel) 13:3810
2. Taghizadeh A et al (2020) Conductive polymers in water treatment: a review. J Mol Liq 312:113447
3. Aigbe UO, Osibote OA (2020) A review of hexavalent chromium removal from aqueous solutions by sorption technique using nanomaterials. J Environ Chem Eng 8:104503
4. Jadoun S, Fuentes JP, Urbano BF, Yáñez J (2023) A review on adsorption of heavy metals from wastewater using conducting polymer-based materials. J Environ Chem Eng 11:109226
5. Oladimeji TE et al (2024) Review on the impact of heavy metals from industrial wastewater effluent and removal technologies. Heliyon 10:e40370
6. Ardila-Leal LD, Poutou-Piñales RA, Pedroza-Rodríguez AM, Quevedo-Hidalgo BE (2021) A brief history of colour, the environmental impact of synthetic dyes and removal by using laccases. Molecules 26:3813
7. Qasem NAA, Mohammed RH, Lawal DU (2021) Removal of heavy metal ions from wastewater: a comprehensive and critical review. NPJ Clean Water 4:36
8. Berber MR (2020) Current advances of polymer composites for water treatment and desalination. J Chem 2020:1–19
9. Muhammad Ekramul Mahmud HN, Huq AKO, Yahya R b (2016) The removal of heavy metal ions from wastewater/aqueous solution using polypyrrole-based adsorbents: a review. RSC Adv 6:14778–14791
10. Awasthi A, Jadhao P, Kumari K (2019) Clay nano-adsorbent: structures, applications and mechanism for water treatment. SN Appl Sci 1:1076
11. Järup L (2003) Hazards of heavy metal contamination. Br Med Bull 68:167–182
12. Bora T, Dutta J (2014) Applications of nanotechnology in wastewater treatment—a review. J Nanosci Nanotechnol 14:613–626
13. Anjum M, Miandad R, Waqas M, Gehany F, Barakat MA (2019) Remediation of wastewater using various nano-materials. Arab J Chem 12:4897–4919
14. Zare EN, Motahari A, Sillanpää M (2018) Nanoadsorbents based on conducting polymer nanocomposites with main focus on polyaniline and its derivatives for removal of heavy metal ions/dyes: a review. Environ Res 162:173–195
15. Le T-H, Kim Y, Yoon H (2017) Electrical and electrochemical properties of conducting polymers. Polymers (Basel). 9:150
16. Singh G, Kumar Y, Husain S (2020) High charge retention and optimization of polyaniline–titanium dioxide nanoparticles composite nanostructures for dominantly stable pseudocapacitive nature. J Energy Storage 31:101660
17. Singh G, Kumar Y, Husain S (2021) Improved electrochemical performance of symmetric polyaniline/activated carbon hybrid for high supercapacitance: comparison with indirect capacitance. Polym Adv Technol 32:4490–4501
18. Singh G, Alam M, Kumar Y, Husain S (2022) Highly capacitive mesoporous polyaniline spheres as scalable and high electrochemical performance supercapacitor electrode. ChemistrySelect 7
19. Singh G, Kumar Y, Husain S (2023) Fabrication of symmetric polyaniline/nano-titanium dioxide/activated carbon supercapacitor device in different electrolytic mediums: role of high surface area of carbon and facile interactions with Nano-titanium dioxide for high-performance supercapacitor. Energ Technol 11
20. Singh G, Kumar Y, Husain S (2023) Fabrication of high energy density symmetric polyaniline/functionalized multiwalled carbon nanotubes supercapacitor device with swift charge transport in different electrolytic mediums. J Energy Storage 65:107328
21. Singh G, Kumar Y, Husain S (2024) Nickel ferrite doped polyaniline composites: synthesis and analysis for a high-performance symmetric pseudo-supercapacitor. Energy Storage 6
22. Singh G et al (2022) Fabrication of polyaniline nanospheres as a good electrode material in supercapacitors. pp 393–400. https://doi.org/10.1007/978-981-16-7691-8_39

23. Singh G, Husain S (2022) Review on polyaniline-based composites with and without binder as advanced supercapacitor electrode materials. pp 551–582. https://doi.org/10.1007/978-981-19-0553-7_13
24. Singh G, Kumar Y, Husain S (2022) Fabrication of polyaniline structures for enhanced supercapacitance: effect of PANI morphologies. ECS Trans 107:15843–15858
25. Shumaila et al (2019) Facile field emission characteristics of polyaniline doped with MgB 2 nanowires. Mater Res Express 6:105315
26. Culebras M, Gómez C, Cantarero A (2014) Review on polymers for thermoelectric applications. Materials (Basel) 7:6701–6732
27. Beygisangchin M, Abdul Rashid S, Shafie S, Sadrolhosseini AR, Lim HN (2021) Preparations, properties, and applications of polyaniline and polyaniline thin films—a review. Polymers (Basel) 13:2003
28. Aksu Z, Kutsal T (1991) A bioseparation process for removing lead(II) ions from waste water by using C. vulgaris. J Chem Technol Biotechnol 52:109–118
29. Deshmukh K et al (2017) Biopolymer composites with high dielectric performance: interface engineering. In: Biopolymer composites in electronics. Elsevier, pp 27–128. https://doi.org/10.1016/B978-0-12-809261-3.00003-6
30. Liu H, Ge J, Ma E, Yang L (2019) Advanced biomaterials for biosensor and theranostics. In: Biomaterials in translational medicine. Elsevier, pp 213–255. https://doi.org/10.1016/B978-0-12-813477-1.00010-4
31. Mehmood U, Al-Ahmed A, Hussein IA (2016) Review on recent advances in polythiophene based photovoltaic devices. Renew Sust Energ Rev 57:550–561
32. Ramesan MT, Suhailath K (2017) Role of nanoparticles on polymer composites. In: Micro and nano fibrillar composites (MFCs and NFCs) from polymer blends. Elsevier, pp 301–326. https://doi.org/10.1016/B978-0-08-101991-7.00013-3
33. Verma A, Gupta R, Verma AS, Kumar T (2023) A review of composite conducting polymer-based sensors for detection of industrial waste gases. Sens Actuators Rep 5:100143
34. Xiaolong W, Jingyi H, Ligang X, Qi Z (2010) Spatial and seasonal variations of the contamination within water body of the Grand Canal, China. Environ Pollut 158:1513–1520
35. Baharin SNA, Rusmin R, Sambasevam KP (2022) Basic concept and application of conducting polymers for environmental protection. Chem Teach Int 4:173–183
36. Ilin I, Kalinina O, Iliashenko O, Levina A (2016) Sustainable urban development as a driver of safety system development of the urban underground. Procedia Eng 165:1673–1682
37. Cao Y et al (2017) Enhancing the soil heavy metals removal efficiency by adding HPMA and PBTCA along with plant washing agents. J Hazard Mater 339:33–42
38. Dixit R et al (2015) Bioremediation of heavy metals from soil and aquatic environment: an overview of principles and criteria of fundamental processes. Sustainability 7:2189–2212
39. Zhang H et al (2024) Contaminant transport modelling of heavy metal pollutants in soil and groundwater: an example at a non-ferrous smelter site. J Cent South Univ 31:1092–1106
40. Asgari Lajayer B, Ghorbanpour M, Nikabadi S (2017) Heavy metals in contaminated environment: destiny of secondary metabolite biosynthesis, oxidative status and phytoextraction in medicinal plants. Ecotoxicol Environ Saf 145:377–390
41. Popa C, Petrus M (2017) Heavy metals impact at plants using photoacoustic spectroscopy technology with tunable CO 2 laser in the quantification of gaseous molecules. Microchem J 134:390–399
42. Sun B et al (2017) Effects of sulforaphane and vitamin E on cognitive disorder and oxidative damage in lead-exposed mice hippocampus at lactation. J Trace Elem Med Biol 44:88–92
43. Wang S, Ang HM, Tadé MO (2008) Novel applications of red mud as coagulant, adsorbent and catalyst for environmentally benign processes. Chemosphere 72:1621–1635
44. Zhang C, Yu K, Li F, Xiang J (2017) Acute toxic effects of zinc and mercury on survival, standard metabolism, and metal accumulation in juvenile ridgetail white prawn, Exopalaemon carinicauda. Ecotoxicol Environ Saf 145:549–556
45. Shao D, Chen C, Wang X (2012) Application of polyaniline and multiwalled carbon nanotube magnetic composites for removal of Pb(II). Chem Eng J 185–186:144–150

46. Kumar PA, Chakraborty S, Ray M (2008) Removal and recovery of chromium from wastewater using short chain polyaniline synthesized on jute fiber. Chem Eng J 141:130–140
47. Ren J et al (2016) Preparation of polyaniline-coated polyacrylonitrile fiber mats and their application to Cr(VI) removal. Synth Met 222:255–266
48. Qiu B et al (2015) Polyaniline coating with various substrates for hexavalent chromium removal. Appl Surf Sci 334:7–14
49. Jiang Y et al (2018) Polyaniline-based adsorbents for removal of hexavalent chromium from aqueous solution: a mini review. Environ Sci Pollut Res 25:6158–6174
50. Karthik R, Meenakshi S (2015) Removal of Cr(VI) ions by adsorption onto sodium alginate-polyaniline nanofibers. Int J Biol Macromol 72:711–717
51. Ghorbani M, Lashkenari MS, Eisazadeh H (2011) Application of polyaniline nanocomposite coated on rice husk ash for removal of Hg(II) from aqueous media. Synth Met 161:1430–1433
52. Li Q et al (2014) Synthesis of SBA-15/polyaniline mesoporous composite for removal of resorcinol from aqueous solution. Appl Surf Sci 290:260–266
53. Jiang N, Xu Y, Dai Y, Luo W, Dai L (2012) Polyaniline nanofibers assembled on alginate microsphere for Cu2+ and Pb2+ uptake. J Hazard Mater 215–216:17–24
54. Karthik R, Meenakshi S (2015) Removal of Pb(II) and Cd(II) ions from aqueous solution using polyaniline grafted chitosan. Chem Eng J 263:168–177
55. Mansour MS, Ossman ME, Farag HA (2011) Removal of Cd (II) ion from waste water by adsorption onto polyaniline coated on sawdust. Desalination 272:301–305
56. Soltani H et al (2019) Copper(II) removal from aqueous solutions by PANI-clay hybrid material: fabrication, characterization, adsorption and kinetics study. J Inorg Organomet Polym Mater 29:841–850
57. Mohammadi Nodeh MK, Gabris MA, Rashidi Nodeh H, Esmaeili Bidhendi M (2018) Efficient removal of arsenic(III) from aqueous media using magnetic polyaniline-doped strontium–titanium nanocomposite. Environ Sci Pollut Res 25:16864–16874
58. Holkar CR, Jadhav AJ, Pinjari DV, Mahamuni NM, Pandit AB (2016) A critical review on textile wastewater treatments: possible approaches. J Environ Manag 182:351–366
59. Singh P, Ullah Mirza A, Ahmad Bhat S, Kareem A, Nishat N (2021) Synthesis, characterization and evaluation of thermal, adsorption and antioxidant studies of amino functionalized poly(methyl methacrylate)/titanium dioxide nanocomposites. Inorg Chem Commun 132:108817
60. Abu Bakar N et al (2020) Physico-chemical water quality parameters analysis on textile. IOP Conf Ser Earth Environ Sci 498:012077
61. Younas F et al (2021) Current and emerging adsorbent technologies for wastewater treatment: trends, limitations, and environmental implications. Water 13:215
62. Jeon S, Haynie T, Chung S, Callmann CE (2024) Bioinspired, carbohydrate-containing polymers efficiently and reversibly sequester heavy metals. ACS Cent Sci 10:1782–1788
63. Kour G et al (2021) Conducting polymer based nanoadsorbents for removal of heavy metal ions/dyes from wastewater. pp 135–157. https://doi.org/10.1007/978-3-030-62090-5_7
64. Alkhaldi H et al (2024) Sustainable polymeric adsorbents for adsorption-based water remediation and pathogen deactivation: a review. RSC Adv 14:33143–33190
65. Kyomuhimbo HD et al (2024) Synthesis and dye adsorption dynamics of chitosan–polyvinylpolypyrrolidone (PVPP) composite. Polymers (Basel) 16:2555
66. Mohamad N, Salleh NM, Mahmud HNME (2024) Synthesis, characterization and analysis of conducting polymer composites toward dye adsorption. Int J Polym Anal Charact 29:127–145
67. Wilson K, Yang H, Seo CW, Marshall WE (2006) Select metal adsorption by activated carbon made from peanut shells. Bioresour Technol 97:2266–2270
68. Mohan D, Pittman CU (2007) Arsenic removal from water/wastewater using adsorbents—a critical review. J Hazard Mater 142:1–53
69. Vijayaraghavan K, Palanivelu K, Velan M (2006) Biosorption of copper(II) and cobalt(II) from aqueous solutions by crab shell particles. Bioresour Technol 97:1411–1419

70. Iftekhar S, Ramasamy DL, Srivastava V, Asif MB, Sillanpää M (2018) Understanding the factors affecting the adsorption of Lanthanum using different adsorbents: a critical review. Chemosphere 204:413–430
71. Salleh MAM, Mahmoud DK, Karim WAWA, Idris A (2011) Cationic and anionic dye adsorption by agricultural solid wastes: a comprehensive review. Desalination 280:1–13
72. Šoštarić TD et al (2018) Study of heavy metals biosorption on native and alkali-treated apricot shells and its application in wastewater treatment. J Mol Liq 259:340–349
73. Firdaus ML, Krisnanto N, Alwi W, Muhammad R, Serunting MA (2017) Adsorption of textile dye by activated carbon made from rice straw and oil palm midrib. Aceh Int J Sci Technol 6:1–7
74. Tan IAW, Chan JC, Hameed BH, Lim LLP (2016) Adsorption behavior of cadmium ions onto phosphoric acid-impregnated microwave-induced mesoporous activated carbon. J Water Process Eng 14:60–70
75. Long J et al (2017) Biosorption of strontium(II) from aqueous solutions by Bacillus cereus isolated from strontium hyperaccumulator Andropogon gayanus. Process Saf Environ Prot 111:23–30
76. Morosanu I, Teodosiu C, Paduraru C, Ibanescu D, Tofan L (2017) Biosorption of lead ions from aqueous effluents by rapeseed biomass. New Biotechnol 39:110–124

# Chapter 7
# Tailored Conducting Polymers for Surface Protection

**Jannatun Zia, Ufana Riaz, and M. S. S. R. Tejaswini**

**Abstract** Conducting polymers (CPs) have garnered significant attention in materials science due to their unique electrical, optical, and mechanical properties. This chapter explores the innovative use of tailored conducting polymers (CPs) for surface protection, highlighting their unique properties and diverse applications. These polymers provide not only functional benefits such as corrosion resistance and anti-fouling properties but also enhance the esthetic appeal of surfaces; conducting polymers, that mainly include polyaniline (PANI), polyacetylene (PA), polypyrrole (PPy), polythiophene (PTh), and PEDOT. These polymers possess distinct properties that make them ideal for protective coatings. This chapter discusses methods for tailoring these polymers through doping, composite formation, and functionalization to enhance their protective capabilities. Specific applications in corrosion resistance, anti-fouling coatings, wear-resistant coatings, and electromagnetic interference shielding are examined, supported by reported studies demonstrating their effectiveness in various industries. Additionally, future directions are considered, including the development of smart coatings and sustainable materials. This chapter underscores the potential of tailored conducting polymers as versatile solutions for modern surface protection challenges.

**Keywords** Conducting polymers · Surface protection · Coatings · Polyaniline · Polypyrrole

---

J. Zia (✉)
Department of Chemistry, Siddhartha Academy of Higher Education, Deemed to be University, Vijayawada, AP, India

U. Riaz
Chemistry and Biochemistry, North Carolina Central University, Durham, NC, USA
e-mail: uriaz@jmi.ac.in

M. S. S. R. Tejaswini
Department of Chemistry, School of Applied Sciences and Humanities, Vignan Foundation for Science, Technology and Research, Guntur, India



S. Jadoun, N. Kumari Jangid (eds.), *Tuning of Conducting Polymers*, Springer Series on Polymer and Composite Materials,
https://doi.org/10.1007/978-3-032-00461-1_7

## 7.1 Introduction

In recent years, the search for innovative materials that offer enhanced surface protection has led to the rise of conducting polymers (CPs) as an up-and-coming solution. These materials combine the beneficial attributes of traditional polymers, such as flexibility, lightweight design, and ease of processing, with the ability to conduct electricity. This unique combination makes conducting polymers attractive for various applications, including corrosion prevention, wear resistance, and anti-fouling measures.

The capacity to customize the properties of conducting polymers through chemical modifications, doping, and composite formation has unlocked new possibilities for their use in surface protection. By fine-tuning their structural and electronic features, researchers can improve adhesion, durability, and environmental stability, rendering them ideal for various industries, from aerospace to marine engineering [1, 2].

This chapter explores the principles and techniques behind developing conducting polymers tailored specifically for surface protection. We will examine their synthesis, the mechanisms they offer protection, and practical applications highlighting their effectiveness. Additionally, we will address the challenges in optimizing these materials for widespread commercial use and discuss the prospects that could further expand their potential.

Through this discussion, we seek to offer a comprehensive understanding of how conducting polymers can be engineered to meet the rigorous demands of surface protection, ultimately advancing materials science and engineering.

## 7.2 Overview of Conducting Polymers

Conducting polymers are those organic materials that primarily include metals and inorganic semiconductors, which could conduct electricity. Conducting polymers can conduct electricity because their backbone contains conjugated systems with alternating sigma (σ) and pi (π) links that permit electrons to flow throughout the structure [3]. The main appeal of conducting polymers lies in their versatility, lightweight nature, and ability to tailor properties by modifying their chemical structure. With their synthesis, doping strategies, and stability advancements, these materials hold great promise for various applications, from sensors to flexible electronics, energy storage devices, actuators, bioelectronics, solar cells, organic light-emitting diodes, genetic engineering, and biomedicine [4, 5].

There has been a noticeable increase in interest in performing polymer research in recent years. The electrical conductivity and distinctive properties of a number of conductive polymers, including polyacetylene (PA), polyaniline compounds (PANI), polydopamine (PDA), poly(3,4-ethylene dioxythiophene) (PEDOT), polystyrene sulfonate (PSS), poly(para-phenylene) (PPP), poly(phenylenevinylene)

**Table 7.1** Common conducting and their properties for surface protection

| Polymer | Conductivity (S cm$^{-1}$) | Type of Doping | Properties | Limitations |
|---|---|---|---|---|
| Polypyrrole | 10–7.5 × $10^3$ | p | High electrical conductivity, ease of preparation and ease of surface modification | Rigid, brittle and insoluble |
| Polyaniline | 30–200 | n,p | Diverse structural forms, environmentally stable, low cost | Hard to process, non-biodegradable, limited to solubility |
| Polythiophene | 10–$10^3$ | p | High electrical conductivity, ease of preparation, good optical properties | Hard to process |
| Poly(3,4-ethylene dioxythiophene) | 0.4–400 | n,p | Transparent conductor, environmentally and electrochemically stable | Limited solubility |

(PPV), polypyrrole (PPy), and polythiophene (PTh), have been thoroughly investigated [2, 3]. Table 7.1 presents an overview of these conducting polymers, including their doping mechanisms, conductivity and properties.

There are two primary methods of making conducting polymers: electrochemical polymerization and oxidative polymerization. Of these, oxidative polymerization is particularly notable for its efficiency and cost-effectiveness, providing higher yields while ensuring enhanced safety and shorter processing times [6]. This chapter primarily focuses on the design, customization, and practical applications of conducting polymers in surface protection. By fine-tuning these materials to meet specific needs, such as corrosion resistance, electromagnetic shielding, anti-static properties, and self-healing abilities, their performance can be optimized for use in challenging or demanding environments.

## 7.3 Properties of Conducting Polymers

The unique characteristics of conductive polymers set them apart from conventional insulating polymers and enable their usage in a wide range of applications. Here are some of the key properties of conducting polymers. [3]

### 7.3.1 *Electrical Conductivity*

Conductive polymers can conduct electricity due to a conjugated π-electron system in their molecular structure. Like metals, the delocalized electrons within this system enable charge flow. Doping, which promotes charge transport via the delocalized electrons, can enhance electrical conductivity. In conducting polymers, charge transport typically occurs through a mechanism involving the doping or tunneling of charge carriers along the conjugated backbone.

### 7.3.2 Flexibility and Lightweight

Conducting polymers are lightweight and flexible, making them ideal for wearable devices and flexible electronics. Additionally, they can be processed into thin films, coatings, and other flexible forms without losing their conductive properties.

### 7.3.3 Optical Properties

Conducting polymers have distinctive optical characteristics, including color shifts and the emission and absorption of visible and near-infrared light. These properties are used in OLEDs, solar cells, and other optoelectronic devices.

### 7.3.4 Chemical Stability

Their chemical stability is a key consideration for practical use, and efforts are being made to enhance their durability through advanced formulations and protective measures.

### 7.3.5 Redox Activity

Conducting polymers exhibit redox (reduction–oxidation reactions) behavior, which is helpful for energy storage devices and sensors.

### 7.3.6 Doping and Conductivity Enhancement

Doping is critical in enhancing the electrical properties of conducting polymers, making them suitable for various applications, from flexible electronics to sensors and energy storage devices. The key to boosting performance while preserving stability and durability is the capacity to regulate and optimize doping levels.

### 7.3.7 Environmental Sensitivity

Conducting polymers' conductivity can be highly sensitive to environmental factors such as humidity, temperature, and gases or ions. This sensitivity can detect and respond to specific environmental conditions in electronics, energy storage, sensors, and so on.

### 7.3.8 Thermal Properties

Conducting polymers generally have lower thermal stability than inorganic conductors like metals. Elevated temperatures can lead to the degradation of the polymer backbone or loss of conductivity due to the breakdown of the conjugated system; however, the thermal stability can be improved with proper material design or by incorporating nanomaterials into the polymer matrix

### 7.3.9 Biocompatibility

Some conducting polymers are biocompatible and can be used in biomedical applications such as neural interfaces, tissue engineering, and drug delivery systems. This is due to their ability to conduct electricity and compatibility with biological tissues.

### 7.3.10 Corrosion Resistance

Unlike metals, conducting polymers generally resist corrosion, especially in certain environmental conditions (e.g., in humid or chemically aggressive environments). This makes them useful for applications where metal corrosion would be problematic, such as in coatings for electronic components or sensors.

## 7.4 Common Conductive Polymers and Their Mechanism

The most common conductive polymers used in surface protection are polyaniline (PANI), polypyrrole (PPy), polythiophene, PEDOT, polyacetylene (PA), polyphenylene, and derivatives.

### 7.4.1 Polyaniline

Polyaniline (PANI) is one of the most widely used conducting polymers due to its ease of synthesis, stability in different environments, and versatile electrochemical properties. PANI is synthesized through chemical polymerization using an oxidizing agent, such as ammonium peroxydisulfate, or through electrochemical polymerization. It is a polymer of aniline, a common organic compound, and can be synthesized in various oxidation states, leading to different electrical conductivities and colors. The polymer's structure consists of conjugated aromatic rings, contributing to its ability to conduct electricity when appropriately doped. PANI is a versatile and practical material for surface protection, offering significant advantages in corrosion inhibition, electrostatic dissipation, and electromagnetic shielding. [7–9] Its ability to form conductive, protective films on metal surfaces makes it an attractive alternative to traditional coating materials.

### 7.4.2 Polypyrrole

Polypyrrole (PPy) is another widely utilized conductive polymer known for its strong mechanical properties and high electrochemical stability, which can be synthesized via the oxidative polymerization process, as shown in Fig. 7.1. PPy finds applications across various fields, including electronics, optics, biology, and medicine. Its conductivity ranges from 40 to 200 $S{\cdot}cm^{-1}$. PPy is a heterocyclic, positively charged, conductive polymer with oxidized nitrogen atoms. Over-oxidation can lead to a loss of conductivity, while aqueous solutions and organic electrolytes enhance its electroactivity. Notable characteristics of PPy include its environmental stability, excellent conductivity, affordability, and redox properties. [10] Chainet et al. [11] proposed using scanning tunneling microscopy (STM) to observe the real-time structure of polypyrrole during doping and dedoping processes. Their results indicated significant structural changes in the conducting polymer, primarily

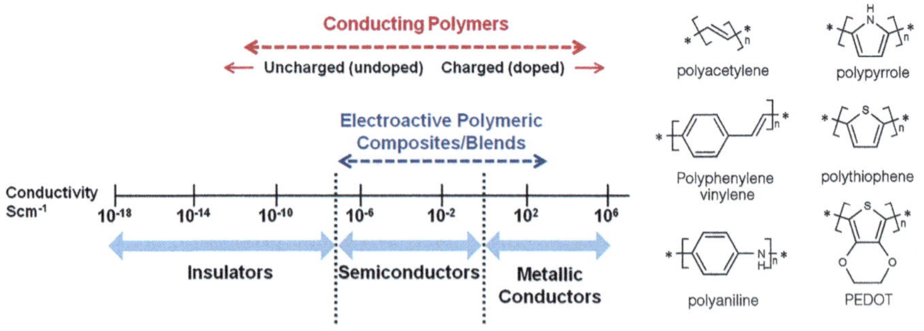


Fig. 7.1 Common conductive polymers' chemical structure

due to the incorporation of anions. PPy is particularly favored in corrosion protection applications, where it can be applied as a thin film to protect metal surfaces from environmental degradation (Fig. 7.2).

### *7.4.3 Polythiophene and Derivatives PEDOT*

Polythiophene (PT) and its derivatives, such as poly (3, 4-ethylenedioxythiophene) (PEDOT), are well-known for their excellent conductivity and stability. These materials are widely used in organic electronics and surface protection, particularly in environments requiring mechanical durability and corrosion resistance [2, 12, 13].

### *7.4.4 Polyacetylene*

Polyacetylene (PA) has become a significant focus of research in functional materials in recent years. It exhibits a wide range of conductivity with reversibly adjustable values between 200 and 1000 S/cm. PA is the simplest conjugated polymer and was the first conductive polymer synthesized by a Ziegler-Natta catalyst [14]. The conductivity of polyacetylene can be enhanced through improvements in the synthesis process and doping techniques. PA has significant applications in material science, particularly in electronics and surface protection. [2, 15]

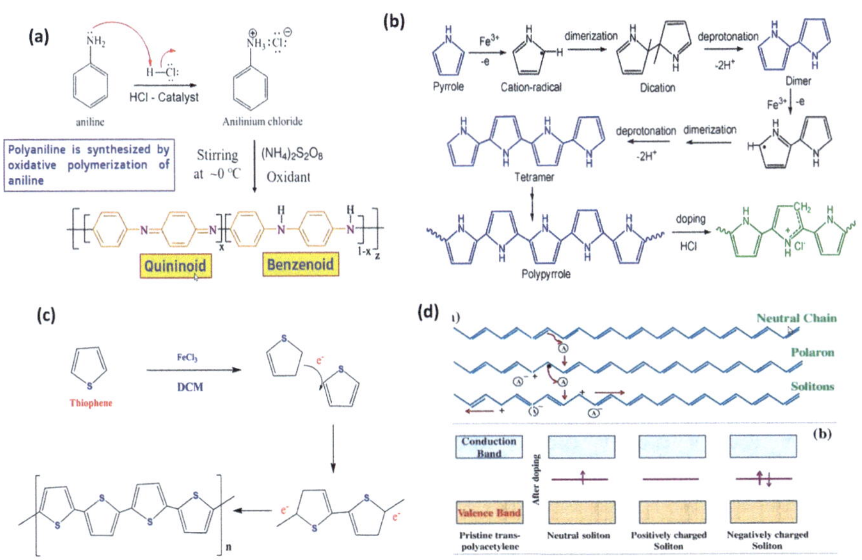


**Fig. 7.2** Polymerization synthesis mechanism of (**a**) PANI, (**b**) PPY, (**c**) PTh, and (**d**) Polyacetylene

### 7.4.5 Polyphenylene and Derivatives

Polyphenylene (PP) and its derivatives are aromatic conductive polymers with a backbone of benzene rings linked by single bonds. These polymers are prized for their exceptional chemical resistance, high thermal stability, and mechanical strength. Variants of polyphenylene, including poly(p-phenylene) (PPP) and polyphenylene sulfide (PPS), have been studied for a variety of applications, especially in high-performance coatings and surface protection [2, 16]. These materials offer notable advantages in demanding environments where other polymers may fail or degrade (Fig. 7.3).

## 7.5 Tailoring of Conducting Polymers for Surface Protection

It has been found that polymers, such as polypyrrole (PPy), polyaniline (PANI), and polyacetylene (PA), and derivatives, can be further tailored through doping, functionalization, and composite formation to meet the specific demands of surface protection applications. This emerging field leverages the unique properties of conducting polymers, such as electrical conductivity, environmental stability, and the ability to form protective coatings. When applied as coatings or films, conducting polymers offer multiple benefits, including corrosion resistance, electrostatic dissipation, self-healing capabilities, and enhanced mechanical properties. Below is a detailed overview of how conducting polymers can be tailored for surface protection.

### 7.5.1 Doping and Functionalization of Conducting Polymer

The tailoring of conducting polymers through doping and functionalization is essential for enhancing their electrical, mechanical, and chemical properties. Doping improves conductivity, while functionalization can optimize solubility, stability, and interaction with other materials.

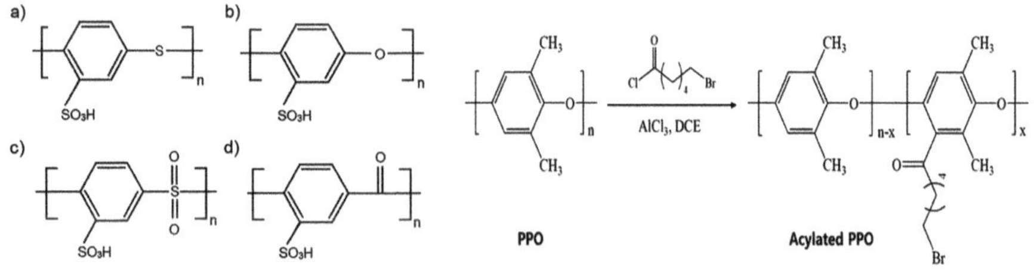


**Fig. 7.3** Polythiophene derivatives

#### 7.5.1.1 Doping

The addition of electron donors or acceptors to intrinsic conducting polymers causes dramatic electronic changes in them, which increases conductivity almost to the metallic range. Conjugated conducting polymers have a low ionization potential and high electron affinities, making them quickly oxidized or reduced. Oxidation or reduction creates positive and negative charges on the polymer backbone, thus increasing conductivity. For example, the conductivity of undoped polyacetylene is $<10^{-7}$ S/cm, while its conductivity after doping is $10–10^5$ S/cm. By analogy with semiconductor technology, doping is referred to as doping and is of two types.

**N-Doping (Reductive doping)** involves introducing electron-donating dopants (e.g., alkali metals, halide salts, or Lewis bases) into the polymer. This increases the electron density of the polymer backbone and thus enhances its conductivity. For example, polypyrrole (PPy) can be doped with sodium or lithium salts to increase electrical conductivity.

**P-Doping (Oxidative doping)** In p-doping, electron-withdrawing dopants such as acids, metal ions, or halide salts create "holes" or missing electrons in the polymer structure. These holes facilitate the movement of positive charge carriers (holes), thereby increasing conductivity. A well-known example is **polyaniline (PANI)**, which undergoes p-doping when treated with an acid such as hydrochloric acid, forming a protonated emeraldine salt form that exhibits high electrical conductivity [17] (Fig. 7.4).

#### 7.5.1.2 Functionalization

Another critical strategy for improving and customizing the performance of conducting polymers is functionalization, which involves chemically modifying the polymer backbone by introducing specific functional groups or side chains. This process enhances the polymer's conductivity, processability, stability, and interactions with other materials, enabling the fine-tuning of its properties to suit specific application requirements. Functionalization can be achieved either during polymerization or through post-polymerization modification. The different methods of functionalization are discussed below.

**Side-Chain Functionalization** Side-chain functionalization is one of the most common methods for enhancing the properties of conducting polymers. Introducing polar or non-polar functional groups can significantly improve the solubility, processability, and interactions of CPs with other materials. For example, **polypyrrole (PPy)** can be functionalized with amine groups to enhance its solubility in water and organic solvents, making it suitable for applications in drug delivery (Lee et al. 2015) [18] shown in Fig. 7.5. A study by **Periyat et al. (2022)** demonstrated the functionalization of PPy with carboxylic acid groups to improve its interaction with metal nanoparticles for catalytic applications. The carboxyl groups on the polymer

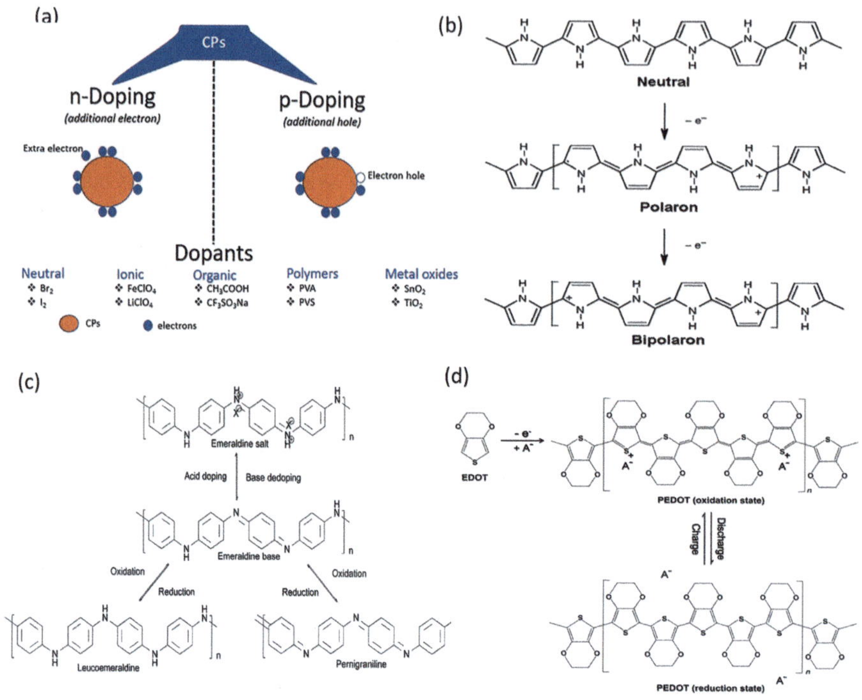


**Fig. 7.4** (**a**) Types Doping with dopant and doping of conduction polymers (**b**) PPY, (**c**) PANI and (**d**) PEDOT

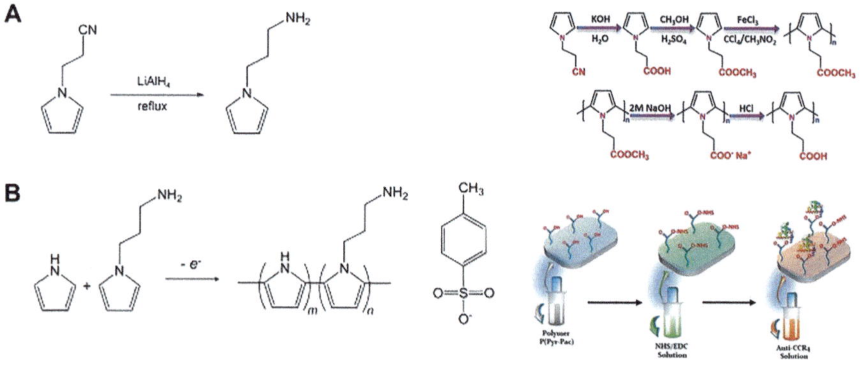


**Fig. 7.5** Functionalization of polypyrrole (PPy) with amine groups and acid

surface provided anchoring sites for nanoparticles, enhancing the catalytic performance of the composite material [19].

**Backbone Functionalization** Backbone functionalization involves incorporating functional groups directly into the polymer backbone. This approach is often used to tune the electronic properties of conducting polymers. For instance, **thiophene-**

**based polymers** can be functionalized with electron-withdrawing groups (e.g., –COOH, $-NO_2$) or electron-donating groups (e.g., $-NH_2$, $-CH_3$) to modify the polymer's bandgap and conductivity [20]. The functionalization of poly(3,4-ethylenedioxythiophene) (PEDOT) with various side groups, such as carboxylates or amines, has enhanced its conductivity and stability in energy storage devices [21]

**Grafting or Copolymerization** Both **grafting** and **copolymerization** are potent strategies for tuning the properties of conducting polymers. Grafting allows for incorporating functional groups or side chains, providing control over processability, stability, and compatibility with other materials. On the other hand, copolymerization offers more flexibility in combining different monomers to create polymers with tailored electronic, mechanical, and chemical properties.

## 7.6 Composite Materials for Conducting Polymers

The tailoring of conducting polymers by composite materials holds significant promise for surface protection applications, offering improved mechanical properties, environmental stability, and anti-corrosive behavior. Recent advancements in composite formulations have demonstrated the potential for conducting polymer-based coatings in various industries, from automotive to marine. Conducting polymer composites (CPCs) can be broadly categorized into three main types based on the filler material used: polymer-polymer composites, polymer-nanoparticles composites, and polymer-inorganic composites [22].

### *7.6.1 Polymer-Polymer Composites*

These composites are formed by combining a conducting polymer with another conducting polymer or a non-conducting polymer to improve processability, mechanical properties, and thermal stability. Some polymer-polymer-based composites reported in Table 7.2 are used for surface protection.

### *7.6.2 Polymer-Nanoparticle Composites*

These composites incorporate nanoparticles (e.g., metal, metal oxide, or carbon-based nanoparticles) to enhance the conducting polymer's mechanical, thermal, and anti-corrosive properties. Conducting coatings filled with carbon nanoparticles (carbon nanotubes, graphene, graphene oxide, etc.) have good corrosion resistance, durability, mechanical properties, wear resistance, adhesion, and barrier properties for metal substrates.

**Table 7.2** shows the information of some polymer-polymer composites in surface protection

| Composite | Substrate | Test medium | Observations | References |
|---|---|---|---|---|
| Terpolymer from 2-pyridylamine (2PA), aniline (AN) and 2,3-xylidine (XY) and poly (2PA-co-AN-co-XY) poly (AN-co-XY), poly (2PA-co-AN) | Mild steel | 0.1 M HCl, 5% NaCl solution, artificial seawater, distilled water and open atmosphere | Synthesized terpolymer exhibited excellent protection against mild steel corrosion, the protection efficiency being in the range of 93%–96% after 30 days of immersion. | [23] |
| PANi/PPy PPy/PAni | Carbon steel and stainless steel | 0.028 M NaCl Solution | Bilayered coatings revealed better protection, especially against pitting corrosion, compared with single Polyaniline coatings for stainless steel. PPy/PAni revealed better performance than PAni/PPy. | [24] |
| Homopolymers and bilayers of PANI/ PPy | Mild steel | 0.5 M NaCl 0.1 M HCl 0.5 M H2SO4 | PPy and PAni/PPy bilayer coatings were a stable host matrix on mild steel against corrosion. PPy exhibited the best corrosion resistance among all coatings. | [25] |
| PBTh as homopolymer and PAni/PBTh | Stainless steel | 0.5 M NaCl and 0.5 M HCl solutions | The corrosion test revealed that among the protective coatings obtained, PBTh as homopolymer and PAni/PBTh as bilayer exhibited the most effective anti-corrosive properties. According to the linear potentiodynamic polarization test, the dissolution current of these coatings at 1.6 V decreased to 99.8% and 99.6% in NaCl solution, respectively, when compared to that of uncoated SS surfaces. | [26] |
| PAni/PNMA PNMA/PAni | Mild steel | 3.5% NaCl | Poly(N-methylaniline) (PNMA)/PAni layers showed higher stability and better Protection than the PAni/ PNMA layers. | [27] |
| PAni/co-polyamide | Carbon steel | Neutral salt spray | Single co-polyamide and PAni on carbon steel both were not sufficient to prevent corrosion in a chloride medium by itself, but the combined coating (a primer of PAni and a top layer of co-polyamide coating) could result in a minimized corrosion. | [28] |

**Table 7.2** (continued)

| Composite | Substrate | Test medium | Observations | References |
|---|---|---|---|---|
| PPy/PAniPPy/ PAni-PW12 PPy-PW12/PAni | Mild steel | 0.1 M HCl | PAni/PPy-PW12 coating exhibited better corrosion resistance for mild steel; the presence of PW12 in the inner Layer of PPy/PAni may hinder the maintenance of passive layer and reduce the corrosion resistance. | [29] |
| PPy/polyphenol | Mild steel | 0.05 M H2SO4 | The polyphenol layer improved the barrier effect of the coating remarkably. The polypyrrole/ polyphenol could provide better and longer corrosion protection for mild steel than single PPy layer. | [30] |
| Poly(pyrrole-co-*N*-methyl pyrrole) | Mild steel | 3.5% NaCl aqueous solution | The copolymerization rate and protective behavior of copolymer were strongly affected by the monomer concentration ratio. The ratio of 1:3 (*N*-methyl pyrrole/pyrrole) gave the most protective copolymer coating and it exhibited better barrier property than single PPY | [31] |
| PANI/PVC blends | Carbon steel | 3% NaCl) and acidic (0.1 M HCl | The PANI/PVC blends demonstrated superior corrosion protection in both environments as compared to pure PANI and pure PVC coatings. The PANI/ PVC blend with a 1/1 proportion offered the most effective corrosion protection in acidic and saline environments based on high charge transfer resistance values and corrosion potential values that approach zero volts. | [32] |
| Polyaniline/chitosan (PANI/CTS) and $TiO_2$-PANI/CTS nanocomposites | 1018 carbon steel | 3.5 wt% NaCl solution | The best performance against corrosion was observed for the coating made by $TiO_2$-PANI/ CTS, since it shows the noblest values in potential, as well as the highest protection efficiencies, even after its immersion for 30 days | [33] |
| PPY/PTh | Mild steel | 3.5% NaCl | Provide anodic protection to MS, an efficiency value of 98.2% was calculated after 220 h of exposure time. | [34] |

### 7.6.3 Polymer-Inorganic Composites

These composites involve the incorporation of inorganic materials such as ceramics or glass to improve the overall stability and mechanical performance of the conducting polymer. Some polymer-nanoparticle (metal, metal oxide, and carbon-based) composites reported in Table 7.3 are used for surface protection.

**Table 7.3** Polymer-based nanoparticle (metal, metal oxide, and carbon-based) composites are used for surface protection

| Composite | Substrate | Test medium | Observations | References |
|---|---|---|---|---|
| Cu-PPy | Carbon steel | 0.5 M $H_2SO_4$ | Cu-PPy exhibits high $NH_3$ yield rate of 0.55 mmol $h^{-1}$ $cm^{-2}$ with $NH_3$ selectivity of 99.99% and FE of 96.0% at high $NO^{3-}$ conversion | [35] |
| PAni/Zn and Ni/PAni | Mild steel and iron | Aqueous oxalic acid medium | PAni/Zn coatings offered the maximum protection, followed by PAni/Ni coatings. Metal/PAni coatings provided the least resistance to corrosion. Zn offered sacrificial protection, and PAni and Ni acted by the barrier mechanism. | [36] |
| PANI/$CeO_2$ | Carbon steel | 3.5% NaCl solution | PANI/$CeO_2$ NPs exhibited excellent corrosion resistance, superior to epoxy coatings added with PANI in NaCl solution. The role of CeO2 NPs and PANI against the diffusion of aggressive ions (e.g., cl−) and the improvement of self-healing protection are attributed to PANI'S redox behavior. | [37] |
| Polyaniline (PANI)-based metal oxides ($SiO_2$, $CeO_2$, and $TiO_2$A) | Commercial 1018 steel | 3.5% NaCl medium | Open circuit potential (OCP) study, Tafel curves, and electrochemical impedance spectroscopy (EIS) showed that coatings made with $TiO_2$A NPs exhibit the best anti-corrosion properties compared to those synthesized with $SiO_2$ and $CeO_2$ NPs. | [38] |
| PANi/$SiO_2$ | | | The coating containing PANi/$SiO_2$ nanocomposite with 17.2 wt% silica presented good mechanical properties (20 L/mil abrasion resistance, 100 kg cm impact strength) and admissible surface and volume resistivity ($1.3 \times 10^{11}$ Ω/$cm^2$ and $6.6 \times 10^{10}$ Ω cm, respectively) for anti-static coatings | [39] |
| PPy-metal oxide ($Al_2O_3$, ZnO, $TiO_2$, $CeO_2$ and $SnO_2$) nanocomposite | Mild steel | 0.5 M $H_2SO_4$. | Results reveal that PPy/$Al_2O_3$ nanocomposite provided the best performance for corrosion protection of the MS by reducing its corrosion current density by 18 times | [40] |

**Table 7.3** (continued)

| Composite | Substrate | Test medium | Observations | References |
|---|---|---|---|---|
| V-$TiO_2$/ polypyrrole | Carbon steels | 0.1 M HCl solution | Results showed that the V-$TiO_2$/PPy composite coatings exhibited better anti-corrosion performance than $TiO_2$/PPy composite coatings | [41] |
| PTh-$TiO_2$-rGO | **Low-carbon steel** | 3.5 wt % NaCl | The salt spray and immersion test results conducted on coated LCS suggest that the PTh-$TiO_2$-rGO nanocomposite coating provides significantly higher corrosion protection than PTh-$TiO_2$ and PTh coatings. | [42] |
| PPy-ZnO | Mild steel | 3.5% NaCl | Electrochemical impedance spectroscopy (EIS) measurements indicated that the coating resistance ($R_{coat}$) and corrosion resistance ($R_{corr}$) values for the PPy-ZnO nanocomposite coating was much higher than that of pure PPy coated electrode. | [43] |
| PPy/$Fe_2O_3$ | Mild steel | Salt spray | After exposure to salt spray test conditions, a lower drop in coating resistance in the case of $Fe_2O_3$/PPy coating as compared to $Fe_2O_3$ coating suggested better corrosion protection is offered by $Fe_2O_3$/PPy coating | [44] |
| PANI–CNTs | Stainless steel | $H_2SO_4$ solution | The mCNTs improved the protective properties of PANI for the SS in aggressive $H_2SO_4$ solution | [45] |
| PANI/GO | Mild steel | NaCl | The amount of the resistance value >106 Ω $cm^2$ was estimated using the composite coatings. | [46] |

## 7.7 Cross-Linking and Nanostructuring

### *7.7.1 Cross-Linking of Conducting Polymers*

Cross-Linking involves creating chemical bonds between polymer chains, resulting in a three-dimensional network. This process can improve conducting polymers' mechanical properties, thermal stability, and chemical resistance. Crosslinked conducting polymers are particularly useful in applications requiring enhanced structural integrity. **Conducting polymer hydrogels**, which are crosslinked networks of hydrophilic conducting polymers, have been extensively studied for biomedical applications, including drug delivery and tissue engineering. These materials combine the electrical conductivity of CPs with the biocompatibility and flexibility of hydrogels. [47, 48]

### 7.7.2 Nanostructuring of CPs

Nanostructuring refers to the manipulation of materials at the nanoscale, typically in the range of 1–100 nm, to create structures with unique properties. In the context of conducting polymers, nanostructuring can be achieved through various techniques such as electrochemical polymerization, template-assisted synthesis, and self-assembly. These methods lead to the formation of nanostructures, including nanoparticles, nanowires, nanotubes, and nanofibers, which significantly enhance the material's properties.

At the nanoscale, materials exhibit a larger surface area to volume ratio, which enhances interactions with the substrate and the surrounding environment. This increased surface area can facilitate faster charged transport and improve conductivity. Additionally, the unique arrangement of nanostructures can lead to enhancements in mechanical strength, flexibility, and wear resistance.

Nanostructured conducting polymers also offer improved performance in surface protection applications. The increased surface area enhances adhesion to substrates, creating more uniform and dense coatings that improve barrier properties against moisture, gases, and corrosive substances. These coatings are more effective at preventing the accumulation of dirt, algae, and other contaminants, especially in marine environments. Furthermore, nanostructured films often exhibit superior wear resistance compared to bulk polymers, making them ideal for coatings exposed to mechanical stress. This approach can lead to more efficient, durable, and multifunctional conducting polymer coatings for various surface protection applications.

A notable example of nanostructured conducting polymers used for surface protection is the development of polyaniline (PANI) and Pyrrole (PPY) nanostructures for corrosion protection. Polyaniline, a conducting polymer with excellent environmental stability, can be fabricated into various nanostructures such as nanofibers and nanorods, significantly improving its protective coating performance. [7, 49]

## 7.8 Applications of Tailored Conducting Polymers in Surface Protection

### 7.8.1 Corrosion Protection

Conducting polymers can form protective coatings on metals like steel, aluminum, and copper, preventing oxidation and corrosion. These coatings can be used in the aerospace, marine, and automotive industries. One of the most common applications of conducting polymers is preventing corrosion, particularly in metal substrates like steel, aluminum, and copper. Coatings made from materials such as PANI, PPy, and PEDOT are applied to metal surfaces to create a conductive barrier that both prevents direct exposure to corrosive agents (e.g., water, oxygen, salts) and reduces the electrochemical reactions that lead to corrosion

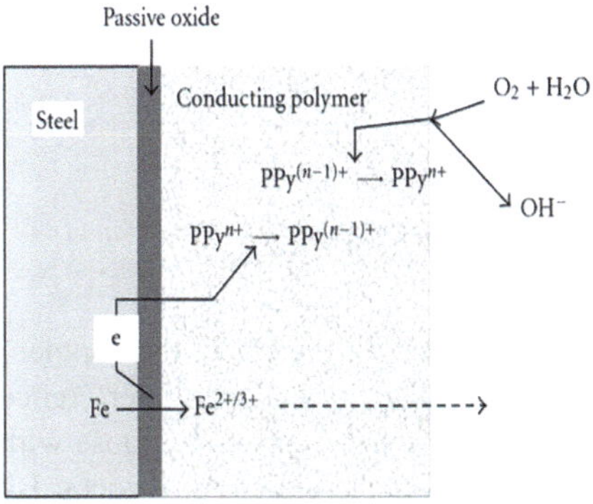


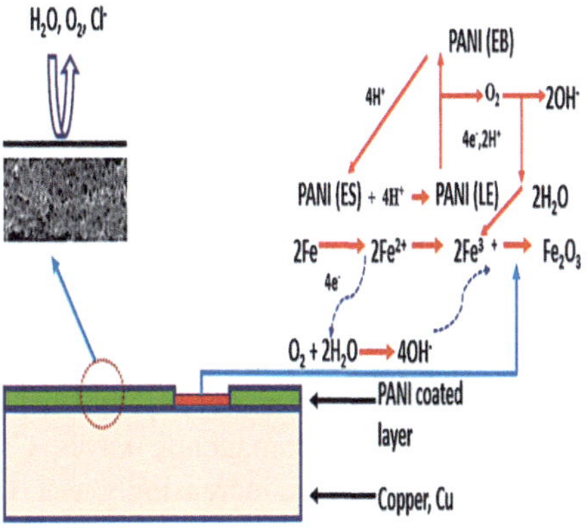


## 7.8.2 Marine and Aerospace Applications

Conducting polymers offers significant benefits for surface protection in marine and aerospace applications. In marine environments, they provide corrosion resistance by forming protective coatings that prevent moisture and oxygen from reaching metal surfaces. In aerospace, conducting polymers are used for lightweight electromagnetic interference shielding and can also prevent ice buildup or melt ice through electro-thermal de-icing. Additionally, their self-healing properties enhance coating longevity by automatically repairing damage. Conducting polymers can also incorporate environmental sensors to monitor and respond to changes, further improving protection and reducing maintenance needs.

## 7.8.3 Anti-static Coatings

Conducting polymers are used in coatings for electronic equipment and sensitive materials to prevent static charge buildup, which can damage components or affect device performance. Anti-static coatings protect sensitive components from electrostatic discharge (ESD) in the electronic and semiconductor industries. Conducting polymers, especially those with low to moderate conductivity like PEDOT: PSS, coat sensitive electronics, preventing static charge accumulation, and ensuring device longevity and reliability.

## 7.8.4 Wear-Resistant Coatings

Conducting polymers can be blended with complex nanomaterials (like carbon nanotubes or graphene) to improve wear resistance while maintaining their conductive properties. The integration of conducting polymers with nanomaterials such as carbon nanotubes (CNTs) or graphene can enhance the wear resistance of the coatings. These hybrid materials offer improved mechanical properties while retaining the conductivity needed for specific applications.

### 7.8.5 Electromagnetic Interference (EMI) Shielding

Conducting polymers are increasingly used in EMI shielding applications for electronic devices and sensitive equipment, where they prevent unwanted interference by dissipating electromagnetic waves. Conducting polymers, especially those with high conductivity, are increasingly used for EMI shielding in electronic devices. By dissipating electromagnetic waves, conducting polymer coatings protect sensitive equipment from interference, improving performance and reliability.

### 7.8.6 Self-Healing Coatings

Self-healing coatings, incorporating conducting polymers and healing agents, are promising for maintaining long-term surface protection. These coatings can autonomously repair damage, such as scratches or cracks, which occur over time, ensuring the protective layer remains intact. When damaged, the conducting polymers used in these coatings are often designed to undergo reversible chemical or physical changes, effectively restoring their functionality. These coatings can detect damage and initiate healing without external intervention by embedding healing agents within the polymer matrix. This self-healing capability not only extends the lifespan of the protective coating but also reduces the need for frequent maintenance and repairs, making them particularly valuable in demanding environments like marine and aerospace applications.

## 7.9 Challenges and Future Directions

The synthesis of high-quality conducting polymers with tailored properties remains complex and expensive, limiting their industrial scalability. Current production methods require costly rearrangements and sophisticated equipment, making large-scale applications economically unfeasible. Prospective research should focus in developing scalable and reasonably priced synthesis methods, like electrochemical polymerization or green chemistry strategies, to navigate through this. A major challenge is improving the long-term stability of conducting polymers, especially in harsh environments exposed to UV light, extreme temperatures, and chemicals. Further research is needed to enhance their molecular design to resist degradation and incorporate stabilizing coatings or additives to extend their lifespan. Balancing conductivity with mechanical strength is another ongoing challenge. Conducting polymers often faces trade-offs between these properties, and achieving a uniform coating with high conductivity and adequate mechanical flexibility remains difficult. Future studies should explore strategies to enhance both simultaneously, possibly through composite materials or molecular design modifications. The increasing demand for environmentally sustainable materials has increased interest in conducting polymers derived from renewable resources. However, challenges like feedstock availability, scalable green synthesis processes, and end-of-life disposal remain. Research should focus on creating bio-based polymers that are recyclable or biodegradable to reduce environmental impact. Despite their potential, conducting polymers still face barriers to large-scale commercial use. They must meet the technical and economic demands of applications like flexible electronics, sensors, and energy storage to achieve broader adoption. Future research should optimize these materials for specific uses, improve their reliability, and reduce costs, with collaboration between researchers and industry stakeholders playing a key role. In conclusion, conducting polymers holds excellent promise but faces challenges that must be addressed regarding synthesis, stability, sustainability, and application. Ongoing advancements in these areas will be essential for unlocking their full potential in next-generation technologies.

## 7.10 Conclusion

Tailoring conducting polymers for surface protection involves fine-tuning of their chemical structure, doping level, morphology, and integration with additives or fillers to meet specific requirements. Conducting polymers can offer unique properties such as corrosion resistance, ESD protection, environmental resistance, self-healing, mechanical durability, and antimicrobial protection. These properties make conducting polymers highly versatile for various surface protection applications, from electronics and coatings to medical devices and industrial machinery. By optimizing the synthesis and processing conditions, conducting polymers can be engineered for maximum performance in specific environments.

## References


1. Guo X, Facchetti A (2020) The journey of conducting polymers from discovery to application. Nat Mater 19:922–928
2. Namsheer K, Chandra SR (2021) Conducting polymers: a comprehensive review on recent advances in synthesis, properties and applications. RSC Adv 11:5659–5697
3. Megha G, Singh K, Bhatnagar N (2024) Conductive polymers: a multipurpose material for protecting coating. Prog Org Coat 187:108083
4. Shahid MA, Rahman MM, Hossain MT, Hossain I, Sheikh MS, Rahman MS, Uddin N, Ul Hoque MI (2025) Advances in conductive polymer-based flexible electronics for multifunctional applications. J Compos Sci 9:42

5. Paramshetti S, Angolkar M, Fatease AA, Alshahrani SM, Hani U, Garg A, Ravi G, Riyaz AM (2023) Revolutionizing drug delivery and therapeutics: the biomedical applications of conductive polymers and composites-based systems. Pharmaceutics 15:1204
6. Magu TO, Agobi AU, Dass P (2019) A review on conducting polymers-based composites for energy storage application. J Chem Rev 1:19–34
7. Beygisangchin M, Rashid SA, Shafie S, Sadrolhosseini AR, Lim HN (2021) Preparations, properties, and applications of polyaniline and polyaniline thin films—a review. Polymers 13:2003
8. Kaushik P, Bharti R, Sharma R, Verma M, Olsson RT, Pandey A (2024) Progress in synthesis and applications of polyaniline-coated nanocomposites: a comprehensive review. Eur Polym J 221:113574
9. Bhadra S, Singha NK, Khastgir D (2011) Polyaniline based anticorrosive and anti-molding coating. J Chem Eng Mater Sci 2:1–11
10. Hao L, Dong C, Zhang L, Zhu K, Yu D (2022) Polypyrrole nanomaterials: structure, preparation and application. Polymers 14:5139
11. E Chaînet, Billon M (1998) In situ study of polypyrrole morphology by STM: effect of the doping state. J Electroanal Chem 451:273–277
12. Wang XS, Tang PH, Li XD, Hua X (2009) Investigations on the mechanical properties of conducting polymer coating-substrate structures and their influencing factors. Int J Mol Sci 10(12):5257–5284
13. AL-Refai HH, Ganash AA, Hussein MS (2021) Polythiophene and its derivatives -based nanocomposites in electrochemical sensing: a mini review. Mater Today Commun 26:101935
14. Poddar A, Patel SS (2021) Synthesis, characterization and applications of conductive polymers: a brief review. Polym Adv Technol 32:1–26
15. Shirakawa H (1989) Synthesis and properties of Polyacetylene. MCLC Incorp Nonlin Optics 171:235–242
16. Rahate AS, Nemade KR, Waghuley SA (2013) Polyphenylene sulfide (PPS): state of the art and applications. Rev Chem Eng 29:209
17. Giri H, Dowell TJ, Almtiri M, Scott CN (2023) Polyaniline derivatives and their applications. IntechOpen. (book)
18. Lee JY, Schmidt CE (2014) Amine-functionalized polypyrrole: inherently cell adhesive conducting polymer. J Biomed Mater Res A 103:2126–2132
19. Thadathil A, Pradeep H, Joshy D, Ismail YA, Periyat P (2022) Polyindole and polypyrrole as a sustainable platform for environmental remediation and sensor applications. Mater Adv 3:2990–3022
20. Abedinifar F, Rezaei EB, Biglar M, Larijani B, Hamedifar H, Ansari S, Mahdavi M (2021) Recent strategies in the synthesis of thiophene derivatives: highlights from the 2012-2020 literature. Mol Divers 25:2571–2604
21. Jeong MK, Lee K, Kang J, Jang J, Jung IJ (2020) Thiophene backbone-based polymers with electron-withdrawing pendant groups for application in organic thin-film transistors. New J Chem 44:9321–9327
22. Bazlia L, Yusuf M, Farahanic A, Kiamarzid M, Seyedhosseinie Z, Nezhadmansarif M, Aliasgharig M, Iranpoorh M (2020) Application of composite conducting polymers for improving the corrosion behavior of various substrates: a review. J Compos Sci 2:228–240
23. Xu H, Zhang Y (2019) A review on conducting polymers and nanopolymer composite coatings for steel corrosion protection. Coatings 9:807
24. Tan CK, Blackwood DJ (2003) Corrosion protection by multilayered conducting polymer coatings. Corros Sci 45:545–557
25. Yagan A, Pekmez NÖ, Yıldız A (2007) Inhibition of corrosion of mild steel by homopolymer and bilayer coatings of polyaniline and polypyrrole. Prog Org Coat 59:297–303
26. Pekmez NÖ, Abacı E, Cınkıllı K, Ya˘gan A (2009) Polybithiophene and its bilayers with polyaniline coatings on stainless steel by electropolymerization in aqueous medium. Prog Org Coat 65:462–468
27. Narayanasamy B, Rajendran S (2010) Electropolymerized bilayer coatings of polyaniline and poly (N-methylaniline) on mild steel and their corrosion protection performance. Prog Org Coat 67:246–254

28. Martyak NM, Mc Andrew P (2007) Corrosion performance of steel coated with co-polyamides and polyaniline. Corros Sci 49:3826–3837
29. Xu J, Zhang Y, Zhang D, Tang Y, Cang H (2015) Electrosynthesis of PAni/PPy coatings doped by phosphotungstate on mild steel and their corrosion resistances. Prog Org Coat 88:84–91
30. Tüken T, Yazıcı B, Erbil M (2004) A new multilayer coating for mild steel protection. Prog Org Coat 50:115–122
31. Tüken T, Tansuğ G, Yazıcı B, Erbil M (2007) Poly (N-methyl pyrrole) and its copolymer with pyrrole for mild steel protection. Surf Coat Technol 202:146–154
32. Bereket G, Hür E (2009) The corrosion protection of mild steel by single layered polypyrrole and multi-layered polypyrrole/poly (5-amino-1-naphthol) coatings. Prog Org Coat 65:116–124
33. Rangel-Olivares FR, Arce-Estrada EM, Cabrera-Sierra R (2024) Development of polyaniline/chitosan (PANI/CTS) and $TiO_2$-PANI/CTS nanocomposites as anti-corrosion coatings: synthesis and characterization. Surf Coat Int 467:130163
34. Tüken T, Yazici B, Erbil M (2004) The use of polythiophene for mild steel protection. Prog Org Coat 51:205–212
35. Chen H, Zhang Z, Li Y, Yu L, Chen F, Li L (2024) Conductive polymer protection strategy to promote electrochemical nitrate reduction to ammonia in highly acidic condition over Cu-based catalyst. Chem Eng J 481:148596
36. Kumar SA, Meenakshi KS, Sankaranarayanan TSN, Srikanth S (2008) Corrosion resistant behaviour of PAni–metal bilayer coatings. Prog Org Coat 62:285–292
37. Lei Y, Qiu Z, Tan N, Du H, Li D, Liu J, Liu T, Zhang W, Chang X (2020) Polyaniline/$CeO_2$ nanocomposites as corrosion inhibitors for improving the corrosive performance of epoxy coating on carbon steel in3.5% NaCl solution. Prog Org Coat 139:105430
38. Francisco R, Arce-Estrada EM, Cabrera-Sierra R (2021) Synthesis and characterization of polyaniline-based polymer nanocomposites as anti-corrosion coatings. Coatings 11:653
39. Shi S, Zhao Y, Zhang Z, Yu L (2019) 1Corrosion protection of a novel SiO2@ PANI coating for Q235 carbon steel. Prog Org Coat 132:227–234
40. Rasoul BS, Parsa JB, Vakili-Azghandi M (2019) Electrodeposition of polypyrrole/metal oxidenano composites for corrosion protection of mild steel—a comparative study. Synth Met 247:183–190
41. Chen Z, Yang W, Xu B, Chen Y, Qian M, Su X, Li Z, Yin X, Liu Y (2019) Corrosion protection of carbon steels by electrochemically synthesized V-TiO/polypyrrole composite coatings in 0.1 M HCl solution. J Alloys Comp 771:857–868
42. Mobin M, Ansar F (2022) Polythiophene (PTh)–TiO–reduced graphene oxide (rGO)nano-composite coating: synthesis, characterization, and corrosion protection performance on low-carbon steel in 3.5wt % NaCl solution, ACS. Omega 50:46717–46730
43. Hosseini MG, Bagheri R, Najjar R (2011) Electropolymerization of polypyrrole and polypyrrole-ZnO nanocomposites onmild steel and its corrosion protection performance. J Appl Polym Sci 121:3159–3166
44. Jadhav N, Kasisomayajula S, Gelling V (2020) Polypyrrole/metal oxides-based composites/nanocomposites for corrosion protection. Front Mater 7:95
45. Hermas AE, Salam MA, Al-Juaid SS (2013) In situ electrochemical preparation of multi-walled carbon nanotubes/polyaniline composite on the stainless steel. Prog Org Coat 76:1810–1813
46. Maurya R, Siddiqui AR, Katiyar PK, Balani K (2019) Mechanical, tribological and anti-corrosive properties of polyaniline/graphene coated mg-9Li-7Al-1Sn and mg-9Li-5Al-3Sn-1Zn alloys. J Mater Sci Technol 3:1767–1778
47. Mouss M, El-Kady MF, Dubal D, Tung TT, Nine MJ, Mohamed N, Kaner RB, Losic D (2024) Self-assembly and cross-linking of conducting polymers into 3D hydrogel electrodes for supercapacitor applications. ACS Appl Energy Mater 3:1–11
48. Horrocks MS, Zhurenkov K, Malmström J (2024) Conducting polymer hydrogels for biomedical application: current status and outstanding challenges. APL Bioeng 8:031503
49. Sood Y, Singh K, Mudila H, Lokhande PE, Singh L, Kumar D, Kumar A, Mubarak NM, Dehghani MH (2024) Insights into properties, synthesis and emerging applications of polypyrrole-based composites, and future prospective: a review. Heliyon 10:33643

# Chapter 8
# Modification of Conducting Polymers for Electronics and Optoelectronics

**Ignacio A. Jessop**

**Abstract** Conducting polymers (CPs), or conjugated polymers, represent a unique class of materials designed for use in electronic applications due to their combination of semiconducting, electrochemical, photophysical, and morphological properties within a single material. These characteristics offer a viable alternative to inorganic materials, enabling the development of low-cost, lightweight, and flexible organic electronic devices. CPs have found applications in organic photovoltaics (OPVs), organic light-emitting diodes (OLEDs), organic field-effect transistors (OFETs), and chemical sensors, among others.

**Keywords** Conducting polymers (CPs) · Organic electronics · Semiconducting properties · Organic photovoltaics (OPVs) · Organic light-emitting diodes (OLEDs) · Organic field-effect transistors (OFETs) · Chemical sensors

## 8.1 Introduction

Conducting polymers (CPs), or conjugated polymers, represent a unique class of materials designed for use in electronic applications due to their combination of semiconducting, electrochemical, photophysical, and morphological properties within a single material. These characteristics offer a viable alternative to inorganic materials, enabling the development of low-cost, lightweight, and flexible organic electronic devices. CPs have found applications in organic photovoltaics (OPVs), organic light-emitting diodes (OLEDs), organic field-effect transistors (OFETs), and chemical sensors, among others.

---

I. A. Jessop (✉)
Organic and Polymeric Materials Research Laboratory, Department of Chemistry, Universidad de Tarapacá, Arica, Chile
e-mail: iajessop@academicos.uta.cl



S. Jadoun, N. Kumari Jangid (eds.), *Tuning of Conducting Polymers*, Springer Series on Polymer and Composite Materials,
https://doi.org/10.1007/978-3-032-00461-1_8

Since the 70s, the chemical structures of CPs have evolved significantly, transitioning from single repeating units—such as polyacetylene, polyaniline (PANI), and polythiophene (PT)—to more complex architectures, as illustrated throughout this chapter. While early research focused on fundamental properties like conductivity, electrochemistry, and photophysics, subsequent studies have emphasized how subtle structural modifications influence performance in specific applications [1, 2]. This structural evolution has mainly been driven by advances in new building blocks and polymerization methods, allowing fine-tuning of CP's electronic and physical properties. Strategies such as the incorporation of heteroatoms, fused rings, side chains, or 2D extensions along the CP backbones have enabled the design of macromolecular materials with high structural precision [1].

This chapter explores the strategic structural modifications of CPs that have significantly enhanced their performance in optoelectronics and electronics. By tuning their molecular structures, researchers have expanded the functional capabilities of CPs, unlocking new possibilities in areas such as renewable energy, flexible electronics, and wearable technologies, among others [3–10].

Section 8.2 provides an overview of the evolution of CP structures and the synthetic methods used to obtain these materials. Section 8.3 examines the practical implications of the structural modifications, highlighting their impact on specific applications such as OPVs, OLEDs, OFETs, and sensors. Other applications will not be discussed here, including electrochromic devices, organic lasers, organic thermoelectrics, photocatalysis, and biomedical uses (e.g., imaging, sensing, and therapeutics). However, readers can refer to several comprehensive reviews for further insights. Also, given the vast structural diversity of CPs, this chapter focuses on selected literature from the last decade, particularly on key macromolecular materials of current interest.

Finally, Sect. 8.4 discusses the challenges and future opportunities in the fields analyzed. While significant progress has been made, critical issues such as scalability and long-term stability remain. Integrating experimental and computational approaches, including machine learning-guided design, offers a promising strategy to accelerate the development of advanced CPs by predicting optimal structural modifications and reducing the time and cost associated with material discovery.

## 8.2 An Overview of Conducting Polymers

### *8.2.1 Evolution of CPs: From Simple Structures to Advanced Functional Materials*

In the early years of CPs, simple repeating unit-based structures such as PANI, PT, and poly(paraphenylene) (PPP) (Fig. 8.1) were extensively studied to understand their metallic states achieved through doping [11]. This process, which involves

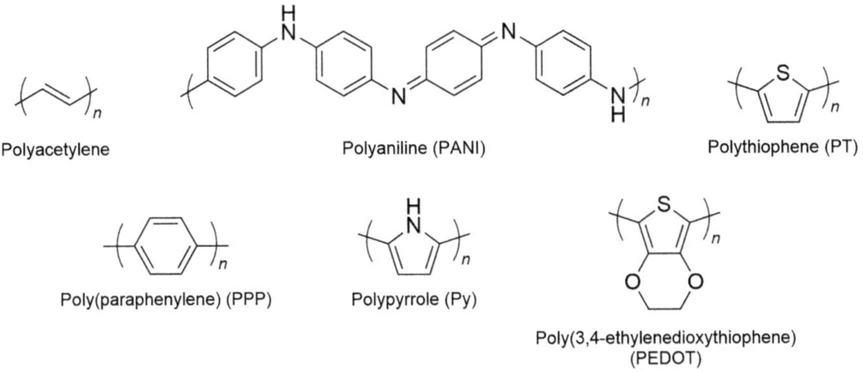


**Fig. 8.1** Chemical structures of early CPs

adding or removing electrons, significantly enhances conductivity. However, the structural simplicity of these early CPs was constrained by limited monomer and polymer synthesis strategies. Consequently, they suffered from poor solubility, low processability, and inadequate environmental stability, which hindered their practical applications [1].

As new synthetic methodologies emerged, CPs became attractive as low-cost alternatives in synthesis and processing, paving the way for their integration into optoelectronic and electronic applications. The introduction of diverse aryl units, alkyl side-chain substituents, and heteroatoms enable precise tuning of key properties such as absorption and emission spectra, frontier molecular orbitals (HOMO/LUMO), bandgap ($E_g$) energy levels, charge transport, solubility, and morphology [1].

A significant breakthrough in CP design was the development of donor–acceptor (DA) copolymers, which incorporated electron-rich and electron-poor aromatic units along the backbone [12]. This innovation allowed researchers to systematically modify electronic properties, including HOMO, LUMO, and $E_g$. Scientists then shifted their focus to understand how copolymer structure, molecular weight, crystallinity, photon absorption and emission, charge transport, and/or structural defects influence device performance [1].

This rational design approach enables fine-tuning electrical, optical, and morphological properties to meet specific application requirements. In OPVs, the goal was to develop materials with optimized low-$E_g$s, strong light absorption, and efficient charge transport. For OLEDs, CPs with high photoluminescence efficiency, tunable emission spectra, and operational stability were highly sought after. FETs benefited from CPs with high crystalline and charge-carrier mobility, while sensors leveraged their environmental sensitivity.

## 8.2.2 Advances in Synthetic Strategies for CPs

In the early stages of CP research, synthetic methodologies were highly limited. Polymerization techniques such as electrochemical, chemical oxidative, and some organometallic methods were the first approaches employed. However, these methods primarily yielded homopolymers with significant structural defects and poor regio- and chemo-selectivity. These limitations negatively impacted the optical and electronic properties of the resulting materials, restricting their practical applications [1].

A significant breakthrough in CP synthesis came with transition metal-catalyzed cross-coupling reactions, including Suzuki-Miyaura, Heck, Yamamoto, and Migita-Stille. These methods revolutionized the field by enabling the synthesis of DA-CPs with improved structural control [1, 13, 14]. Suzuki and Stille's polycondensations have been the most widely used due to their reliability in producing structurally diverse CPs with controlled molecular weights and crystalline and minimized structural defects. Over time, numerous protocols have been developed and optimized to fine-tune the polymer structure and enhance performance [1] (Fig. 8.2).

A more recent approach is Direct Arylation Polymerization (DArP) [15, 16], which leverages selective C-H bond activation. Unlike Suzuki and Stille polymerizations, which require boronic acid/ester or trialkylstannane functionalized monomers, DArP eliminates the need for these pre-functionalized reagents [1]. An even more advanced version of this strategy, Oxidative DArP (Oxi-DArP) [17] achieves polymerization via dehydrogenative coupling, bypassing the need for all pre-functionalized monomers, including brominated monomers (Fig. 8.2). These approaches significantly reduce the number of synthetic steps and eliminate hazardous reagents, marking a significant step toward easier, cheaper, and eco-friendly CP synthesis [1, 17].

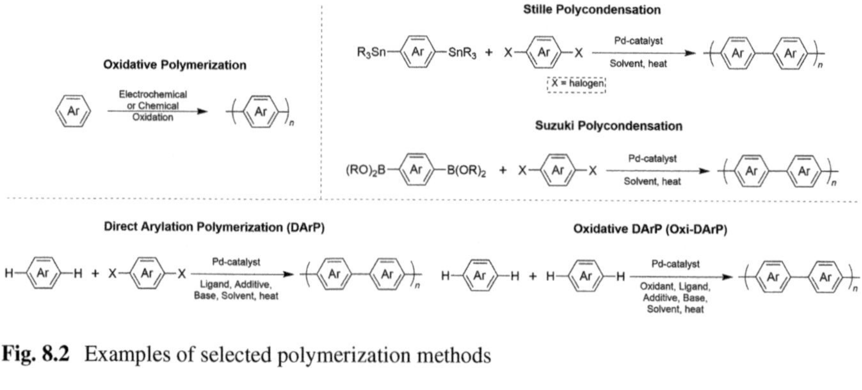


**Fig. 8.2** Examples of selected polymerization methods

# 8.3 Tailoring CPs for Specific Applications

Over the last decades, CPs have gained the scientific community's attention for their versatility in optoelectronic and electronic applications. These materials can be tailored to meet specific application requirements through specific structural modifications, enhancing their performance in cutting-edge technologies. This section explores their use in optoelectronics and electronics, emphasizing organic photovoltaics, light-emitting diodes, field-effect transistors, and sensors (Fig. 8.3).

## *8.3.1 Optoelectronic Applications*

### 8.3.1.1 Organic Solar Cells (OSCs)

Solar energy conversion is essential in transitioning from fossil energy to a more sustainable energy system. Compared with silicon, perovskite, and other inorganic materials, organic polymeric materials exhibit lightweight, semi-transparency, flexibility, and roll-to-roll processability, significantly increasing production efficiency [3, 18, 19]. Due to the breakthrough emergence of non-fullerene acceptors (NFAs), like ITIC and Y6 series, the PCE of polymer solar cells (PSCs) has gradually reached 19% in recent years [18, 20–22]. This device mainly uses the CPs as donor materials within the photoactive layers, promoting light absorption, charge

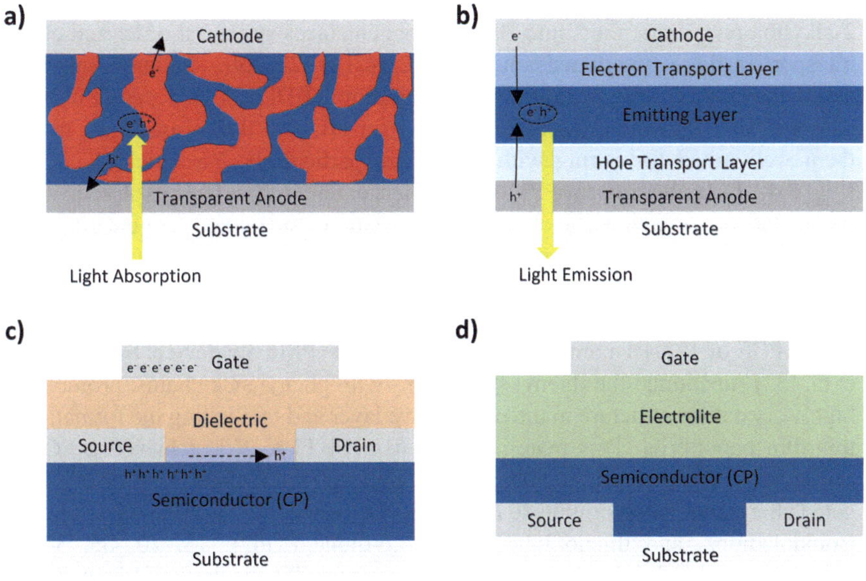


**Fig. 8.3** Schematic representation of some CP-based device architectures: (**a**) OPV, (**b**) OLED, (**c**) OFET, and (**d**) OECT

diffusion, and separation. In this sense, the structural modifications have been focused on optimizing their optical, electronic, and morphological properties.

In a typical bulk heterojunction (BHJ) device (Fig. 8.3a), the donor and acceptor materials are blended to form the photoactive layer. When incident photons are absorbed by the donor material, typically a CP, electrons are excited from the highest occupied molecular orbital (HOMO) to the lowest unoccupied molecular orbital (LUMO), generating excitons (electron–hole pairs bound by electrostatic forces). These excitons diffuse to the donor–acceptor interface, where charge separation occurs, driven by the energy difference between the LUMO levels of the donor and acceptor materials. This energy difference must exceed the exciton binding energy to facilitate dissociation. The resulting free electrons and holes then migrate to their respective metallic electrodes.

Meticulous tuning of the electronic and physical properties of the CPs through the addition of heteroatoms, fused rings, or extended 2D-conjugation along the π-conjugated backbone or the introduction of side chains for improving solubility have pushed the performance of OPV devices over the past decade [1, 18, 23, 24]. Traditionally, research on CPs for OPV applications has focused on developing low-$E_g$ materials ($E_g$ < 1.6 eV) to enhance absorption in the near-infrared (NIR) region, aligning with the fullerene-based acceptors that dominate early OPV architectures [25]. However, the advent of NFAs has shifted the focus toward medium (1.6 eV < $E_g$ < 1.8 eV) and wide- $E_g$ ($E_g$ > 1.8 eV) CPs, which provide complementary absorption and well-aligned energy levels, leading to more efficient devices [26, 27].

High-performance CPs in OPVs share a standard structural building block: the electron–donor monomer benzo[1,2-*b*:4,5-*b*']dithiophene (BDT). Incorporating 2-alkylthiophene side units into BDT donors enhances structural order and charge transport in these two-dimensional CPs (2D-CPs) [28, 29]. Among them, PM6, based on benzo[1,2-*c*:4,5-*c*']dithiophene-4,8-dione (BDD) as the acceptor unit, and its copolymer derivatives meet key criteria for efficient donor materials, making them well-suited for paring with the NFAs, particularly, Y6 series [1, 21, 30] (Fig. 8.4). To further improve PM6 performance, researchers have explored various molecular modifications, including halogenation, side-chain engineering, and acceptor substitution. For example, Ma et al. [31] replaced fluorine atoms with less electronegative chlorine atoms in the lateral thienyl groups of BDT units. This modification led to the development of PM7, which, when blended with Y6, achieved a higher PCE of 17.04% compared to 15.94% for the PM6:Y6 device. Similarly, Xie et al. [32] substituted the thienyl side chains with phenyl side chains, promoting a highly crystalline structure in the photoactive layer and facilitating the formation of thin-film nanofibrils. This material, known as PBT1-C-2Cl, achieved a PCE of 16.1% with Y6. Beyond side-chain modifications, researchers have also investigated alternative acceptor units to replace BDD, such as quinoxaline (Qx), triazoloisoindoledione, and thieno[3,4-*c*]pyrrole-4,6-dione (TPD) [18, 20, 33, 34]. A notable example of an alternative acceptor is presented in the work of Liu et al. [35], who conducted an insightful study on the building block dithieno[3′,2′:3,4:2″,3″:5,6]benzo[1,2-*c*][1,2,5]thiadiazole (DTBT). This fused-ring

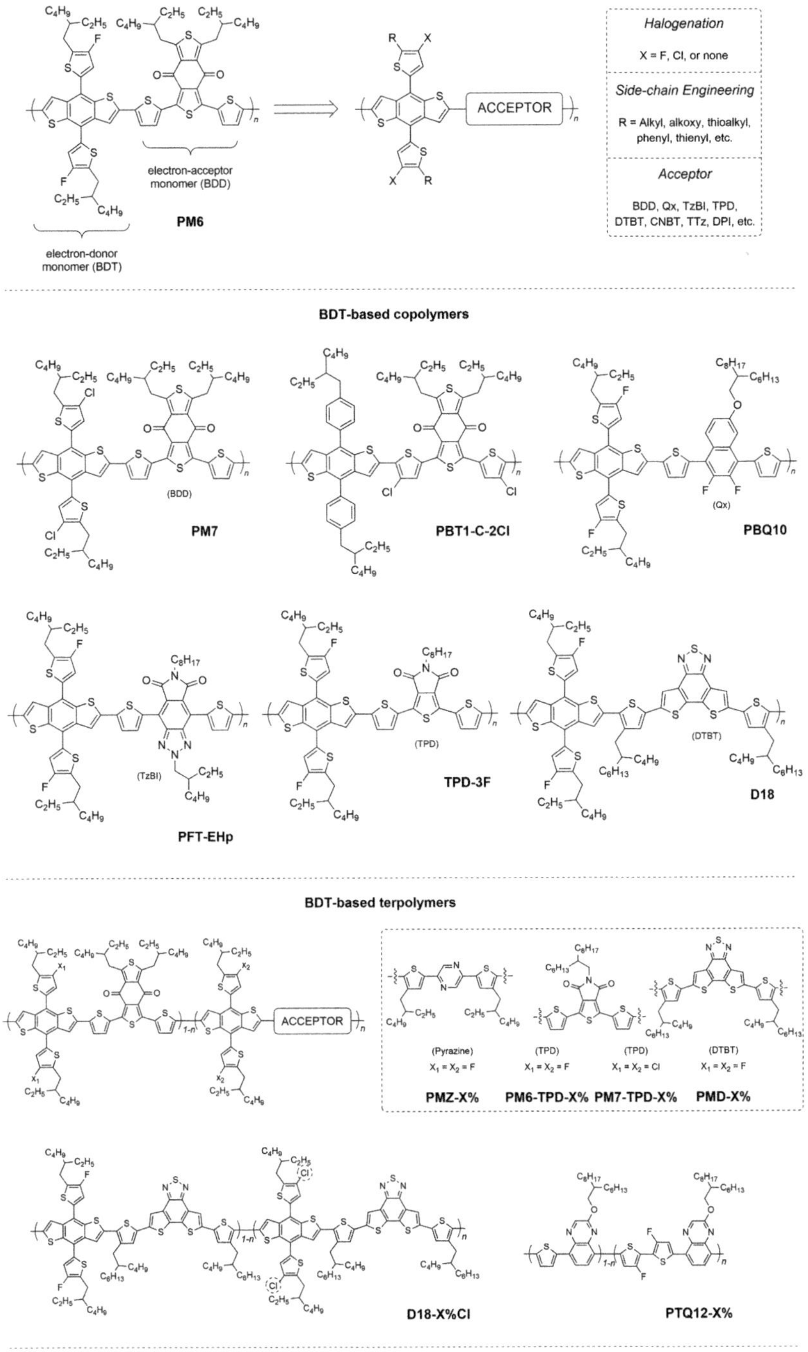


**Fig. 8.4** Chemical structures of current CPs used for OPVs

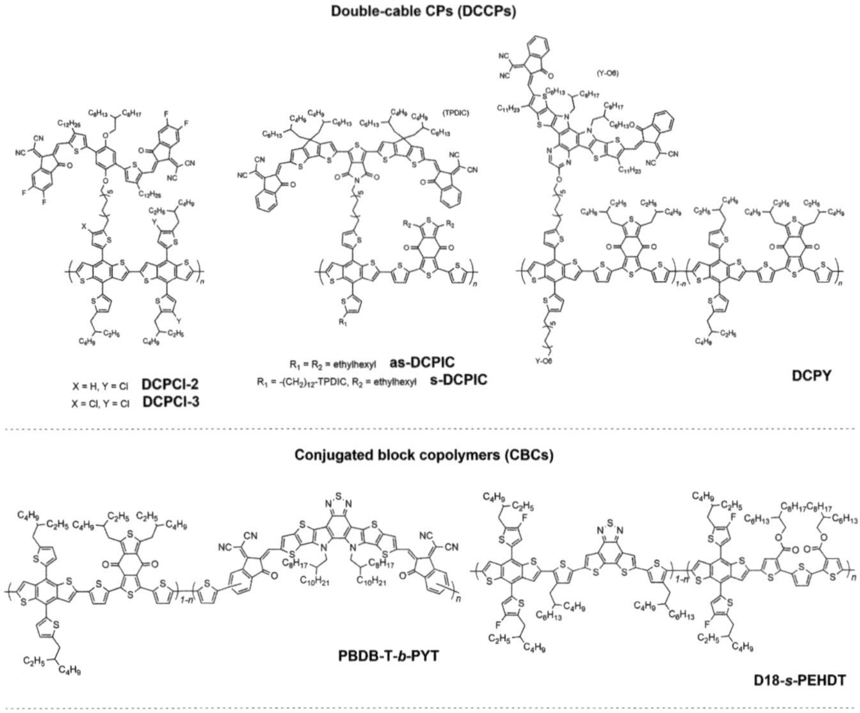


**Fig. 8.4** (continued)

acceptor unit exhibits slightly weaker electron-withdrawing properties than BDD, leading to the development of a CP called D18. D18 features a wide-$E_g$ (1.98 eV) and a blue-shifted absorption spectrum compared to PM6. Despite this, its high hole mobility enables the fabrication of a device that achieves a remarkable PCE of 18.22% when paired with Y6. These advancements highlight the ongoing efforts to optimize polymer structures for improved photovoltaic performance.

Over the past decade, the increased synthetic and structural complexity of top-performing CPs has led to more significant challenges in synthesis and purification, along with the associated waste and higher material costs. To address these issues, there has been growing interest in developing terpolymers, copolymers incorporating a third monomer unit into their backbones at an optimized ratio (Fig. 8.4). This approach offers a straightforward method to fine-tune the physical and electronic properties of CPs, particularly broadening photon absorption, modulating HOMO, LUMO, and $E_g$ energies, and optimizing morphology [18, 21, 36, 37].

Several studies have demonstrated the effectiveness of the terpolymer strategy in enhancing the performance of BDT-based CPs, such as those by Zhou et al. [38] who added a low-cost pyrazine derivative as the third monomer into the PM6 backbone. With just 10% incorporation, the resulting device achieved a PCE of 18.23%, surpassing the PCE of 16.63% for a PM6:Y6-based device. The pyrazine unit influenced key properties such as HOMO energy level, absorption coefficient, aggregation, crystallinity, and charge-carrier mobility of PM6. Similar improvements have been observed when other acceptor monomers, such as TPD, DTBT, dicyanobenzothiadiazole (CNBT), thiazolothiadiazole (TTz), and dithienophthalimide (DPI) [39–42], are incorporated into the CP backbone in controlled amounts. Even minor structural modifications can significantly enhance their properties, such as introducing a chlorine-substituted BDT unit into the D18 backbone. For example, incorporating 20% of this third monomer enabled modulation of D18's physical and optoelectronic properties, resulting in a PCE of 18.28%, higher than the 17.50% achieved with a D18:Y6 blend [43]. The proportion of the third monomer is crucial for achieving high-performance terpolymers, highlighting the need for systematic studies to optimize its incorporation. For instance, Zhang et al. [44] synthesized a series of PM7 derivatives—like PM6 but with chlorinated BDT instead of fluorinated BDT units—gradually incorporating TPD units to enhance crystallization. The resulting material, PM7-TPD, when blended with the Y-series molecule L8-BO, formed a uniform fibrous morphology. This structural arrangement improved charge generation, suppressed charge recombination, and achieved a PCE of 18.35% with just 10% of TPD, outperforming PM6-TPD:Y6 [31]and other PM7-based devices [44].

Given the complex and time-consuming synthesis, isolation, and purification processes of PM6 and its derivatives, which significantly increase production costs, there is a strong drive to develop low-cost, high-performance materials. Wu et al.'s work is a notable example [45]., who designed structurally simple terpolymers based on the well-known wide-$E_g$ quinoxaline-thiophene material PTQ10. When blended with the small molecule K2 (an L8-BO derivative), these terpolymers demonstrated improved excitation, enhanced charge transfer and extraction, suppressed charge recombination, and more efficient, balanced charge transport, ultimately achieving a PCE of 18.77%. Studies like this encourage the transition from PM6 to other wide-bandgap CPs with lower costs.

Standard CP-based OPV devices typically employ a binary blend layer as photoactive material. Researchers have extensively explored ternary systems incorporating a third component, either a donor or an acceptor material, to advance toward a new generation of devices. These ternary blends offer multiple advantages, including enhanced photon utilization, broader spectral absorption, optimized morphology, improved exciton separation, and more balanced charge transport, ultimately leading to a synergistic enhancement of photovoltaic performance [46–49]. The selection of the third component is critical, as it must align with both the processing conditions and the chemical and physical properties of the ternary blend. Specifically, the chosen material should complement the absorption profile of the CP and the acceptor, exhibit good compatibility with the donor or acceptor in the binary host system, induce controlled modification of energy levels, and contribute to improved crystallinity and device stability [46–49]. Despite the significant PCEs achieved with ternary OPVs [50–52], identifying a material that simultaneously satisfies all these criteria remains a considerable challenge.

In binary or ternary systems, the photoactive layer often suffers from thermodynamic instability, leading to the spontaneous formation of self-aggregated structures [53, 54]. This results in reduced thermal and light stability over long-term operation. One strategy to address this issue is the covalent linkage of donor and acceptor units within a single material, forming double-cable CPs (DCCPs) for single-component organic solar cells (SCOSCs) [55–58]. DCCPs typically consist of a CP backbone acting as the electron donor material, with conjugated side units functioning as electron acceptors [59] (Fig. 8.4). Traditionally, fullerenes and rylene diimides have been widely employed as side units, yielding devices with PCEs approaching 8% [56, 60, 61]. More recently, NFAs have been incorporated into DCCPs to enhance performance. For example, Liu et al. [62] successfully integrated a thienyl-phenyl-thienyl core with 2-(3-oxo-2,3-dihydroinden-1-ylidene)malonitrile (IC) as a side group into a poly(BDT) backbone, producing a DCCP named DCPY. By varying the number of chlorine atoms in the BDT units, they achieved a PCE of 8.23%. Similarly, Liang et al. [63] developed a DCCP based on a non-fused ring electron acceptor featuring a thienopyrroledione (TPDIC) core attached to a PBDB-T backbone (a fluorine-free derivative of PM6). In this study, the authors incorporated one or two acceptor molecules per BDT unit, resulting in symmetric (s-DCPIC) and asymmetric (as-DCPIC) materials. While s-DCPIC exhibited a low PCE of 2.99%, the as-DCPIC achieved a significantly higher PCE of 8.12% (10.09% if DIO is used as additive). Both CPs exhibited low voltage losses and excellent thermal and light stability. The length of the alkyl linkers connecting the CP backbone to the acceptor side units also plays a crucial role in determining device performance. as-DCPIC demonstrated its highest PCE when a dodecyl ($C_{12}H_{24}$) chain was employed [64]. Liang et al. [65] introduced a highly crystalline Y-series acceptor featuring a Qx core as a side chain within a PBDB-T backbone. The resulting DCCP (DCPY1) exhibited a PCE of 2.75% when blended with a Y-series molecule with Qx core (Y-O6). However, when a BDT unit was incorporated as a third monomer, the resulting DA-terpolymer (DCPY2) achieved a remarkable PCE of 13.02%.

Another class of CPs used in SCOSCs is conjugated block copolymers (CBCs), which incorporate oligomeric donor and acceptor blocks along the polymer backbone [66]. For instance, Wu et al. [67] synthesized a CBC named PBDB-T-*b*-PYT, where the PBDB-T donor block was linked to an acceptor block composed of thiophene and L8-BO [22]. This material was obtained via a one-pot polymerization approach: First, the donor polymer block was prepared via Stille polycondensation, and after a fixed time, the monomers of the acceptor block were rapidly introduced into the reaction mixture. The reaction was then allowed to proceed for a designated period before being finished and processed like any other method to obtain CPs. This CBC exhibited a PCE of 11.32% and enhanced storage stability compared to the corresponding PBDB-T:PYT BHJ. More recently, Lee et al. [68] synthesized a CBC of 80% D18 and 20% of a BDT-terthiophene-based CP (PEHDT). This sequentially prepared material combined the superior electrical properties of the D18 block with the excellent mechanical stretchability of PEHDT. When used as the donor material in conjunction with L8-BO as the small-molecule acceptor, the resulting D18-*s*-PEHDT: L8-BO-based OPV device achieved a PCE of 19.0% and a crack-onset strain (COS), a measure of polymer stretchability, of 17.2%, surpassing both D18:L8-BO and PEHDT: L8-BO blends. Although DCCP and CBC provide a promising strategy to improve exciton dissociation, charge generation, environmental stability, and device fabrication simplicity, the PCEs of SCOSCs remain less competitive compared to other systems discussed in this section.

#### 8.3.1.2 Organic Light-Emitting Diodes (OLEDs)

OLEDs have been established as essential light sources in the display industry, finding applications in OLED TVs, mobile phones, laptops, and more [69, 70]. These multilayer devices generate charge carriers at the cathode and anode under an applied voltage. The charge carriers are injected into the electron and hole injection layers, migrating through transport layers into the emissive layer. Here, the recombination of electrons and holes leads to the formation of excitons. Electroluminescence (EL) occurs when these excitons decay radiatively to their ground state, emitting photons [71, 72] (Fig. 8.3b).

Early OLEDs relied on fluorescent small molecules or polymers to generate photoluminescence. However, these materials utilized only 25% of the generated excitons due to the singlet-to-singlet transitions, resulting in low external quantum efficiencies (EQEs), which is related to the efficiency of converting electricity into light, of approximately 5%. The remaining 75% of excitons formed in the triplet state were initially inaccessible. To overcome this limitation, phosphorescent heavy-metal complexes were introduced, taking advantage of the triple state to achieve EQEs approaching 100%. While effective, these materials are costly and environmentally harmful. Then, researchers focused on the thermally activated delayed fluorescence (TADF) mechanism to enhance OLED efficiencies. TADF molecules feature small $E_g s$ between the lowest singlet state ($S_1$) and triplet state ($T_1$), enabling thermally activated intersystem crossing from $T_1$ to $S_1$. This process allows triplet excitons to contribute to EL, theoretically achieving EQEs of 100%. However, TADF-based devices still face challenges, including achieving high efficiency, color purity, luminance, brightness, and long operational lifetimes. "Hyperfluorescence" (or thermally assisted fluorescence) was introduced to improve their performance. In these systems, excitons are formed in TDAF molecules and subsequently transferred to a fluorescent emitter via long-range Förster resonance energy transfer (FRET) or Dexter energy transfer (DET) processes, resulting in EL. Building on this approach, hybrid systems combining TADF materials with fluorescent molecules have been explored to enhance device efficiency and stability [72–75].

While much of the research into new materials for OLEDs has focused on small molecules, polymer-based OLEDs offer a unique combination of electrical and mechanical properties, including flexibility, ease of processing, low fabrication costs, moldability, and lightweight characteristics [74]. In the case of CP-based devices, the rational design of the emissive layer is the most crucial factor for achieving high efficiency. In this sense, CPs must possess well-tuned $E_g s$ that align with other layers, a rigid chemical structure, good thermal stability, and excellent charge transport properties [72, 76].

Several strategies have been employed to improve the efficiency of OLEDs based on CPs with TDAF units grafted along their backbones. These strategies include increasing the dihedral angle between donor and acceptor units, incorporating branched side chains, and introducing large steric substituents. These modifications reduce the overlap of HOMO and LUMO energy levels, thereby minimizing the singlet-triplet energy gap ($\Delta E_{ST}$) and weakening intermolecular π-π coupling interactions. Donor units such as tetramethylbenzene, dibenzothiophene sulfone, and carbazole, which provide good charge transport properties, are polymerized together or with acceptor units at varying molar ratios to produce highly twisted backbones. This approach enables OLEDs to emit in the visible spectrum, ranging from blue-green to orange-red, with EQE values reaching up to 20% [76–83] (Fig. 8.5).

Despite their high efficiency, TADF CPs often exhibit broad emission spectra, compromising color purity. Due to their narrowband emission properties, multi-resonance TADF (MR-TADF) emitters have recently emerged as promising candidates for high-performance OLEDs. In these materials, the MR effect arises from the interaction between electron-withdrawing boron (B) and electron-donating atoms (N, O, or S) within a rigid polycyclic aromatic hydrocarbon unit. This interaction enhances HOMO-LUMO separation due to opposing resonance effects on the same π-conjugated plane, reducing the $\Delta E_{ST}$ energy gap and minimizing vibration relaxation. Consequently, these materials exhibit narrowband emission [84, 85]. MR-TADF emitters can be incorporated as pendant groups or integrated directly into the CP backbone [86, 87].

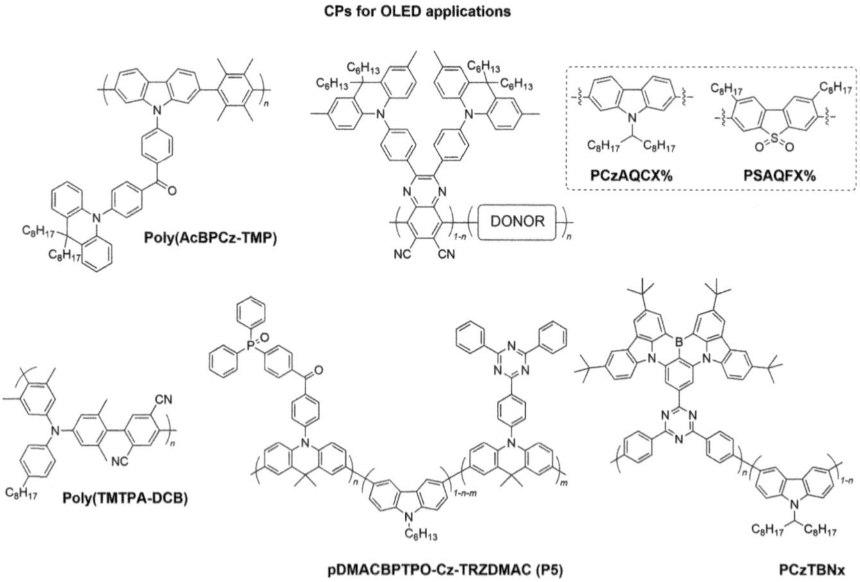


**Fig. 8.5** Chemical structures of current CPs used for OLEDs

## 8.3.2 Electronics Applications

### 8.3.2.1 Organic Thin-Film Transistors (OTFTs)

CPs-based OTFTs offer several advantages over inorganic semiconductor-based transistors, including solution processability for large-area fabrication, intrinsic flexibility and stretchability, and biocompatibility [88, 89]. These features make them particularly attractive for commercial applications [90].

OTFTs can be categorized into organic field-effect transistors (OFETs) and organic electrochemical transistors (OECTs). The performance of these devices is significantly influenced by charge-carrier mobility, molecular alignment, and film morphology, which depend on the CP structure. A typical polymer-based OFET comprises three metallic terminals: the source, drain, and gate electrodes. The channel, made of a CP thin film, is deposited on a substrate, with an insulating layer separating the gate electrode from the channel and the source and drain electrodes. Field-effect doping occurs in the CP when a voltage is applied between the gate and the channel. This induces charge accumulation at the dielectric interface, modulating the channel's conductivity. If the charge density is high enough, current flows along the channel between the source and drain electrodes (Fig. 8.3c) [91, 92]. The gate voltage thus controls the channel's charge density, enabling the device to function as an electrically controlled switch or amplifier [71, 91]. In the OECTs, the current is regulated by ion exchange between an electrolyte, which replaces the dielectric in OFETs, and the channel. The ions are injected into the channel, altering the charge density throughout its entirely [74, 93, 94] (Fig. 8.3d).

The key performance parameter for CP-based OTFTs is the charge-carrier mobility within the channel, which depends on various factors such as molecular weight, backbone planarity and rigidity, and close intermolecular packing to promote ordered aggregates [71, 90, 95, 96]. The charge-carrier mobility of conducting polymers is inferior to semiconductors because organic molecules have more considerable intermolecular distances than inorganic molecules. This increased spacing and the complex organization of polymers results in disordered regions that introduce electronic traps that hinder charge transport. However, significant progress has been made in recent years. Strategies focusing on the selection of appropriate electron-donating and acceptor-withdrawing monomers have produced D-A polymers with considerable π-π orbital overlap, reduced stacking distance, and optimized HOMO/LUMO energy levels that allow efficient charge injection and extraction from the electrodes. Additionally, the addition of suitable heteroatoms, substituents, and/or side chains has facilitated dense molecular packing, improved solubility, and improved charge-carrier mobility [76, 90].

These advancements have resulted in OFETs with charge-carrier mobilities exceeding 10 $cm^2\ V^{-1}\ s^{-1}$ for p-type (hole-dominant transport) devices. As stated before, such improvements are driven by molecular design strategies, including side-chain engineering for solubility, chemical substitution to lock conformations (e.g., F···H or S···O interactions), and the incorporation of donor-acceptor building blocks to tune molecular energy levels [88, 90, 95, 97–99]. Notable CPs, including indacenodithieno[3,2-*b*]thiophene (IDTT) [100], cyclopentadithiophene (CDT) [101, 102], dithienosilole (DTS) [103, 104], diketopyrrolopyrrole (DPP) [105–108], bis(oxoindolinylidene)benzodifurandione (BDOPV) [109], naphthalenedimide (NDI) [110], among others, have shown good performance (Fig. 8.6). These materials exhibit highly rigid backbones with small torsion angles, reducing intra-chain energetic disorder and maximizing charge transport. Their well-ordered microstructures and optimized HOMO and LUMO energy levels contribute to achieving stable and high hole or electron mobilities across various gate voltages [88].

For OECTs, CPs functionalized with ethylene glycol side chains demonstrate effective ion uptake throughout the bulk and high electronic mobility, resulting in significant mobility-capacitance products (μC*) [111–113]. However, compared to OFETs, OECT research remains in its early stages, with several challenges still to be addressed [91, 94].

#### 8.3.2.2 Organic Sensors

Chemical sensors are widely used in food, pharmacy, and environmental and industrial safety monitoring [114]. CPs exhibit sensitivity to environmental stimuli such as light, gas molecules, ions, solvents, or biomolecules. When incorporated into sensor devices, CPs can convert these external stimuli into electrical signals or other output information [115]. This property has positioned CPs as crucial components in chemical sensors, where their versatility, ease of synthesis, large surface area, and high sensitivity make them highly desirable. Chemical sensors are mainly composed of three parts: the sensitive material, the transduction element, and the transduction circuit, where the sensitive material is the core of this technology [116].

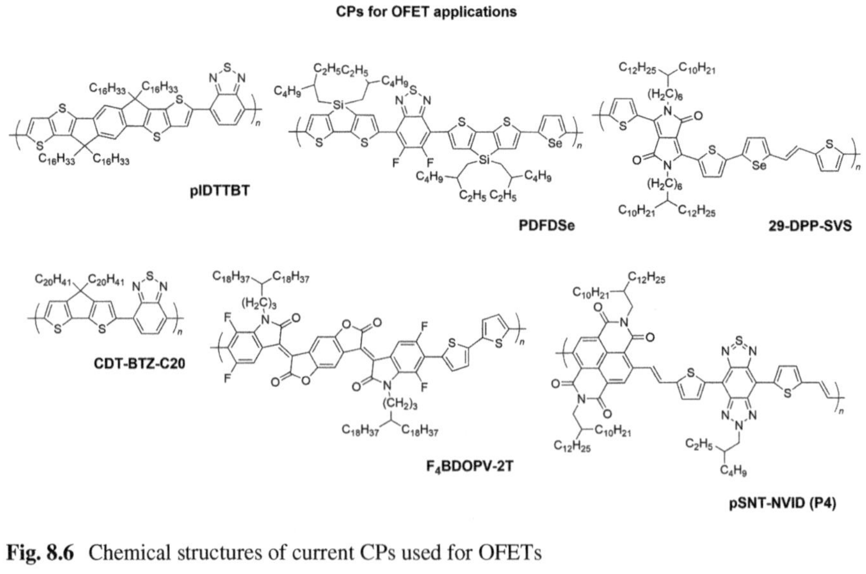


**Fig. 8.6** Chemical structures of current CPs used for OFETs

OTFTs have demonstrated high environmental sensitivity, selectivity, and mechanical flexibility, making them promising candidates for various sensor applications (Fig. 8.3c, d) [12, 92].

Common and simple CPs such as PANI, polypyrrole (Py), poly(3,4-ethylenedioxythiophene) (PEDOT), and their derivatives have been extensively employed in these applications [117] (Fig. 8.1). To enhance the sensing capabilities of these CPs, they are often combined with nanomaterials, such as carbon-based materials or metal nanoparticles, through chemical or electrochemical polymerization. The resulting CP nanocomposites (CPNs) [118] are hybrid materials that integrate the distinct properties of each component, leading to improved electrical sensing activity, thermal stability, mechanical strength, and chemical properties. By modifying the structure, proportion, or size of CPNs, their sensing performance can be fine-tuned to achieve highly sensitive detection with low detection limits for specific analytes [119].

In chemical sensor applications, CP-based sensors operate by converting the concentration of an analyte into measurable physical signals such as changes in absorption, emission, pH, or conductivity. For example, CPs are widely used for detecting environmentally hazardous gases produced by natural and human activities. When gas molecules interact with a CP, they modify their doping level, promoting changes in their physical and chemical properties, including electric conductivity and morphology.

The CP may undergo oxidation, reduction, doping, or dedoping depending on the analyte. Studies have shown that CP or CPN-based sensors utilizing PANI, PPy, PEDOT, PT, and their derivatives are effective for monitoring gases such as ammonia ($NH_3$), carbon monoxide (CO), nitrogen dioxide ($NO_2$), and hydrogen sulfide ($H_2S$) [120, 121]. Additionally, these sensors have shown potential for detecting volatile organic compounds such as 1,3-butadiene, formaldehyde, ethylene glycol, and tetrachloroethylene [117, 122, 123]. For advanced gas sensing, conjugated microporous polymers (CMPs), a class of porous organic polymers (POPs), have gained growing interest owing to their high chemical/thermal stability, high surface area, microporosity, efficient host-guest interactions with the analyte, efficient exciton migration along the backbone, and reversible redox properties. CMPs are synthesized through polycondensation reactions using building blocks that possess simple phenyl units, extended aromatic hydrocarbons, heteroatoms, and/or large macrocyclic compounds. The pore sizes and specific surface area are controlled by carefully selecting building blocks and the reaction conditions. The major drawbacks of these CMP materials are the time-consuming fabrication process, short duration, and high cost for large-scale production [114, 116].

CPs incorporating hydrophilic metal-receptor units have also gained attention as optical sensors. Various CP backbones with distinct fluorescent properties and side chains with specific functional groups have been explored for metal ion detection. For instance, derivatives of poly(p-phenyleneethynylene) (PPE), polyfluorene (PF), or PT have been successfully employed as colorimetric and/or fluorimetric probes for detecting metal ions such as sodium ($Na^+$), potassium ($K^+$), mercury ($Hg^{2+}$), copper ($Cu^{2+}$), lead ($Pb^{2+}$), iron ($Fe^{3+}$), aluminum ($Al^{3+}$), and other ions [124–126]. By adjusting the water solubility CP through side-chain functionalization, some hydrophilic CP-based probes have been designed to determine metal ions in aqueous solutions, expanding their applicability.

This section shows that CPs can be adapted to meet the unique demands of diverse optoelectronic and electronic applications. PSCs and OLEDs see efficiency improvements through bandgap engineering and the design of new advanced materials. OFETs benefit from their molecular alignment and morphological control, while sensors gain improved selectivity and sensitivity from functionalization. Continued innovation in CP design and processing is necessary for unlocking new possibilities for sustainable, high-performance, and commercially viable technologies.

## 8.4 Challenges and Future Perspectives

CPs have revolutionized the fields of optoelectronics and electronics due to their unique combination of electrical conductivity, light-weight structure, and mechanical flexibility. However, intrinsic properties often require strategic modifications to meet the demands of advanced applications. Chemical modification, including functional group substitution, side-chain engineering, or co/terpolymerization, has proven effective in tuning the $E_g$, improving solubility, and enhancing performance. Physical strategies such as blending and nanocompositing have allowed the creation of hybrid materials that leverage the strengths of both polymers and inorganic components.

Despite significant advancements, several challenges remain in optimizing and innovating high-performance CPs. Overcoming these challenges is essential to enhancing their performance and enabling their widespread adoption in commercial applications.

A primary challenge in material development is novel high-performance CPs tailored for specific applications. Achieving high efficiency and highly stable blue-emissive materials in OLEDs remains a critical hurdle. In OTFTs, the charge mobility and overall performance of n-type CPs still lag their p-type counterparts, limiting their practical implementation. Similarly, the demand for CPs with enhanced sensitivity, selectivity, and low detection limits continues growing in organic sensors. Additionally, a deeper understanding of the structure-property relationship is essential for designing materials with superior performance. Integrating theoretical calculations, machine learning techniques, and experimental data presents an opportunity to accelerate molecular design. Furthermore, developing low-cost and environmentally friendly synthetic methods is also essential for transitioning CPs from laboratory research to commercial-scale production. Another key challenge is the precise control and optimization of microstructure-property relationships, which directly influence morphology and device performance. Advancing our ability to manipulate these relationships will be instrumental in achieving superior material properties.

From a device fabrication perspective, ensuring the large-scale fabrication of uniform, high-quality films with repeatable performance are also essential for scalable fabrication processes. Producing cost-effective, energy-efficient, and environmental-friendly devices is also a major goal. A critical barrier to the commercialization of CP-based electronic devices is their stability. Many CPs and devices degrade under exposure to high temperatures, humidity, and atmospheric oxygen, compromising long-term functionality. Improving the mechanical resilience and operational stability of CP-based devices is, therefore, imperative.

CPs hold immense potential for next-generation applications, particularly in wearable electronics and bioelectronics. Continuous innovation in molecular structures and device fabrication techniques will be key drivers of progress. As research efforts intensify, the convergence of material science, artificial intelligence, and advanced synthetic and manufacturing methods will shape the future of CP-based techonologies, bringing it closer to widespread commercialization and societal impact.

**Acknowledgments** The author acknowledges the financial support of the Agencia Nacional de Investigación y Desarrollo (ANID) through the FONDECYT N° 1231028 research grant.

## References

1. Pankow RM, Thompson BC (2020) The development of conjugated polymers as the cornerstone of organic electronics. Polymer 207
2. Rasmussen SC (2020) Conjugated and conducting organic polymers: the first 150 years. ChemPlusChem 85(7):1412–1429
3. Subudhi P, Punetha D (2023) Progress, challenges, and perspectives on polymer substrates for emerging flexible solar cells: a holistic panoramic review. Prog Photovolt Res Appl 31(8):753–789
4. Gualandi I, Tessarolo M, Mariani F, Possanzini L, Scavetta E, Fraboni B (2021) Textile chemical sensors based on conductive polymers for the analysis of sweat. Polymers (Basel) 13(6)
5. Lee KW, Wan Y, Huang Z, Zhao Q, Li S, Lee CS (2024) Organic optoelectronic materials: a rising star of bioimaging and phototherapy. Adv Mater 36(17):e2306492
6. Xu X, Zhao Y, Liu Y (2023) Wearable electronics based on stretchable organic semiconductors. Small 19(20):e2206309
7. Hwang S, Yasuda T (2022) Indoor photovoltaic energy harvesting based on semiconducting π-conjugated polymers and oligomeric materials toward future IoT applications. Polym J 55(4):297–316
8. Dai C, Pan Y, Liu B (2020) Conjugated polymer nanomaterials for solar water splitting. Adv Energy Mater 10(42)
9. Liu Y, Feig VR, Bao Z (2021) Conjugated polymer for implantable electronics toward clinical application. Adv Healthc Mater 10(17):e2001916
10. Wang S, Zuo G, Kim J, Sirringhaus H (2022) Progress of conjugated polymers as emerging thermoelectric materials. Prog Polym Sci 129
11. Poddar AK, Patel SS, Patel HD (2021) Synthesis, characterization and applications of conductive polymers: a brief review. Polym Adv Technol 32(12):4616–4641
12. Müllen K, Scherf U (2022) Conjugated polymers: where we come from, where we stand, and where we might go. Macromol Chem Phys 224(3)

13. Krishna A, Lunchev AV, Grimsdale AC (2017) Suzuki polycondensation. In: Synthetic methods for conjugated polymers and carbon materials, pp 59–95
14. Liang Z, Neshchadin A, Zhang Z, Zhao F-G, Liu X, Yu L (2023) Stille polycondensation: a multifaceted approach towards the synthesis of polymers with semiconducting properties. Polym Chem 14(40):4611–4625
15. Gobalasingham NS, Thompson BC (2018) Direct arylation polymerization: a guide to optimal conditions for effective conjugated polymers. Prog Polym Sci 83:135–201
16. Ye L, Thompson BC (2022) Improving the efficiency and sustainability of catalysts for direct arylation polymerization (DArP). J Polym Sci 60(3):393–428
17. Chakraborty B, Luscombe CK (2023) Cross-Dehydrogenative coupling polymerization via C-H activation for the synthesis of conjugated polymers. Angew Chem Int Ed Engl 62(21):e202301247
18. Jin J, Wang Q, Ma K, Shen W, Belfiore LA, Bao X, Tang J (2023) Recent developments of polymer solar cells with photovoltaic performance over 17%. Adv Funct Mater 33(14)
19. Li S, Li Z, Wan X, Chen Y (2023) Recent progress in flexible organic solar cells. eScience 3(1)
20. Luo D, Jang W, Babu DD, Kim MS, Wang DH, Kyaw AKK (2022) Recent progress in organic solar cells based on non-fullerene acceptors: materials to devices. J Mater Chem A 10(7):3255–3295
21. Guo Q, Guo Q, Geng Y, Tang A, Zhang M, Du M, Sun X, Zhou E (2021) Recent advances in PM6:Y6-based organic solar cells. Mater Chem Front 5(8):3257–3280
22. Tian L, Liu C, Huang F (2023) Recent progress in side chain engineering of Y-series non-fullerene molecule and polymer acceptors. SCIENCE CHINA Chem 67(3):788–805
23. Tian H, Zhao M, Ma X, Xu C, Xu W, Liu Z, Zhang M, Zhang F (2023) Critical progress of polymer solar cells with a power conversion efficiency over 18%. Energies 16(11)
24. Luscombe CK, Maitra U, Walter M, Wiedmer SK (2021) Theoretical background on semiconducting polymers and their applications to OSCs and OLEDs. Chem Teach Int 3(2):169–183
25. Scharber MC, Sariciftci NS (2021) Low band gap conjugated semiconducting polymers. Adv Mater Technol 6(4)
26. Xu X, Yu L, Peng Q (2021) Recent advances in wide bandgap polymer donors and their applications in organic solar cells. Chin J Chem 39(2):243–254
27. Sharma VV, Landep A, Lee S-Y, Park S-J, Kim Y-H, Kim G-H (2024) Recent advances in polymeric and small molecule donor materials for Y6 based organic solar cells. Next Energy 2
28. Zhou D, Wang Y, Yang S, Quan J, Deng J, Wang J, Li Y, Tong Y, Wang Q, Chen L (2024) Recent advances of benzodithiophene-based donor materials for organic solar cells. Small 20(8):e2306854
29. Angel FA, Camarada MB, Jessop IA (2023) Computational chemistry advances on benzodithiophene-based organic photovoltaic materials. Crit Rev Solid State Mater Sci 48(3):333–360
30. Shoaee S, Luong HM, Song J, Zou Y, Nguyen T-Q, Neher D (2024) What we have learnt from PM6:Y6. Adv Mater 36(20):2302005
31. Ma R, Liu T, Luo Z, Guo Q, Xiao Y, Chen Y, Li X, Luo S, Lu X, Zhang M, Li Y, Yan H (2020) Improving open-circuit voltage by a chlorinated polymer donor endows binary organic solar cells efficiencies over 17%. SCIENCE CHINA Chem 63(3):325–330
32. Xie Y, Cai Y, Zhu L, Xia R, Ye L, Feng X, Yip HL, Liu F, Lu G, Tan S, Sun Y (2020) Fibril network strategy enables high-performance semitransparent organic solar cells. Adv Funct Mater 30(28)
33. Li M, Zeng Z, Fan B, Zhong W, Zhang D, Zhang X, Zhang Y, Ying L, Huang F, Cao Y (2020) Tailoring the side chain of imide-functional benzotriazole based polymers to achieve internal quantum efficiency approaching 100%. J Mater Chem A 8(44):23519–23525
34. Wu J, Liao C-Y, Chen Y, Jacobberger RM, Huang W, Zheng D, Tsai K-W, Li W-L, Lu Z, Huang Y, Wasielewski MR, Chang Y-M, Marks TJ, Facchetti A (2021) To fluorinate or not to fluorinate in organic solar cells: achieving a higher PCE of 15.2% when the donor polymer is halogen-free. Adv Energy Mater 11(47):2102648

35. Liu Q, Jiang Y, Jin K, Qin J, Xu J, Li W, Xiong J, Liu J, Xiao Z, Sun K, Yang S, Zhang X, Ding L (2020) 18% Efficiency organic solar cells. Sci Bull (Beijing) 65(4):272–275
36. Cheng F, Cui Y, Ding F, Chen Z, Xie Q, Xia X, Zhu P, Lu X, Zhu H, Liao X, Chen Y (2023) Terpolymerization and regioisomerization strategy to construct efficient terpolymer donors enabling high-performance organic solar cells. Adv Mater 35(30):e2300820
37. Li L, Que M, Fang Y, Wang X, Yue Y, Jeong SY, Woo HY, Chen S, Liu J, Huang B (2025) Fine-tuning intra/inter-molecular interaction via ternary copolymerization strategy to obtain efficient polymer donors. Small:e2410198
38. Zhou L, Meng L, Zhang J, Zhu C, Qin S, Angunawela I, Wan Y, Ade H, Li Y (2021) Introducing low-cost pyrazine unit into terpolymer enables high-performance polymer solar cells with efficiency of 18.23%. Adv Funct Mater 32(8)
39. Yang N, Cheng Y, Kim S, Huang B, Liu Z, Deng J, Wang J, Yang C, Wu F, Chen L (2022) Random copolymerization strategy for host polymer donor PM6 enables improved efficiency both in binary and ternary organic solar cells. ChemSusChem 15(8):e202200138
40. Shao J, Liao C, Xu X, Deng M, Yu L, Li R, Peng Q (2022) Synergistically improving the absorption, energy level, and crystallization of PM6 by a dicyanobenzothiadiazole block-based terpolymer strategy. Chem Mater 34(17):7971–7981
41. Chen X, Liao C, Deng M, Xu X, Yu L, Li R, Peng Q (2023) Improving the performance of PM6 donor polymer by random ternary copolymerization of BDD and DTBT segments. Chem Eng J 451
42. Zhou L, Meng L, Zhang J, Qin S, Zhang J, Li X, Li J, Wei Z, Li Y (2022) Terpolymer donor with inside alkyl substituents on thiophene pi-bridges toward thiazolothiazole A(2) -unit enables 18.21% efficiency of polymer solar cells. Adv Sci (Weinh) 9(34):e2203513
43. Bai H-R, An Q, Zhi H-F, Jiang M, Mahmood A, Yan L, Liu M-Q, Liu Y-Q, Wang Y, Wang J-L (2022) A random terpolymer donor with similar monomers enables 18.28% efficiency binary organic solar cells with well polymer batch reproducibility. ACS Energy Lett 7(9):3045–3057
44. Zhang G, Wu Q, Duan Y, Liu W, Jeong SY, Woo HY, Zhao Q, Zhou H (2024) Achieving 19.4% efficiency polymer solar cells by reducing backbone disorder in donor terpolymers. Adv Funct Mater 34(48)
45. Wu X, Zhang X, Zhang J, Wu Y, Li J, Kong X, Li Z, Zhang X, Li X, Li A, Yang G, Sun C (2024) 19.36% Efficiency organic solar cells based on low-cost terpolymer donors with simple molecular structures. Adv Funct Mater 34(44)
46. Zhao C, Wang J, Zhao X, Du Z, Yang R, Tang J (2021) Recent advances, challenges and prospects in ternary organic solar cells. Nanoscale 13(4):2181–2208
47. Doumon NY, Yang L, Rosei F (2022) Ternary organic solar cells: a review of the role of the third element. Nano Energy 94
48. Xu X, Li Y, Peng Q (2022) Ternary blend organic solar cells: understanding the morphology from recent progress. Adv Mater 34(46):e2107476
49. Liu J, Liu X, Xin J, Zhang Y, Wen L, Liang Q, Miao Z (2024) Dual function of the third component in ternary organic solar cells: broaden the spectrum and optimize the morphology. Small 20(24):e2308863
50. Sun R, Wu Y, Yang X, Gao Y, Chen Z, Li K, Qiao J, Wang T, Guo J, Liu C, Hao X, Zhu H, Min J (2022) Single-junction organic solar cells with 19.17% efficiency enabled by introducing one asymmetric guest acceptor. Adv Mater 34(26):e2110147
51. Peng W, Lin Y, Jeong SY, Genene Z, Magomedov A, Woo HY, Chen C, Wahyudi W, Tao Q, Deng J, Han Y, Getautis V, Zhu W, Anthopoulos TD, Wang E (2022) Over 18% ternary polymer solar cells enabled by a terpolymer as the third component. Nano Energy 92
52. Liu S, Xue Z, Liang Z, Zhao B, Wang W, Cong Z, Wu H, Lu G, Zheng J, Gao C (2024) High-performance PM6:Y6-based ternary solar cells with enhanced open circuit voltage and balanced mobilities via doping a wide-band-gap amorphous acceptor. ACS Appl Mater Interfaces 16(28):36705–36714
53. Li Y, Pacalaj RA, Luo Y, Ai K, Hai Y, Liang S, Fan K, Sergeev AA, Ma R, Dela Pena TA, Muller JS, Jin Z, Tuladhar PS, Jia T, Wang J, Li G, Wong KS, Li W, Durrant JR, Wu J (2024)

Molecular control of the donor/acceptor interface suppresses charge recombination enabling high-efficiency single-component organic solar cells. Adv Mater:e2409212

54. Yang X, Gao Y, Sun R, Chen M, Wang Y, Wang S, Shao Y, Xu L-Y, Zhang M, Fu Y, Lu X, Ashraf RS, Luponosov YN, Ponomarenko SA, Min J (2024) The application of Y series acceptor-based double-cable polymers in single-material organic solar cells. Macromolecules 57(3):1011–1020
55. Yao H, Hou J (2022) Recent advances in single-junction organic solar cells. Angew Chem Int Ed Engl 61(37):e202209021
56. Liang S, Li W, Ding L (2023) Single-component organic solar cells. J Semicond 44(3)
57. Liang S, Wang J, Ouyang Y, Tan WL, McNeill CR, Chen Q, Tang Z, Li W (2022) Double-cable conjugated polymers with rigid phenyl linkers for single-component organic solar cells. Macromolecules 55(7):2517–2523
58. Wang T, Sun R, Wu Y, Wang W, Zhang M, Min J (2022) Constructing a double-cable polymer acceptor for efficient all-polymer solar cells with a non-radiative recombination energy loss of 0.16 eV. Chem Mater 34(22):9970–9981
59. Zhang Z, Chen Q, Wang J, Xiao C, Tang Z, McNeill CR, Li W (2024) Correlating crystallinity and performance in single-component organic solar cells based on double-cable conjugated polymers. Giant 19
60. Liang S, Jiang X, Xiao C, Li C, Chen Q, Li W (2021) Double-cable conjugated polymers with pendant rylene diimides for single-component organic solar cells. Acc Chem Res 54(9):2227–2237
61. Lai W, Karuthedath S, Xiao C, Meng L, Laquai F, Li W, Li Y (2024) Alkyl-thiophene-alkyl linkers to construct double-cable conjugated polymers for single-component organic solar cells. Chin Chem Lett 35(1)
62. Liu B, Liang S, Karuthedath S, He Y, Wang J, Tan WL, Li H, Xu Y, Laquai F, Brabec CJ, McNeill CR, Xiao C, Tang Z, Hou J, Yang F, Li W (2023) Double-cable conjugated polymers based on simple non-fused electron acceptors for single-component organic solar cells. Macromolecules 56(3):1154–1164
63. Liang S, Liu B, Karuthedath S, Wang J, He Y, Tan WL, Li H, Xu Y, Li N, Hou J, Tang Z, Laquai F, McNeill CR, Brabec CJ, Li W (2022) Double-cable conjugated polymers with pendent near-infrared electron acceptors for single-component organic solar cells. Angew Chem Int Ed Engl 61(35):e202209316
64. Wang Y, Chen Q, Liang S, Xia D, Zhao C, McNeill CR, Li W (2024) Near-infrared double-cable conjugated polymers based on alkyl linkers with tunable length for single-component organic solar cells. Chin Chem Lett 35(4)
65. Liang S, Xiao C, Xie C, Liu B, Fang H, Li W (2023) 13% single-component organic solar cells based on double-cable conjugated polymers with pendent Y-series acceptors. Adv Mater 35(18):e2300629
66. Cheng Y, Huang B, Mao Q, Huang X, Liu J, Zhou C, Zhou W, Ren X, Kim S, Kim W, Sun Z, Wu F, Yang C, Chen L (2024) Three-in-one strategy enables single-component organic solar cells with record efficiency and high stability. Adv Mater 36(19):e2312938
67. Wu Y, Guo J, Wang W, Chen Z, Chen Z, Sun R, Wu Q, Wang T, Hao X, Zhu H, Min J (2021) A conjugated donor-acceptor block copolymer enables over 11% efficiency for single-component polymer solar cells. Joule 5(7):1800–1815
68. Lee J-W, Lee H-G, Oh ES, Lee S-W, Phan TN-L, Li S, Kim T-S, Kim BJ (2024) Rigid- and soft-block-copolymerized conjugated polymers enable high-performance intrinsically stretchable organic solar cells. Joule 8(1):204–223
69. Ren A, Wang H, Zhang W, Wu J, Wang Z, Penty RV, White IH (2021) Emerging light-emitting diodes for next-generation data communications. Nat Electron 4(8):559–572
70. Guo K, Tang Z, Chou X, Pan S, Wan C, Xue T, Ding L, Wang X, Huang J, Zhang F, Wei B (2023) Printable organic light-emitting diodes for next-generation visible light communications: a review. Adv Photonics Nexus 2(04)
71. Blom PWM (2020) Polymer electronics: to be or not to be? Adv Mater Technol 5(6)

72. Hong G, Gan X, Leonhardt C, Zhang Z, Seibert J, Busch JM, Brase S (2021) A brief history of OLEDs-emitter development and industry milestones. Adv Mater 33(9):e2005630
73. Bauri J, Choudhary RB, Mandal G (2021) Recent advances in efficient emissive materials-based OLED applications: a review. J Mater Sci 56(34):18837–18866
74. Chauhan AK, Jha P, Aswal DK, Yakhmi JV (2021) Organic devices: fabrication, applications, and challenges. J Electron Mater 51(2):447–485
75. Liu W, Zhang C, Alessandri R, Diroll BT, Li Y, Liang H, Fan X, Wang K, Cho H, Liu Y, Dai Y, Su Q, Li N, Li S, Wai S, Li Q, Shao S, Wang L, Xu J, Zhang X, Talapin DV, de Pablo JJ, Wang S (2023) High-efficiency stretchable light-emitting polymers from thermally activated delayed fluorescence. Nat Mater 22(6):737–745
76. Xie Z, Liu D, Gao C, Zhang X, Dong H, Hu W (2025) High mobility emissive organic semiconductors for optoelectronic devices. J Am Chem Soc 147:2239–2256
77. Liu Y, Xie Y, Hua L, Tong X, Ying S, Ren Z, Yan S (2023) Realizing external quantum efficiency over 25% with low efficiency roll-off in polymer-based light-emitting diodes synergistically utilizing intramolecular sensitization and bipolar thermally activated delayed fluorescence monomer. CCS Chem 5(4):1005–1017
78. Wang T, Yao B, Li K, Chen Y, Zhan H, Yi X, Xie Z, Cheng Y (2021) Backbone-acceptor/pendant-donor strategy for efficient thermally activated delayed fluorescence conjugated polymers with external quantum efficiency close to 25% and emission peak at 608 nm. Adv Opt Mater 9(8)
79. Rao J, Yang L, Li X, Zhao L, Wang S, Tian H, Ding J, Wang L (2021) Sterically-locked donor-acceptor conjugated polymers showing efficient thermally activated delayed fluorescence. Angew Chem Int Ed Engl 60(17):9635–9641
80. Zhan H, Wang Y, Li K, Chen Y, Yi X, Bai K, Xie G, Cheng Y (2020) Saturated red electroluminescence from thermally activated delayed fluorescence conjugated polymers. Front Chem 8:332
81. Wang T, Li K, Yao B, Chen Y, Zhan H, Xie Z, Xie G, Yi X, Cheng Y (2020) Rigidity and polymerization amplified red thermally activated delayed fluorescence polymers for constructing red and single-emissive-layer white OLEDs. Adv Funct Mater 30(46)
82. Li M, Hua L, Liu J, Ren Z (2023) Recent advances in regulating the excited states of conjugated thermally activated delayed fluorescence polymers for high-efficiency OLEDs. Mater Chem Front 7(23):6141–6153
83. Liu Y, Xie Y, Hua L, Li S, Tong X, Ying S, Yan S, Ren Z (2023) High-efficiency TADF polymers with a spatially confined conjugated backbone enable solution-processable blue OLEDs realizing over 20% EQE. Adv Opt Mater 12(5)
84. Situ Z, Li X, Gao H, Zhang J, Li Y, Zhao F, Kong J, Zhao H, Zhou M, Wang Y, Kuang Z, Xia A (2024) Accelerating intersystem crossing in multiresonance thermally activated delayed fluorescence emitters via long-range charge transfer. J Phys Chem Lett 15(15):4197–4205
85. Wang T, Huang Z, Zhang H, Miao J, Yang C (2024) Multi-resonance TADF conjugated polymers toward highly efficient solution-processible narrowband green OLEDs. Adv Funct Mater 34(48)
86. Luo W, Wang T, Huang Z, Huang H, Li N, Yang C (2023) Blue TADF conjugated polymers with multi-resonance feature toward solution-processable narrowband blue OLEDs. Adv Funct Mater 34(6)
87. Xu K, Li N, Lin H, Ye Z, Jiang P, Miao J, Cao X, Yang C (2024) Multi-resonance thermally activated delayed fluorescence polymers for high-efficiency and narrowband solution-processed green OLEDs. Chem Commun 60(75):10318–10321
88. Xiao M, Ren X, Ji K, Chung S, Shi X, Han J, Yao Z, Tao X, Zelewski SJ, Nikolka M, Zhang Y, Zhang Z, Wang Z, Jay N, Jacobs I, Wu W, Yu H, Abdul Samad Y, Stranks SD, Kang B, Cho K, Xie J, Yan H, Chen S, Sirringhaus H (2023) Achieving ideal transistor characteristics in conjugated polymer semiconductors. Sci Adv 9(22):eadg8659
89. Song J, Liu H, Zhao Z, Lin P, Yan F (2024) Flexible organic transistors for biosensing: devices and applications. Adv Mater 36(20):e2300034

90. Kim M, Ryu SU, Park SA, Choi K, Kim T, Chung D, Park T (2019) Donor–acceptor-conjugated polymer for high-performance organic field-effect transistors: a progress report. Adv Funct Mater 30(20)
91. Zhao W, Fu GE, Yang H, Zhang T (2024) Two-dimensional conjugated polymers: a new choice for organic thin-film transistors. Chem Asian J 19(3):e202301076
92. Raj B, Kaur P, Kumar P, Gill SS (2021) Comparative analysis of OFETs materials and devices for sensor applications. SILICON 14(9):4463–4471
93. Raza A, Farooq U, Naseem K, Alam S, Ehtisham Khan M, Mohammad A, Zakri W, Yasir Khan M (2024) A focused review on organic electrochemical transistors: a potential futuristic technological application in microelectronics. Microchem J 207
94. Wang Y, Liu Y (2023) Insight into conjugated polymers for organic electrochemical transistors. Trends Chem 5(4):279–294
95. Fratini S, Nikolka M, Salleo A, Schweicher G, Sirringhaus H (2020) Charge transport in high-mobility conjugated polymers and molecular semiconductors. Nat Mater 19(5):491–502
96. Pei D, Wang Z, Peng Z, Zhang J, Deng Y, Han Y, Ye L, Geng Y (2020) Impact of molecular weight on the mechanical and electrical properties of a high-mobility diketopyrrolopyrrole-based conjugated polymer. Macromolecules 53(11):4490–4500
97. Sui Y, Deng Y, Du T, Shi Y, Geng Y (2019) Design strategies of n-type conjugated polymers for organic thin-film transistors. Mater Chem Front 3(10):1932–1951
98. He Y, Kukhta NA, Marks A, Luscombe CK (2022) The effect of side chain engineering on conjugated polymers in organic electrochemical transistors for bioelectronic applications. J Mater Chem C Mater 10(7):2314–2332
99. Cao X, Han Y (2024) Control over the aggregated structure of donor–acceptor conjugated polymer films for high-mobility organic field-effect transistors. Aggregate 5(3)
100. Zhang W, Han Y, Zhu X, Fei Z, Feng Y, Treat ND, Faber H, Stingelin N, McCulloch I, Anthopoulos TD, Heeney M (2016) A novel alkylated indacenodithieno[3,2-b]thiophene-based polymer for high-performance field-effect transistors. Adv Mater 28(20):3922–3927
101. Luo C, Kyaw AK, Perez LA, Patel S, Wang M, Grimm B, Bazan GC, Kramer EJ, Heeger AJ (2014) General strategy for self-assembly of highly oriented nanocrystalline semiconducting polymers with high mobility. Nano Lett 14(5):2764–2771
102. Yamashita Y, Hinkel F, Marszalek T, Zajaczkowski W, Pisula W, Baumgarten M, Matsui H, Müllen K, Takeya J (2016) Mobility exceeding 10 cm2/(V·s) in donor–acceptor polymer transistors with band-like charge transport. Chem Mater 28(2):420–424
103. Nketia-Yawson B, Lee HS, Seo D, Yoon Y, Park WT, Kwak K, Son HJ, Kim B, Noh YY (2015) A highly planar fluorinated benzothiadiazole-based conjugated polymer for high-performance organic thin-film transistors. Adv Mater 27(19):3045–3052
104. Nketia-Yawson B, Jung AR, Nguyen HD, Lee KK, Kim B, Noh YY (2018) Difluorobenzothiadiazole and selenophene-based conjugated polymer demonstrating an effective hole mobility exceeding 5 cm(2) V(−1) s(−1) with solid-state electrolyte dielectric. ACS Appl Mater Interfaces 10(38):32492–32500
105. Han AR, Dutta GK, Lee J, Lee HR, Lee SM, Ahn H, Shin TJ, Oh JH, Yang C (2014) E-Branched flexible side chain substituted diketopyrrolopyrrole-containing polymers designed for high hole and electron mobilities. Adv Funct Mater 25(2):247–254
106. Back JY, Yu H, Song I, Kang I, Ahn H, Shin TJ, Kwon S-K, Oh JH, Kim Y-H (2015) Investigation of structure–property relationships in diketopyrrolopyrrole-based polymer semiconductors via side-chain engineering. Chem Mater 27(5):1732–1739
107. Ali M, Ewenike RB, Manion JG, Lessard BH (2025) Two is better than one: how the addition of multiple biodegradable polymers can improve organic thin-film transistor performance. ACS Appl Mater Interfaces 17(1):1734–1742
108. Yi Z, Wang S, Liu Y (2015) Design of high-mobility diketopyrrolopyrrole-based pi-conjugated copolymers for organic thin-film transistors. Adv Mater 27(24):3589–3606
109. Zheng YQ, Lei T, Dou JH, Xia X, Wang JY, Liu CJ, Pei J (2016) Strong electron-deficient polymers lead to high electron mobility in air and their morphology-dependent transport behaviors. Adv Mater 28(33):7213–7219

110. Wang Y, Hasegawa T, Matsumoto H, Michinobu T (2019) Significant improvement of unipolar n-type transistor performances by manipulating the coplanar backbone conformation of electron-deficient polymers via hydrogen bonding. J Am Chem Soc 141(8):3566–3575
111. Hallani RK, Paulsen BD, Petty II AJ, Sheelamanthula R, Moser M, Thorley KJ, Sohn W, Rashid RB, Savva A, Moro S, Parker JP, Drury O, Alsufyani M, Neophytou M, Kosco J, Inal S, Costantini G, Rivnay J, McCulloch I (2021) Regiochemistry-driven organic electrochemical transistor performance enhancement in ethylene glycol-functionalized polythiophenes. J Am Chem Soc 143(29):11007–11018
112. Moser M, Savva A, Thorley K, Paulsen BD, Hidalgo TC, Ohayon D, Chen H, Giovannitti A, Marks A, Gasparini N, Wadsworth A, Rivnay J, Inal S, McCulloch I (2021) Polaron delocalization in donor–acceptor polymers and its impact on organic electrochemical transistor performance. Angew Chem Int Ed 60(14):7777–7785
113. Luo X, Shen H, Perera K, Tran DT, Boudouris BW, Mei J (2021) Designing donor–acceptor copolymers for stable and high-performance organic electrochemical transistors. ACS Macro Lett 10(8):1061–1067
114. Rajput SK, Mothika VS (2024) Powders to thin films: advances in conjugated microporous polymer chemical sensors. Macromol Rapid Commun 45(10):e2300730
115. Lu H, Li X, Lei Q (2021) Conjugated conductive polymer materials and its applications: a mini-review. Front Chem 9:732132
116. He W, Duan J, Liu H, Qian C, Zhu M, Zhang W, Liao Y (2024) Conjugated microporous polymers for advanced chemical sensing applications. Prog Polym Sci 148
117. Verma A, Gupta R, Verma AS, Kumar T (2023) A review of composite conducting polymer-based sensors for detection of industrial waste gases. Sens Actuators Rep 5
118. Raza S, Li X, Soyekwo F, Liao D, Xiang Y, Liu C (2021) A comprehensive overview of common conducting polymer-based nanocomposites; recent advances in design and applications. Eur Polym J 160
119. Kaur G, Kaur A, Kaur H (2020) Review on nanomaterials/conducting polymer based nanocomposites for the development of biosensors and electrochemical sensors. Polym-Plast Technol Mater 60(5):504–521
120. Farea MA, Mohammed HY, Shirsat SM, Sayyad PW, Ingle NN, Al-Gahouari T, Mahadik MM, Bodkhe GA, Shirsat MD (2021) Hazardous gases sensors based on conducting polymer composites: review. Chem Phys Lett 776
121. Andre RS, Mercante LA, Facure MHM, Sanfelice RC, Fugikawa-Santos L, Swager TM, Correa DS (2022) Recent progress in amine gas sensors for food quality monitoring: novel architectures for sensing materials and systems. ACS Sens 7(8):2104–2131
122. Miah MR, Yang M, Khandaker S, Bashar MM, Alsukaibi AKD, Hassan HMA, Znad H, Awual MR (2022) Polypyrrole-based sensors for volatile organic compounds (VOCs) sensing and capturing: a comprehensive review. Sensors Actuators A Phys 347
123. Li T, Zhu X, Hai X, Bi S, Zhang X (2023) Recent progress in sensor arrays: from construction principles of sensing elements to applications. ACS Sens 8(3):994–1016
124. Choudhury N, Saha B, De P (2021) Recent progress in polymer-based optical chemosensors for Cu2+ and Hg2+ ions: a comprehensive review. Eur Polym J 145
125. Pak YL, Wang Y, Xu Q (2021) Conjugated polymer based fluorescent probes for metal ions. Coord Chem Rev 433
126. Gezahegn TF, Ambaye AD, Mekoyete TM, Malefane ME, Oyedotun KO, Mokrani T (2024) Breakthroughs in nanostructured-based chemical sensors for the detection of toxic metals. Talanta Open 10

# Chapter 9
# Tailored Conducting Polymers as Energy Storage Devices

**Shagufta Jabin, Priti Gupta, Shilpa Sharma, and Anupama Rajput**

**Abstract** The challenges and quest for a sustainable and greener world are some of the most important goals of the present decade. Due to substantial energy requirements, energy storage devices are immensely required for a clean and greener world. So, it is essential to have workable and ecological devices for energy storage. Tailored conducting polymers have been synthesized and designed as energy storage devices in recent years and be promising and efficient for energy storage, which may surpass conventional devices soon because of extraordinary performance in this field. More efforts are being made to design and synthesize conducting polymers, which can have good electrical conductivity and ionic transport properties. Tailored conducting polymers are light weight, highly stable, resistant to corrosion, and have good electrical conductivity. Depending upon the types of synthesis technologies, they can be synthesized in different forms such as polymer composites, thin films, and powdered forms. These materials can be synthesized so that their mechanical, physical, and chemical properties can be modified per various requirements in diverse fields. The present research compiles different state-of-the-artwork in synthesis techniques, characteristics, and multiple applications as energy storage materials for tailored conducting polymers.

**Keywords** Tailored conducting polymers · Polymer composites · Conventional devices · Energy storage devices · Ionic transport properties

---

S. Jabin (✉) · A. Rajput
Department of Applied Sciences, School of Engineering & Technology, Manav Rachna International Institute of Research & Studies, Faridabad, India
e-mail: shagufta.set@mriu.edu.in

P. Gupta · S. Sharma
Department of Chemistry, Manav Rachna University, Faridabad, India
e-mail: priti@mru.edu.in; shilpa@mru.edu.in



S. Jadoun, N. Kumari Jangid (eds.), *Tuning of Conducting Polymers*, Springer Series on Polymer and Composite Materials,
https://doi.org/10.1007/978-3-032-00461-1_9

## 9.1 Introduction

Conducting polymers (CPs) are a unique class of organic compounds that are recognized by their extraordinary power to conduct electricity due to the presence of conjugated π-electron systems along with their polymer backbones [1]. This property allows them to exhibit metallic or semiconducting behavior, making them attractive for various applications. The development of CPs began in the nineteenth century, but significant advancements were made in the 1970s with the discovery of conductive properties in polyacetylene, which spurred extensive research into other CPs like polypyrrole (PPy), polyaniline (PANI) and poly(3,4-ethylenedioxythiophene) (PEDOT) [2–4].

Research relevant to conducting polymers has got much attraction in last decade, and their applications vary from material sciences to optoelectronics [5–6]. This property enables them to be utilized in a wide range of applications across various industries. Here are some of the most common uses of conducting polymers:

Conducting polymers are organic compounds that conduct electricity due to conjugation along their backbone. The high charge density, fast charge- discharge kinetics, low cost, quick redox reaction, ease of synthesis, high power capability, morphology, and superior intrinsic conductivity of conducting polymers make them relevant materials in energy storage. Recent developments in CPs and their composites are examined in Conducting Polymers-Based Energy Storage Materials, which focuses on using these materials as electrodes in electrochemical power sources [7]. They are widely used in energy storage technologies, particularly super capacitors and batteries.

### *9.1.1 Organic Light-Emitting Diodes (OLEDs)*

Since the late 1980s, conducting polymers have been integral to developing OLEDs [4]. Due to prominent nonlinear optical figure of merit, energy band gap, electronic structure, ultrafast optical responses, architectural flexibility, and high optical damage thresholds in addition to the mechanical properties them promising for light-emitting devices.

**Sensors** Conducting polymers are extensively utilized in various sensors, including chemical sensors, biosensors, and gas sensors [8]. CP-based biosensors are ideal for portable devices because of their many benefits, which include easy fabrication, mass production, adaptability, and low weight. Simply adjusting the CP-based electrodes' surface shape and nanostructure can improve their performance, as can combining them with other nanomaterials to boost conductivity, which will raise signal sensitivity and reduce the LOD [9].

### 9.1.2 Corrosion Protection Materials

CPs like polypyrrole (PPy) are used as protective coatings for metals to prevent corrosion [10]. CPs have been applied to the surface of metals and are one of the most studied alternatives for corrosion protection materials [11].

### 9.1.3 Dye-Sensitized Solar Cells (DSSCs)

Conducting polymers are commonly used as hole conductors in dye-sensitized solar cells (DSSCs). In these devices, CPs transport holes generated during the photo electrochemical conversion [12].

### 9.1.4 Proton Exchange Membrane Fuel Cells (PEMFCs)

In PEMFCs, conducting polymers are used as proton-conducting membranes. These membranes facilitate the transport of protons from the anode to the cathode while preventing the passage of gases like hydrogen and oxygen [13]. Unique materials used to enhance CPs are dopants and main enhancers to create tailored polymers. Dopants are crucial in improving the properties and functionalities of conducting polymers (CPs). These materials, which include polypyrrole (PPy), polyaniline (PANI), and polyacetylene, typically exhibit low electrical conductivity in their pristine state. However, the introduction of dopants significantly alters their electrical, optical, and mechanical properties [14].

Dopants are essential for increasing the electrical conductivity of conducting polymers. Doping involves either removing electrons from the polymer (p-type) or adding electrons to it (n-type), generating charge carriers (polarons and bipolarons) that enhance electrical conductivity [15]. Different dopants modify the redox potential and improve the stability of CPs, allowing for tailored electrochemical behavior suitable for specific applications such as batteries and sensors [16].

Due to their unique properties and versatility, tailored conducting polymers (CPs) are increasingly crucial for various applications in human healthcare and technology [17–18]. Their customization enables the development of sensitive biosensors, effective drug delivery systems, and innovative tissue engineering solutions [19]. Different kinds of tailored CPs have been discussed in detail.

## 9.2 Types of Tailored Conducting Polymer

### *9.2.1 Polyaniline (PANI)*

#### 9.2.1.1 Structure of PANI

Polyaniline (PANI) is an economical, easily synthesized conducting material and poses no threat to the environment. Due to these properties, it has grabbed many researchers' attention [20]. Conductivity is essential for efficient charge transport in energy storage applications. They are majorly utilized in electrochemical energy storage devices. PANI is characterized by a conjugated structure formed from repeating aniline units. This conjugation facilitates the delocalization of π-electrons, essential for their electrical conductivity. PANI is a compound with three oxidation states: insulation (Leucoemeraldine), conductive (Emeraldine), and fully oxidized (Pernigraniline). They are commonly and significantly used in various applications. Polyaniline contains both benzenoid and quinonoid repeating units in the polyaniline polymeric chain. The efficiency of PANI is closely associated with its molecular and morphological structure. The presence or absence of the repeating units creates a difference between three distinct oxidation states that affects conjugation. All three forms are derived from the same aniline and can be interconverted through oxidation and reduction, which affects conductivity.

**Structural Properties of PANI**
Polyaniline's function in energy storage devices largely depends on its distinct structural properties, including its oxidation states, capacity to form composites, and charge storage processes. PANI can exist in three primary oxidation states: leucoemeraldine, emeraldine, and pernigraniline (blue or violet). The polyaniline polymeric chain has repeating units that are both benzenoid and quinonoid [21]. The presence of repeating units creates three different oxidation states of PANI, which influence conjugation.

**Leucoemeraldine** is the non-conductive and reduced form of PANI containing benzenoid rings, lacking any quinonoid structures leading to a lack of conjugation. Due to its non-conductive nature, it has limited use but can be made conductive by oxidation.

**Emeraldine** It is another form of PANI, partially oxidized, containing an equal ratio of benzenoid to quinonoid units, which results in extended conjugation. It exhibits a conductivity range of 10–10 s cm to $10^{-1}$ s cm, which is why it is the most stable and conductive form [21]. The conductivity usually depends on protonation and environmental conditions like pH and doping level. The conductivity of emeraldine increases when doped with acids. As it is partially oxidized, it can be converted into leucoemeraldine and pernigraniline through reduction and oxidation respectively. Therefore, it is widely used in sensors, electrochromic devices, and as a conducting polymer in various electronic applications.

**Pernigraniline** It is fully oxidized and is less conductive than Emeraldine. But it gets stabilized under certain conditions. It represents a higher oxidation state due to which all benzenoid groups have been oxidized and only quinonoid structures are left. It is used in various applications where high oxidation states are required, though its practical use is limited compared to emeraldine due to lower conductivity [22].

#### 9.2.1.2 Tailored Polyaniline (PANI)

Polyaniline (PANI) is a flexible conducting polymer that has gained popularity, especially in energy storage devices like batteries and supercapacitors. Numerous tailoring techniques, such as doping, structural alterations, and composite synthesis, can significantly improve its qualities.

##### Electrical Conductivity of Tailored PANI

High electrical conductivity, which is essential for effective charge transfer in energy storage devices, is exhibited by tailored PANI. By doping PANI with different chemicals, its conductivity may be significantly increased.

Doping Substances: For instance, doping with para-toluene sulfonic acid (PTSA) produces different conductivity values depending on the doping concentration. PANI doped with PTSA at a 0.5 molar ratio exhibits the highest conductivity of around $10^{-2}$ S/cm, whereas lesser concentrations result in $10^{-3}$ and $10^{-5}$ S/cm, respectively [23]. Also, co-doping PANI with HCl and PA exhibits relevant results in achieving high specific capacitance and stability in super capacitor applications. Recent breakthroughs include the fabrication of PANI nanocomposites, including nitrogen-doped reduced graphene oxide (N-rGO). This method combines the advantages of PANI with those of graphene oxide, improving overall electrochemical performance [24]. The co-doped N-rGO-PANI composites exhibit high specific capacitance (346. 3 $Fg^{-1}$) and outstanding cycle stability, exhibiting a remarkable rate capability [25].

##### Electrochemical Performance of Tailored PANI

The electrochemical characteristics of PANI are essential for its use in supercapacitors.

Capacitance: Customized PANI composites may yield high specific capacitance values. Mesoporous carbon nanosheets coated with polyaniline, for example, exhibit a capacitance of 107.8 F/g at a current density of 0.5 A/g. PANI's redox processes increase the total energy storage capacity by producing more pseudo-capacitance.

Fast ion transport in specially designed structures makes high-rate capability possible, preserving sizable capacitance retention even at higher current densities (e.g., 69.6% retention at 5 A/g) [26].

### Structural Characteristics of Tailored PANI

PANI's morphology and structure can be modified to maximize performance by varying surface area, porosity, and nanocomposite. Customized PANI frequently has a porous structure that improves charge storage capacity and makes electrolyte access easier. Porosity also affects the electrochemical performance. Morphological designs with porosity help improve ion diffusion channels, which are required for quick charge/discharge operations. For example, PANI composites with mesoporous architectures have increased electrochemical performance due to higher ion accessibility. The morphology influences the distribution of protonated and non-protonated sites inside PANI. Depending on their shape, the total charge storage capacity can be impacted by how ions interact with these sites during redox reactions. These morphological designs improve the responsiveness and efficiency of energy storage devices by enabling shorter ion diffusion routes, which can result in quicker charge/discharge kinetics [27]. Pseudo- capacitances, which includes quick redox reactions at the polymer's surface, is the primary way PANI stores energy. This technique makes fast charge and discharge cycles possible, which is crucial for supercapacitor applications. The total energy storage capacity is increased by the capability to go through reversible oxidation and reduction processes. Research is ongoing to improve the performance of PANI in various applications such as supercapacitors and hybrid energy storage devices by maximizing these qualities using various synthesis processes and composite forms. These technologies enable the development of nanostructures such as nanofibers and nanotubes, increasing the surface area accessible for electrochemical reactions. Increased surface area correlates with higher capacitance values in super capacitor applications [21].

### Formation of Nanocomposite from Tailored PANI

PANI can be combined with metal oxides or carbon materials to increase mechanical strength and stability while preserving a large surface area for electrolyte interaction. Conductivity can be further improved by adding elements like titanium dioxide (TiO2) to PANI; for a given composition, values of up to 5.6 S/cm can be achieved [28]. Various materials like nanoparticle composites (Silicon, Silver), phase change materials, and conductive polymer blends have been integrated with polyaniline to enhance its performance in super capacitors and batteries [29–30]. The morphology of PANI can be modified using a variety of synthesis processes, including chemical oxidative polymerization and electrochemical deposition. These technologies enable the development of nanostructures such as nanofibers and nanotubes, increasing the surface area accessible for electrochemical reactions.

## 9.2.2 *Polyindole (PIN)*

### 9.2.2.1 Structure of Polyindole (PIN)

Polyindoles are becoming necessary materials for energy storage devices, notably supercapacitors. The current study focuses on them because of their unique qualities and the possibility of enhancement through composites. The primary causes of polyindole's (PIN) conductivity are its π-conjugated structure and the polarons that develop during doping. Polyindole acts as an insulator in its neutral state due to the presence of localized electrons [31]. The physical structure of polyindole affects its conductivity. PIN's non-planar and loosely packed form may result in weak interactions among polymer chains and lesser conductivity than more organized structures, but controlling the synthesis conditions to produce a more crystalline form can increase conductivity because higher crystallinity in polyindole can improve its electrical characteristics [32].

Conducting polymers such as polypyrrole (PPY) and polyaniline (PANI) have intrinsically poor electrical and electrochemical conductivity. However, polyindoles have good thermal stability, controllable electric conductivity, and vigorous redox activity. Problems like mechanical instability or gradual degradation exist in PIN, but they are still suitable for energy storage materials [5].

### 9.2.2.2 Properties of the Tailored Polyindole (PIN)

Electrical Conductivity

Polyindole (PIN) usually has low electrical conductivity, so its application in energy storage devices is restrictive. Still, it is being used as an energy storage device because of significant redox activity, which allows for better charge transfer when modified or coupled with conductive materials. Integrating polyindole with nitrogen-doped porous carbon (NPC) significantly improves electrical conductivity, resulting in a specific capacitance of 286.2 F $g^{-1}$ at a current density of 0.5 A $g^{-1}$ in acidic electrolytes [33]. Metal oxides and nanoparticles in conjunction with polyindole are used to enhance the electrical performance and the mechanical properties of tailored polyindole.

Thermal Stability of Tailored PIN

Polyindole is known for its good thermal stability and low degradation rate. These properties are essential for consistent performance, functioning effectively without significant loss of capacity over extended periods, and changing operational conditions. A study reported by Vannathan et al. explain the synthesis of two new PIn/$PVMo_{11}$ and PIn/$PV_2Mo_{10}$ composite electrodes [34]. The electrodes have a potential application in supercapacitors. This explains that combining polyoxometalates and polyindole, improves thermal stability and capacitance since polyoxometalates have excellent physical and chemical properties [34].

#### Morphology and Structure of PIN

Polyindole's morphology is pivotal in enhancing its performance in various fields. During synthesis, morphological optimization helps increase the conductivity of polyindole. This can be done by creating nanocomposites or hybrid structures that optimize ion transport within the material, resulting in improved electrical pathways inside the polymer matrix. By increasing surface area and electroactive sites, nano/microspheres may be produced using techniques such as emulsion polymerization, which improves capacitance and charge–discharge efficiency. Doping and composite formation can also rectify the structural imperfections in polyindole, resulting in more resilient and conductive networks inside the material. Doping by oxidation removes an electron from the π-system, resulting in a radical cation (or polaron) that enhances electron mobility down the chain. The role of doping is explained best in supercapacitor applications; for instance, doping with vanadium-based polyoxometalates enhanced capacitance and energy density, suggesting that the interactions between the dopants and polyindole boosted conductivity. By taking electrons out of the π-system, doping adds charge carriers to the polymer. Polarons are mobile charge carriers and allow easy electron travel along the polymer chain, which improves conductivity. Polyindole doping can be enhanced using organic dopants with surfactants like sodium dodecyl sulfate. This improves ionic conductivity and the overall electrical performance of polyindol [35].

#### Electrochemical Performance of Tailored PIN

The electrochemical performance of customized polyindole composites is assessed using various methods, including cyclic voltammetry and galvanostatic charge/discharge experiments. Tailored polyindole composites exhibit minimal degradation over extended use. In the case of supercapacitor applications, polyindole composites containing zinc oxide (ZnO) and nickel oxide (NiO) have demonstrated encouraging outcomes. With the use of the oxides mentioned above, a few changes have been observed in the properties of polyindole, attaining high specific capacitance values and strong cycle stability. Supercapacitors made of NPC-PIn (NPC—nitrogen doped porous carbon) composites have demonstrated excellent cycling stability, with 88.9% capacitance retention after 3000 cycles. This shows during lengthy operation lifetimes, the tailored polyindole can sustain its performance [33].

## 9.2.3 *Polypyrrole (PPy)*

### 9.2.3.1 Structure of Polypyrrole (PPy)

Polypyrrole is a conducting polymer. The backbone of the polymeric chain consists of alternating single and double bonds. The conjugation and overlapping p-orbitals boost the conductivity in polypyrrole, enabling efficient electron transit along the polymer chain. This increased conductivity makes PPy a superior conductor over non-conductive polymers [35]. Apart from being used in supercapacitors, PPy has also found its application in wearable technologies and biomedical devices because of its biocompatible properties.

### 9.2.3.2 Properties of Tailored PPy

Because of its special characteristics, the conducting polymer polypyrrole (PPy) has drawn researchers' interest in the energy storage space. The improved electrochemical activities, conductivity, and structural stability make it a potent applicant for use in a variety of energy storage systems, such as batteries and supercapacitors. Customized PPy is important in improving its electrochemical performance by changing its structure and content. This can be done in different ways, like doping and composite formation, enhancing its performance in supercapacitors and batteries. It is also being studied as a possible alternative for noble metal catalysts in fuel cells because of its exceptional oxygen-reduction activity, which combines affordability and good performance.

#### Electrical Conductivity of Tailored PPy

PPy can quickly transfer charge in energy storage devices, so it can easily attain high electrical conductivity. This high electrical conductivity of polypyrrole is due to its conjugated structure. Apart from its conjugated structure, effective doping mechanisms, favorable morphologies (nanofibers), dynamic charge carrier behavior, and stability under diverse conditions are significant factors for high electrical conductivity [36]. The various materials that enhance the properties of PPy for better energy storage capabilities are doping agents, composite materials, structural changes and surfactants [37–39]. The electrical conductivity of polypyrrole is primarily attributed to its conjugated structure, effective doping mechanisms, favorable morphologies such as nanofibers, dynamic charge carrier behavior, and stability under diverse conditions. One recent discovery is the use of PPy-coated textiles to create flexible supercapacitors. Compared to usual materials, these devices' specific capacitance (343.20 F $g^{-1}$) was much higher. This demonstrates that the customized PPy may outperform conventional electrodes in supercapacitors.

### 9.2.3.3 Electrochemical Performance of Tailored PPy

The electrochemical performance of tailored polypyrrole is marked by its capacitance and energy density. For instance, in supercapacitor applications, polypyrrole composites have shown capacitance values above 198 F/g with energy densities of around 10.19 Wh/kg. Compared to conventional materials, PPy's redox-active sites enable pseudocapacitance, which raises the material's capacity for energy storage [34]. Another type of material, named hydrogels, can give both a greater electrode surface area and better flexibility, as deformation happens through the flow of water in hydrogel holes rather than the movement of conducting polymer chains. Research shows that PPy hydrogels may be made stiffer and perform electrochemically better by adding cross-linking agents. This results in a maximum specific capacitance of 217.07 F $g^{-1}$ is 140% higher than hydrogels that have not been altered [40]. Two-dimensional materials, such as MXenes, when mixed with PPy, create a nanocomposite with enhanced electrochemical characteristics. These composites use the high surface area and conductivity of both materials, resulting in increased energy density and power output [41]. MXenes are two-dimensional inorganic compounds composed of thick layers of transition metal carbides, carbonitrides, or nitrides containing a few atoms. They have a general formula of Mn + 1XnTx ($n$ = 1, 2, 3), where M is an early transition metal such as Ti or Sc, and X is either carbon or nitrogen. Tx represents the surface terminations, such as –O, –OH, and/or –F. Apart from this, the performance and dependability of energy storage devices are adversely affected by the high contact resistance caused by the frequently insufficient adherence of PPy films to current collectors, which can result in delamination during electrochemical processes [42]. Tailored PPy is more versatile than supercapacitors because of its high conductivity and electrochemical activities. When used as a cathode material, PPy improves lithium-ion batteries' stability and capacity retention performance.

#### Thermal Stability of Tailored PPy

Polypyrrole features excellent thermal stability, critical in maintaining performance in different operating environments. This stability leads to better heat resistance among PPy-based materials. The polymer's high temperature resistance leads to slight degradation, making it ideal for use in various energy storage systems.

#### Morphology and Structure of Tailored PPy

The shape of polypyrrole can vary during its synthesis due to hydrothermal or electrochemical polymerization thus improving the ion uptake and electrochemical performance. Improvement in the electrochemical performance and ion accessibility is associated with the area and the porosity. The presence of polypyrrole around manganese oxide composites has improved their conductivity and structural stability. It simply enhances the mechanical properties and charge storage capabilities of PPy-based material [43]. For instance, nanofibers with coaxial structure, consisting of metal oxides and PPy, have demonstrated enhanced cycle stability and specific capacitance owing to their distinctive structural features. By encasing active materials with mechanical strength, PPy enables more stable structures during cycling, critical for performance over time.

## *9.2.4 Poly(3,4-ethylenedioxythiophene) (PEDOT)*

### 9.2.4.1 Structure of PEDOT

PEDOT is a polymer with a conjugated structure. This conjugated structure is formed by repeating units of ethylene dioxythiophene, which facilitates the delocalization of π-electrons. It plays a vital role in its high electrical conductivity. Repeated ethylene dioxythiophene (EDOT) units give PEDOT a conjugated structure.

### 9.2.4.2 Properties

Electrical Conductivity of Tailored PEDOT

PEDOT's remarkable electrical conductivity is essential for effective charge conveyance in energy storage devices. Different structural features such as conjugation, morphological variations, crystallinity, and doping affect the electrical conductivity of PEDOT via a charge transfer mechanism, intermolecular interactions, and electrochemical properties. Conjugated backbone, morphological variations, crystallinity, and doping are the structural features that affect the electrical conductivity of PEDOT via charge transfer mechanism, intermolecular interactions, and electrochemical properties. Electrical conductivity depends on the delocalization of π-electrons, which is made possible by this conjugation. In supercapacitors, the performance relies mainly on the overall conductivity of PEDOT-based films. Hence, the applications of various additives, such as reduced graphene oxide, generally help to improve the electrochemical activity of PEDOT. The unique properties of PEDOT are explored through different ways to improve its performance, cycling stability, and widespread availability in energy storage devices. The use of several materials enhances the performance of PEDOT in energy storage devices. Various metal organic frameworks (MOFs) such as ZIF-671 and nanomaterials like Carbon nanotubes and borophene are utilized to improve the electrical conductivity of PEDOT [44].

### Electrochemical Performance of Tailored PEDOT

The relevant electrochemical features of PEDOT can be tailored by synthesizing PEDOT with specific conductance, energy density, and cycling stability. PEDOT nanofibril electrodes have been synthesized with particular capacitance and cycling stability. They could achieve an energy density of approximately 14.96 Wh $kg^{-1}$ and power densities of about 22.2 kW $kg^{-1}$ for fast charge/discharge applications [45]. As per the literature, tailored PEDOT electrodes often demonstrate high cycling stability with retention rates higher than 90% [46].

### Thermal Stability of Tailored PEDOT

PEDOT is known for its good thermal stability and ability to maintain performance across various temperatures. This stability benefits diverse applications that may experience temperature fluctuations during operation.

### Morphology and Structure of Tailored PEDOT

Several techniques exist for the synthesis of PEDOT, which allow for morphology control through electro-polymerization or vapor deposition. These methods afford control over the microstructure, which affects surface area and porosity. Nanostructured Form: It increases the specific surface area and ion accessibility and provides space to accommodate volume change.

### Composite Structures

PEDOT in conjunction with other conductive materials such as multi-walled carbon nanotubes can have higher conductivity and capacitance [44]. PEDOT's electrical properties are primarily determined by its structural morphology. Furthermore, different synthesis techniques can improve their features. PEDOT's shape (coral and sheet-like structure) can vary greatly depending upon the synthesis, especially the oxidant-to-monomer ratio during polymerization. PEDOT has a coral-like shape at lower oxidant ratios (e. g., 3:1), which increases its surface area and capacity to store charge. Higher oxidant ratios (such as 9:1) result in sheet-like configurations, which might enhance conductivity by improving inter chain contacts [47–48]. The crystallinity correlates with better conductivity due to more efficient charged transport pathways. The easier charge of transport is due to well-aligned EDOT units and strong intermolecular interactions. Different PEDOT geometries show differing stability and charge storage capabilities during electrochemical cycling. The structural variations brought about by synthesis techniques are connected to this variability.

## 9.3 Diverse Applications of TCP in Various Fields

Conducting polymers have diverse and potential usage. They are utilized significantly in the industry, producing batteries and electronic devices. They are also used as antistatic materials. Previous studies also reveal that conducting polymers are tailored to use potentially as diodes, solar cells, supercapacitors, chemical sensors, biosensors, translucent conductors, and actuators [49–50]. They are also gaining attention due to improved electrical and physical properties and low cost. Tailored conducting polymers also have vast applications in medical science, optoelectronics, and energy storage [51]. As per the study, PANI-based conducting polymers are widely used in manufacturing printed circuit boards and also help protect copper from corrosion. However, polyindole-based conducting polymers are gaining interest due to their thermal stability and high oxidation-reduction properties [52]. PPy –Chitosan-tailored polymer composites are used efficiently in food packaging and the medical field [53]. Polythiophene (PTh) derivative-polyurethane conducting polymer composites are potentially utilized in wound curing and tissue engineering [18]. Furthermore, PANI nanofiber-collagen-based tailored conducting polymers are used as scaffold materials. Scaffolds offer a platform for tissue development that can stimulate the original tissues. Scaffolds with bioactive elements act as reservoirs and transport bioactive agents to the tissues [54]. Conducting polymers are tailored to be utilized as redox mediators thus enhancing electron transfer into the catalytic cycle. In addition to the other applications, they are used as suitable catalysts for various redox reactions [55–56]. Among all the applications of conducting polymers, energy storage is the most significant due to the enormous energy requirement from fossil fuel exhaustion and environmental degradation.

### *9.3.1 Role of Tailored Conducting Polymers as Energy Storage Devices*

The basic requirements for energy storage conducting polymers are good conductivity and maximum power transfer. The CPs have a diverse range of conductivity, which ranges from semiconductors to metals. Their properties can be tailored by monitoring temperature, pH, and doping agents [57]. CPs exhibit extraordinary current density and have a low cost compared to carbonaceous materials. As per literature, tailored CPs are significantly used in energy storage due to high specific capacitance, high conductivity, high surface area, and high charge- discharge rate [58]. Because of extraordinary properties, they are proven to be next-generation energy storage materials [59].

#### 9.3.1.1 Supercapacitors

They are helpful for energy storage and have high capacitance values compared to conventional capacitors and batteries. Previous studies considered them better options with excellent power density and quicker chargeable capabilities. They can be tailored with high discharge–charge rate, high energy density, and high conducting properties to be used efficiently in energy storage. As per previous research, metal oxides and carbon-based materials have been utilized to make flexible and efficient supercapacitors and tailored conducting polymers such as PCZ, PANI, and others [60]. Researchers have concluded a simple, fast, and flexible technique for synthesis of conducting and versatile conducting polymers [59]. They could synthesize 3,4-ethylene dioxythiophene-based conducting polymers for energy storage. Researchers also explored various conducting polymers such as PANI, poly(pyrrole-3-carboxylic acid), polycarbazole (PCZ), polythiophene (PTh), and its derivatives for numerous applications in the fields of energy storage because of strong properties of electrical conductivity [61]. Chemists prepared a poly-methylaniline derived porous nitrogen-doped-carbon microsphere polymer with high electro-catalytic properties for energy storage. Poly-methylaniline polymer was found to be thermally stable up to 900 °C and had a surface area of 727 $m^2g^{-1}$ with high amounts of quaternary nitrogen [62]. Literature also revealed that a nitrogen doped porous polymer was synthesized with nano-fiber-like structure by simply carbonization of poly (aniline-*co*-*p*-phenylene-diamine) at a temperature range of 600–900 °C in the $N_2$ atmosphere. High heat energy, high surface area (about 1500 $m^2g^{-1}$), and thermal stability up to 700 °C make the polymer ideal for energy storage applications [63].

Previous research concluded the synthesis of a PPy-based conducting composite with pristine carbon. Lithium perchlorate was used as a dopant material, and sodium dodecyl-benzene sulphonate was used as a surfactant. This sodium benzenesulfonate was specially tailored for use in energy storage.

##### Solar Cells

The first breakthrough in using solar cells as conducting polymers for energy storage was made in 2000. Three chemists, Alan Macdiarmid, Alan J. Heeger, and Hideki Shirakawa, were awarded the Nobel prize for developing the first polymer-based solar cell for energy storage [64]. The first solar cell was synthesized by the macromolecule "poly-2-methoxy-5-(2′-ethylhexyloxy)-p-phenylenevinylene," along with different variants having significant energy conversion efficiency.

Various tailored CPs such as PPy, PANI, PTh, poly (m-phenylene-diamine), polyvinyl pyrrolidone (PVP) and their nanocomposites are used significantly as energy storage devices for solar cells [65]. Anand et al. synthesized and designed a novel sulphonated PANI polymer of a thiol group and assessed its role in solar cells for energy storage [66]. Sulphonated PANI with graphene nano-platelets (GNP) buffer layer was used to potentially increase the energy storage power for the newly prepared solar cells. A strong near field of GNP was equally disseminated on the active layer to improve its diverse applications. GNP helps to improve surface area, significant hole extraction, and low interfacial resistance of these polymers, which allow them to have diverse applications [66].

A previous study in 2018 concluded the synthesis of poly-N-9-heptadecanyl-2,7-carbazolet-5,5-4′,7′-di-2-thienyl-2′,1′,3′-benzothiadiazole conducting polymer based on photovoltaic cells. The flexographic printing technique has synthesized them, and the synthesis method was economical and cost-effective. This novel polymer has uniformity on its surface and contains the development of a low pinhole surface. The flexographic layer of the polymer acts as a photovoltaic cell and has 4% efficiency [67].

### Lithium-Ion Batteries (LIBs)

LIBs are considered as primary energy storage devices for electrical vehicles and consumer electronics. Lithium behaves as an anode and oxidizes as $Li^+$ and moves toward the cathode. The LIBs are lightweight and have high efficiency as energy storage materials and high-power density, which make them easy to transport from one place to another [68]. LIBs have specific energy in the range of 70–250 (Wh-$Kg^{-1}$), round trip efficiency of 80 to 95 (%), self-discharge rate of 0.1 to 0.3 (%) and specific power of 150 to 300 [69–70]. They are comparatively new devices for energy storage as compared to conventional devices such as Ni-Cd and lead–acid batteries.

Tailored CPs have emerged as potential flexible electrodes for LIBs for stability, versatility, low cost, and low density. They are globally used as energy storage devices because of their durability, fast charge capabilities, and high specific energy. Furthermore, because of the unique characteristics of a long lifeline and high efficiency, they are utilized as stationary energy storage appliances. However, the synthesis of electrodes for LIBs is often a challenging task that results in poor performance in terms of energy storage. Hence, tailored CPs such as PPy, PTh, and PANI have been used as active electrode materials previously [71]. These macromolecules have high thermal capacity, good electrical conductivity, and a good oxidation-reduction rate. Non-conducting polymers are also used to make electrodes for LIBs. But CP-based electrodes have good electrochemical stability [71].

The LIBs have broad applications in stationary energy storage, complementing renewable energy resources. They help to enhance the grid penetration power of renewable energy resources and play a major role in downward and upward frequency regulation, improve the power quality, and reduce voltage fluctuation [70]. LIBs have significant applications in the automotive sector. Their utilization as energy storage devices is a promising field that help to reduce fuel consumption and offers alternatives to internal combustion engines. Among various storage techniques, LIBs represent the best choice for automotive fields in terms of compact size, reliability, and high energy density. Due to high market demands, industries are manufacturing hybrid and fully electrical vehicles, so the battery requirement has also increased multiple folds and LIBs are playing major contributions in it.

### Fuel Cells

Fuel cells are electrochemical-based devices that transform chemical energy to electrical energy by the process of electrochemical oxidation of small organic-based molecules such as ethanol, methanol, and hydrogen. Fuel cells produce electricity from the anode and cathode sides, which react in the presence of electrolytes. Fuel cells work in the presence of $H_2$-based fuels with low emissions of $CO_2$ [70]. Among all fuel cells, proton exchange membrane fuel cells achieved much attention due to better characteristics like high energy conversion quality and effective electro-oxidation of organic compounds such as methanol, ethanol, and formic acid at low temperatures [72]. On the other side, microbial-based fuel cells help to convert organic waste into electricity via microbial cathodic and microbially catalyzed anodic electrochemical reactions. The limitations of microbial fuel cells are their high cost and low power output. So, metallic nanoparticles as anodic and platinum-based metals are used as cathodic materials to improve the efficiency of MFC. Fuel cells include the following tailored conducting polymers:

#### Conducting Polymer-Supported Nanohybrid (CPNH) Electrode Materials

Conducting polymers-supported nanomaterials have been tailored uniquely for applications in energy storage. Conducting polymers with high conductivity and high surface area integrated with nanoporous metal oxides play a significant role in stabilizing the performance of fuel cells. An effort has been made to tailor them in such a way that a catalyst (free from Pt) can be deposited on their surface for better use [13]. Few studies have been done on conducting polymers-supported hybrid catalysts, including multi-metallic nanopolymers. The various components of conducting polymers-supported nanomaterials in fuel cells are shown in Fig. 9.1.

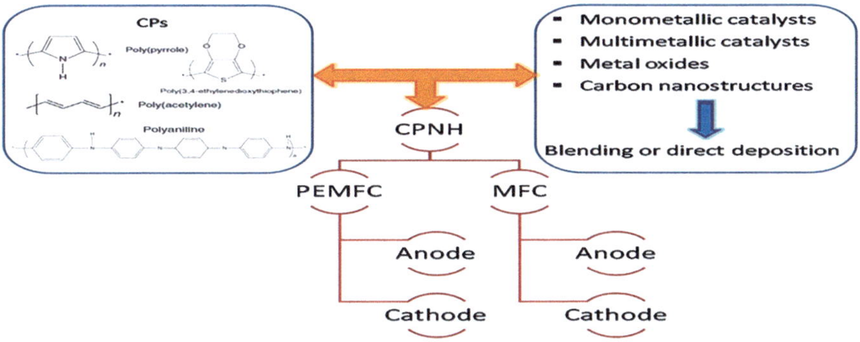


**Fig. 9.1** Various components of conducting polymers [13]

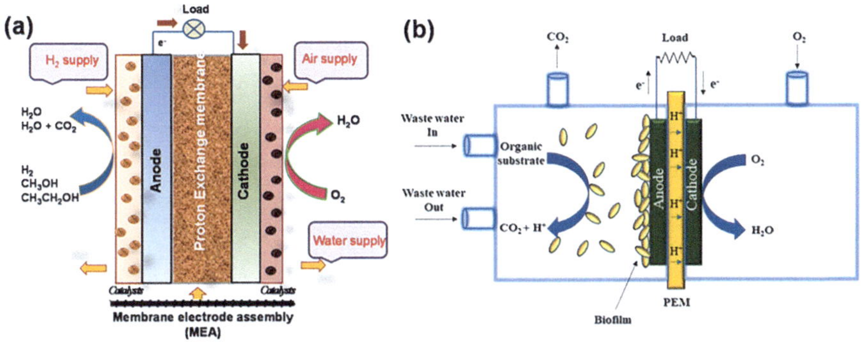


**Fig. 9.2** Working of alcohol fuel cells and MFC [13]

## CPNH-Based Electrode for Alcohol Fuel Cells

CPNH-based electrodes are useful for alcohol fuel cells for the generation of electricity by the oxidation of alcohol in liquid form in combination with the reduction of $O_2$. They transform chemical energy into electrical energy in alcohol in hybrid electro-catalyst based conducting polymers. However, in MFCs, microorganisms play the role of catalysts to produce electricity from inorganic and organic biodegradable compounds. MFC consists of cathode such as graphite, carbon paper, and carbon cloth anode and electrolyte as well as polymer electrolyte membrane, viz.; polyethylene, polystyrene, ultrex, and many more [73]. The working of alcohol fuel cells and MFC has been shown in Fig. 9.2.

The different kinds of tailored polymers, along with their different properties, applications, advantages, and limitations, have been summarized in Table 9.1.

**Table 9.1** Different types of tailored polymers along with their properties, advantages, and limitations

| Types of tailored polymer | Electrical conductivity | Thermal stability | Tailored materials | Applications | Advantages | Limitations | References |
|---|---|---|---|---|---|---|---|
| Polypyrrole (PPy) | Excellent ($10^{-2}$ to $10^{-1}$ S /cm to $10^{-1}$S/ cm) Specific Capacitance- around 300F/g | Medium thermal stability | Doping agents- $ClO_4^-$, P-toluenesulfonate; Composite materials- Polyaniline-polypyrrole, Polybenzopyrrole, Multiwalled carbon nanotubes | Supercapacitors, lithium-sulfur batteries, fuel cells | Ease of synthesis; High Charge storage capacity; Environmental stability | Low cycling stability; Adhesion issues; Self-discharge characteristics; Environmental stability; Complex synthesis process | [34, 37, 39–40, 74] |
| Polythiophene (PTh) | Varies widely, good conductivity | Good thermal stability till 200 °C | Nanoparticle composites- $BaTiO_3$ Hybrid structures- Graphene oxide Conductive fillers- Carbon nanotubes Electrolyte enhancement- ionic liquids Morphological enhancement- 3D architectures | Organic electronics, energy storage systems | High charge storage capacities By structural modifications and doping. it is suitable for energy storage devices | Processing challenges Limited conductivity variability Degradation under cycling | [75–77] |

| | | | | | | | |
|---|---|---|---|---|---|---|---|
| poly(3,4-ethylenedioxythiophene) (PEDOT) | Excellent conductivity-up to 6. 23 × $10^{-6}$ S/cm; Specific Capacitance-105. 35 mF/$cm^2$ | Excellent thermal stability till 200 °C | Metal Organic Frameworks- ZIF-67 Nanomaterials- BoropheneNanoplatelets. Multi-walled carbon nanotubes | Flexible electronics, supercapacitors | High electrical conductivity Retains flexibility Excellent thermal and chemical stability High-charge storage capacities | Rigidity surface area limitations cost of production | [41, 44–48] |
| Polyaniline | High Conductivity (6. 28 x $10^{-9}$ S m) | High thermal stability till 200 °C | Composites- Graphene oxide; Nanoparticle composites- Silicon nanoparticles, Silver nanoparticles; Phase change materials- Paraffin Conductive polymer blends- polypyrrole | Pseudocapacitive applications, energy storage devices | Cost Effectiveness; Good Electrochemical properties; Versatile doping mechanism | Doping dependence; environmental sensitivity; limited cycle life | [21–22, 29] |
| Polyindole | Electrical conductivity—around $10^{-3}$ S/cm Specific Capacitance- <198 F/g (improved when doped) | | Polyoxometalates- Vanadium based Metal oxides- $MnO_2$, $SnO_2$ Nanoparticles- silver, carbon nanotubes, transition metal **($Co_3O_4$)** Composites- Graphene/Pin | Supercapacitors, hybrid electrode materials | High conductivity after modification Good electrochemical stability and flexibility | Low initial conductivity mechanical instability difficult processing | [5, 31–35, 78–80] |

## 9.4 Future Work

Conducting polymers (CPs) hold immense promise for revolutionizing energy storage technologies like supercapacitors and batteries. However, realizing their full potential requires addressing key challenges and exploring promising future directions.

Future research should focus on synthesizing CPs with precisely tailored electrical, mechanical, and thermal properties to enhance performance in energy storage applications. This includes exploring various doping methods to optimize conductivity and charge storage capacity. Combining CPs with other nanomaterials, such as graphene or metal oxides, can create composite materials that leverage the strengths of each component. This approach can improve electrochemical performance and stability, benefiting from the high surface area of nanomaterials and the conductivity of CPs.

Nanostructuring techniques can significantly enhance the surface area and electrochemical activity of CPs. Research should explore various nanofabrication methods to create porous structures that facilitate ion transport and increase capacitance. Investigating three-dimensional architectures for CP electrodes can improve performance by providing more active sites for charge storage and enhancing mechanical stability during cycling.

### *9.4.1 Challenges*

Due to volumetric changes, CPs often suffer degradation during repeated charge–discharge cycles, leading to reduced performance. Research is needed to improve cycling stability.

CPs are often sensitive to environmental factors like humidity and temperature, affecting their conductivity and overall performance. Enhancing environmental stability is crucial. Scaling up the synthesis of CPs for industrial applications remains a challenge. Developing cost-effective and scalable synthesis techniques is essential for commercial viability.

Addressing these challenges and pursuing these future directions will pave the way for the widespread adoption of CPs in energy storage applications, leading to more efficient, sustainable, and innovative energy solutions.

## References

1. Kuznetsova LS, Arlyapov VA, Plekhanova YV, Tarasov SE, Kharkova AS, Saverina EA, Reshetilov AN (2023) Conductive polymers and their nanocomposites: application features in biosensors and biofuel cells. Polym 15(18):3783. https://doi.org/10.3390/polym15183783
2. Kim J, Lee J, You J, Park M-S, Hossain MSA, Yamauchi Y, Kim JH (2016) Conductive polymers for next-generation energy storage systems: recent progress and new functions. Mater Horiz 3(6):517–535. https://doi.org/10.1039/c6mh00165c

3. Madhushani KAU, Gupta RK (2023) Conducting polymers for electrochemical energy storage applications. In: Recent advancements in polymeric materials for electrochemical energy storage. Springer, Singapore, pp 49–70. https://doi.org/10.1007/978-981-99-4193-3_4
4. Magu TO,Agobi AU, Hitler L, Dass PM (2019a) A review on conducting polymers-based composites for energy storage application. J Chem Rev 1(1):19–34. https://doi.org/10.33945/sami/jcr.2019.1.1934
5. Nezakati T, Seifalian A, Tan A, Seifalian AM (2018) Conductive polymers: opportunities and challenges in biomedical applications. Chem Rev 118(14):6766–6843. https://doi.org/10.1021/acs.chemrev.6b00275
6. Khokhar D, Jadoun S, Arif R, Jabin S, Rathore DS (2022) Facile synthesis of the chemically oxidative grafted copolymer of 2, 6-diaminopyridine (DAP) and thiophene (Th) for optoelectronic and antioxidant studies. J Mol Struct 1248:131453. https://doi.org/10.1016/j.molstruc.2021.131453
7. Boddula R, Ahmer MF, Asiri AM (eds) (2019) Conducting polymers-based energy storage materials. CRC Press. https://doi.org/10.1201/9780429202261
8. Ravichandran R, Sundarrajan S, Venugopal JR, Mukherjee S, Ramakrishna S (2010) Applications of conducting polymers and their issues in biomedical engineering. J Royal Soc Interface 7(suppl_5). https://doi.org/10.1098/rsif.2010.0120.focus
9. Lin C-H, Lin J-H, Chen C-F, Ito Y, Luo S-C (2021) Conducting polymer-based sensors for food and drug analysis. J Food Drug Anal, 29(4):544–558. https://doi.org/10.38212/2224-6614.3374
10. Jadhav N, Kasisomayajula S, Gelling VJ (2020) Polypyrrole/metal oxides-based composites/nanocomposites for corrosion protection. Front Mater 7:95. https://doi.org/10.3389/fmats.2020.00095
11. Kumar AN (2023) Role of conducting polymers in corrosion protection. WAJRR 17(2):045–047. https://doi.org/10.30574/wjarr.2023.17.2.0238
12. Vlachopoulos N, Grätzel M, Hagfeldt A (2021) Solid-state dye-sensitized solar cells using polymeric hole conductors. RSC Advan 11(62):39570–39581. https://doi.org/10.1039/d1ra05911d
13. Ghosh S, Das S, Mosquera MEG (2020) Conducting polymer-based nanohybrids for fuel cell application. Polym 12(12):2993. https://doi.org/10.3390/polym12122993
14. Namsheer K, Rout CS (2021) Conducting polymers: a comprehensive review on recent advances in synthesis, properties and applications. RSC Adv 11(10):5659–5697. https://doi.org/10.1039/d0ra07800j
15. Park J-S, Kim JY, Song H-K (2013) Redox-active charge carriers of conducting polymers as a tuner of conductivity and its potential window. Sci Rep 3(1). https://doi.org/10.1038/srep02454
16. Le T-H, Kim Y, Yoon H (2017) Electrical and electrochemical properties of conducting polymers. Polym 9(4):150. https://doi.org/10.3390/polym9040150
17. Le CV, Yoon H (2024) Advances in the use of conducting polymers for healthcare monitoring. Int J Mol Sci 25(3):1564. https://doi.org/10.3390/ijms25031564
18. Guo B, Ma PX (2018) Conducting polymers for tissue engineering. Biomacromoleules 19(6):1764–1782. https://doi.org/10.1021/acs.biomac.8b00276
19. Chapman CAR, Cuttaz EA, Goding JA, Green RA (2020) Actively controlled local drug delivery using conductive polymer-based devices. Appl Phys Lett 116(1). https://doi.org/10.1063/1.5138587
20. Sharma N, Singh N, Kumar N, Tiwari A, Lal M, Arya S (2024) A review on polyaniline and its composites: from synthesis to properties and progressive applications. J Mater Sci:1–39. https://doi.org/10.1007/s10853-024-09562-z
21. Beygisangchin M, Abdul Rashid S, Shafie S, Sadrolhosseini AR, Lim HN (2021) Preparations, properties, and applications of polyaniline and polyaniline thin films—a review. Polym 13(12):2003. https://doi.org/10.3390/polym13122003
22. Zoshki A, Rahmani MB, Masdarolomoor F, Pilehrood SH (2017) Room temperature gas sensing properties of polyaniline/ZnO nanocomposite thin films. J Nanoelect Optoe 12(5):465–471. https://doi.org/10.1166/jno.2017.2031
23. Rathees R, Viswanathan K (2014) Charge transport study in para-toluene sulfonic acid doped polyaniline. Int J Innov ResSci Eng Technol 3(3):10646–10658

24. Kaushik P, Bharti R, Sharma R, Verma M, Olsson RT, Pandey A (2024) Progress in synthesis and applications of polyaniline-coated nanocomposites: a comprehensive review. Europ Polym J 113574. https://doi.org/10.1016/j.eurpolymj.2024.113574
25. Altınışık H, Getiren B, Çıplak Z, Soysal F, Yıldız N (2022) Energy storage performance of nitrogen doped reduced graphene oxide/co-doped polyaniline nanocomposites. J Inorg Organomet Polym Mat 33(2):353–367. https://doi.org/10.1007/s10904-022-02456-0
26. Elkais A, Gvozdenovic M, Jugovic B, Trisovic T, Maksimovic M, Grgur B (2011) Electrochemical synthesis and corrosion behavior of thin polyaniline film on mild steel, copper and aluminum. Hem Ind 65(1):15–21. https://doi.org/10.2298/hemind100901069e
27. Diggikar RS, Deshmukh SP, Thopate TS, Kshirsagar SR (2019) Performance of polyaniline nanofibers (PANI NFs) as PANI NFs-Silver (Ag) nanocomposites (NCs) for energy storage and antibacterial applications. ACS Omega 4(3):5741–5749. https://doi.org/10.1021/acsomega.8b02834
28. Badi N, Roy AS, Al-Aoh HA, Motawea MS, Alghamdi SAM, Alsharari A, Albaqami AS, Ignatiev A (2023) Enhanced and proficient soft template array of polyaniline—TiO2 nanocomposites fibers prepared using anionic surfactant for fuel cell hydrogen storage. Polym 15(20):4186. https://doi.org/10.3390/polym15204186
29. Bora N, Joshi DP, Aulakh JS (2023) Influence of polyaniline conducting polymer on thermal properties of phase change material for thermal energy storage. Polym Bull 81(2):1597–1621. https://doi.org/10.1007/s00289-023-04778-6
30. Fan D, Meng Y, Jiang Y, Qian S, Liu J, Xu Y, Xiong D, Cao Y (2021) Silver/polypyrrole-functionalized polyurethane foam embedded phase change materials for thermal energy harvesting. J Nanomat 11(11):3011. https://doi.org/10.3390/nano11113011
31. Agobi AU, Louis H, Magu TO, Dass PM (2019) A review on conducting polymers-based composites for energy storage application. J Chem Rev 1(1):19–34. https://doi.org/10.33945/SAMI/JCR.2019.1.1934
32. Mudila H, Prasher P, Kumar M, Kumar A, Zaidi MGH, Kumar A (2019) Critical analysis of polyindole and its composites in supercapacitor application. Mater Renew Sustain Energy 8(2). https://doi.org/10.1007/s40243-019-0149-9
33. Zhou P, Xiao F, Weng R, Huang Q, Wang L, He Q, Tang W, Yang P, Su R, He P, Jia B, Bian L (2022) PVP derived nitrogen-doped porous carbon integrated with polyindole: nano/microspheres assembled by emulsion polymerization for asymmetric supercapacitors. J Mat Chem A10(19):10514–10524. https://doi.org/10.1039/d2ta00102k
34. Vannathan AA, Kella T, Shee D, Mal SS (2021) One-pot synthesis of polyoxometalate decorated polyindole for energy storage supercapacitors. ACS Omega (6, 17):11199–11208. https://doi.org/10.1021/acsomega.0c05967
35. Joshi P, Mehtab S, Zaidi MGH, Tyagi T, Bisht A (2019) Development of polyindole/tungsten carbide nanocomposite-modified electrodes for electrochemical quantification of chlorpyrifos. J Nanostruct Chem 10(1):33–45. https://doi.org/10.1007/s40097-019-00326-9
36. Sood Y, Singh K, Mudila H, Lokhande PE, Singh L, Kumar D, Dehghani MH (2024) Insights into properties, synthesis and emerging applications of polypyrrole-based composites, and future prospective: a review. Heliyon 10(13):e33643. https://doi.org/10.1016/j.heliyon
37. Pérez-Torres AF, González-Hernández M, Ortiz P, Cortés MT (2022) Statistical study of the influence of electrosynthesis conditions on the capacitance of polypyrrole. ACS Omega 7(18):15580–15595. https://doi.org/10.1021/acsomega.1c06843
38. Begum B, Bilal S, AUHA S, Roese P (2021) Physical, chemical, and electrochemical properties of redox-responsive polybenzopyrrole as electrode material for faradaic energy storage. Polym 13(17):2883. https://doi.org/10.3390/polym13172883
39. Attia NF, Geckeler KE (2013) Polyaniline–polypyrrole composites with enhanced hydrogen storage capacities. Macromol Rapid Commun 34(11):931–937. https://doi.org/10.1002/marc.201300060
40. Chalmers E, Li Y, Liu X (2019) Molecular tailoring to improve polypyrrole hydrogels' stiffness and electrochemical energy storage capacity. Front Chem Eng 13(4):684–694. https://doi.org/10.1007/s11705-019-1817-0

41. Adekoya GJ, Adekoya OC, Sadiku RE, Hamam Y, Ray SS (2022a) Applications of MXene-containing polypyrrole nanocomposites in electrochemical energy storage and conversion. ACS Omega 7(44):39498–39519. https://doi.org/10.1021/acsomega.2c02706
42. Tahir M, He L, Li L, Cao Y, Yu X, Lu Z, Liao X, Ma Z, Song Y (2023) Pushing the electrochemical performance limits of polypyrrole toward stable microelectronic devices. Nano-Micro Lett 15(1). https://doi.org/10.1007/s40820-023-01027-3
43. Ratsameetammajak N, Autthawong T, Khunpakdee K, Haruta M, Chairuangsri T, Sarakonsri T (2023) Insight Into the Role of Conductive Polypyrrole Coated on Rice Husk-Derived Nanosilica-Reduced Graphene Oxide as the Anodes: Electrochemical Improvement in Sustainable Lithium-Ion Batteries. Polychandanm. 15(24):4638. https://doi.org/10.3390/polym15244638
44. Chandran ACS, Schneider J, Nair R, Bill B, Gadegaard N, Hogg R, Kumar S, Manjakkal L (2024) Enhancing supercapacitor electrochemical performance With 3D printed cellular PEEK/MWCNT electrodes coated with PEDOT: PSS. ACS Omega 9(31):33998–34007. https://doi.org/10.1021/acsomega.4c04576
45. Wustoni S, Nikiforidis G, Inal S, Indartono YS, Suendo V, Yuliarto B (2022) Hydroxymethyl PEDOT microstructure-based electrodes for high-performance supercapacitors. APL Mater 10(6). https://doi.org/10.1063/5.0088452
46. Guan X, Pan L, Fan Z (2021) Flexible, transparent and highly conductive polymer film electrodes for all-solid-state transparent supercapacitor applications. Membr J 11(10):788. https://doi.org/10.3390/membranes11100788
47. Zhao Q, Jamal R, Zhang L, Wang M, Abdiryim T (2014) The structure and properties of PEDOT synthesized by template-free solution method. Nanoscale Res Lett 9(1). https://doi.org/10.1186/1556-276x-9-557
48. Bhagwat N, Kiick KL, Martin DC (2014) Electrochemical deposition and characterization of carboxylic acid functionalized PEDOT copolymers. J Mat Res 29(23):2835–2844. https://doi.org/10.1557/jmr.2014.314
49. Deshmukh K, Ahamed MB, Deshmukh RR, Pasha SK, Bhagat PR, Chidambaram K (2017) Biopolymer composites with high dielectric performance: interface engineering. In: Biopolymer composites electronics. Elsevier, pp 27–128. https://doi.org/10.1016/B978-0-12-809261-3.00003-6
50. Li C, Zhang K, Cheng X, Li J, Jiang Y, Li P, Peng H (2023) Polymers for flexible energy storage devices. Prog Polym Sci 101714. https://doi.org/10.1016/j.progpolymsci.2023.101714
51. Tajik S, Beitollahi H, Nejad FG, Shoaie IS, Khalilzadeh MA, Asl MS, Shokouhimehr M (2020) Recent developments in conducting polymers: applications for electrochemistry. RSC Adv 10(62):37834–37856. https://doi.org/10.1039/D0RA06160C
52. Marriam I, Wang Y, Tebyetekerwa M (2020) Polyindole batteries and supercapacitors. Energy Storage Mater 33:336–359. https://doi.org/10.1016/j.ensm.2020.08.010
53. Pal K, Bharti D, Sarkar P, Anis A, Kim D, Chałas R, Jarzębski M (2021). Selected applications of chitosan composites. Inter J Mol Sci 22(20): 10968 https://doi.org/10.3390/ijms222010968
54. Gupta P, Sharma S, Jabin S, Jadoun S (2024) Chitosan nanocomposite for tissue engineering and regenerative medicine: a review. Int J Biol Macromol 254:127660
55. Bullock RM, Das AK, Appel AM (2017) Surface immobilization of molecular electrocatalysts for energy conversion. Chem Eur J 23(32):7626–7641. https://doi.org/10.1002/chem.201605066
56. Beley M, Gros PC (2014) Ruthenium polypyridine complexes bearing pyrroles and π-extended analogues. Synthesis, spectroelectronic, electrochemical, and photovoltaic properties. Organometallics 33(18):4590–4606. https://doi.org/10.1021/om5002183
57. Maziz A, Özgür E, Bergaud C, Uzun L (2021) Progress in conducting polymers for biointerfacing and biorecognition applications. Sens Actuators Rep 3:100035. https://doi.org/10.1016/j.snr.2021.100035
58. Yang J, Liu Y, Liu S, Li L, Zhang C,Liu T (2017) Xu P, Han X, Zhang B, Du Y, Wang HL. Multifunctional polymer–metal nanocomposites via direct chemical reduction by conjugated polymers. Chem Soc Rev. 43: 1349–1360 https://doi.org/10.1039/C3CS60380F

59. Ibanez JG, Rincón ME, Gutierrez-Granados S, Chahma MH, Jaramillo-Quintero OA, Frontana-Uribe BA (2018) Conducting polymers in the fields of energy, environmental remediation, and chemical–chiral sensors. Chem Rev 118(9):4731–4816
60. Ge Y, Jalili R, Wang C, Zheng T, Chao Y, Wallace GG (2017) A robust free-standing MoS2/ poly (3, 4-ethylenedioxythiophene): poly (styrenesulfonate) film for supercapacitor applications. Electrochim Acta 235:348–355. https://doi.org/10.1016/j.electacta.2017.03.069
61. Li K, Liang M, Wang H, Wang X, Huang Y, Coelho J, Pinilla S, Zhang Y, Qi F, Nicolosi V, Xu Y (2020) 3D MXene architectures for efficient energy storage and conversion. Adv Funct Mater 30(47). https://doi.org/10.1002/adfm.202000842
62. He Y, Han X, Du Y, Song B, Xu P, Zhang B (2016) Bifunctional nitrogen-doped microporous carbon microspheres derived from poly (o-methylaniline) for oxygen reduction and supercapacitors. ACS Appl. Mat. Interfaces 8(6):3601–3608. https://doi.org/10.1021/acsami.5b07865
63. Zhu D, Cheng K, Wang Y, Sun D, Gan L, Chen T, Liu M (2017) Nitrogen-doped porous carbons with nanofiber-like structure derived from poly (aniline-co-p-phenylenediamine) for supercapacitors. Electrochim Acta 224:17–24. https://doi.org/10.1016/j.electacta.2016.12.023
64. Govindarajan D, Chinnakutti KK (2022) Chapter 3: Fundamentals, basic components and performance evaluation of energy storage and conversion devices. In: Oxide free nanomaterials for energy storage and conversion applications. Elsevier, Amsterdam, pp 51–74
65. Jadoun S, Rathore DS, Riaz UNPS (2021) Tailoring of conducting polymers via polymerization-a review. Eur Polym J 155:110561. https://doi.org/10.1016/j.eurpolymj.2021.110561
66. Anand SG, Gopalan AI, Lee KP, Venkatesan S, Kang BH, Lee SW, Kang SW (2015) A futuristic strategy to influence the solar cell performance using fixed and mobile dopants incorporated sulfonated polyaniline based buffer layer. Sol Energ Mater Sol C 141:275–290. https://doi.org/10.1016/j.solmat.2015.05.035
67. Alem S, Graddage N, Lu J, Kololuoma T, Movileanu R, Tao Y (2018) Flexographicprinting of polycarbazole-based inverted solar cells. Org Electron 52:146–152
68. Cole W, Frazier AW, Augustine C (2021) Cost projections for utility-scale battery storage update (No. NREL/TP-6A20-79236). National Renewable Energy Lab. (NREL), Golden
69. Salkuti SR (2020) Energy storage technologies for smart grid: a comprehensive review. Majlesi J Electrical Eng 14(1):39–48. https://doi.org/10.1016/j.apenergy.2014.09.081
70. Kebede AA, Kalogiannis T, Van Mierlo J, Berecibar M (2022) A comprehensive review of stationary energy storage devices for large scale renewable energy sources grid integration. Renew Sustain Energ Rev 159:112213. https://doi.org/10.1016/j.rser.2022.112213
71. Islam J, Chowdhury FI, Raza W, Zobed HM (2021) Towards polymer composites based and architectural engineering induced flexible electrodes for lithium-ion batteries. Renew Sustain Energy Rev 148:111302. https://doi.org/10.1016/j.rser.2021.111302
72. Saleem F, Ni B, Yong Y, Gu L, Wang X (2016) Ultra-small Tetrametallic Pt-Pd-Rh-Ag Nanoframes with tunablebehavior for direct formic acid/methanol oxidation. Small 12:5261–5268. https://doi.org/10.1002/smll.201601299
73. Reddy ALM, Rajalakshmi N, Ramaprabhu S (2008) Cobalt-polypyrrole-multiwalled carbon nanotube catalysts for hydrogen and alcohol fuel cells. Carbon 46:2–11
74. Li Y, Niu J, Xue T, Duan X, Tian Q, Wen Y,Chen Y (2020). Multifunctional porous nanohybrid based on graphene-like tungsten disulfide on poly (3, 4-ethoxylenedioxythiophene) for supercapacitor and electrochemical nanosensing of quercetin. J Electrochem Soc 167(4): 047512. https://doi.org/10.1149/1945-7111/ab721e
75. Shindalkar SS, Redd M, Singh R, Nainar MAM, Kandasubramanian B (2023) Polythiophene blends and composites as potential energy storage materials. Synth Met 299:117467. https://doi.org/10.1016/j.synthmet.2023.117467
76. Tadesse MG, Ahmmed AS, Lübben JF (2024) Review on conductive polymer composites for supercapacitor applications. J Compos Sci 8(2):53
77. AL-Refai HH, Ganash AA, Hussein MA (2021) Polythiophene and its derivatives–based nanocomposites in electrochemical sensing: a mini review. Mater Today Commun 26:101935. https://doi.org/10.1016/j.mtcomm.2020.101935

78. Thadathil A, Pradeep H, Joshy D, Ismail YA, Periyat P (2022) Polyindole and polypyrrole as a sustainable platform for environmental remediation and sensor applications. Mater Adv 3(7):2990–3022
79. Sharma S, Verma A, Rangappa SM, Siengchin S, Ogata S (2023) Recent progressive developments in conductive-fillers based polymer nanocomposites (CFPNC's) and conducting polymeric nanocomposites (CPNC's) for multifaceted sensing applications. J Mater Res Technol 26:5921–5974
80. Koiry SP, Saxena V, Sutar D, Bhattacharya S, Aswal DK, Gupta SK, Yakhmi JV (2007) Interfacial synthesis of long polyindole fibers. J Appl Polym Sci 103(1):595–599

# Chapter 10
# Hybrid Conducting Polymers as Sensors and Biosensors

**Sahid SK, Manda Sathish, Suman Chirra, B. V. Bhaskara Rao, and Nasseb Singh**

**Abstract** The field of hybrid conducting polymers (HCPs) has become a major material category. The combination of conducting polymers with other components generates synergistic effects that improve performance in sensor and biosensor applications. The integration of conducting polymers with different materials by researchers has resulted in better sensing performance. The integration of bioorganic molecules into HCPs becomes possible through combining conducting polymers with inorganic nanoparticles and carbon materials which ultimately delivers enhanced electrical conductivity along with tunable optical properties and biocompatibility as well as stability. These hybrid systems operate as chemical, physical, and optical sensors which detect multiple properties including pH levels, temperature changes, humidity variations, gas presence, ion concentrations, and biomolecular interactions. The distinct properties of HCPs make them effective tools for environmental monitoring tasks as well as healthcare diagnostic procedures and biomedical research applications. This chapter highlights new developments in sensors and biosensors based on HCPs to demonstrate their capabilities for precise

S. SK (✉)
Department of Chemistry, Indian Institute of Technology Kanpur (IIT-K), Kanpur, Uttar Pradesh, India

M. Sathish (✉)
Centro de Investigación de Estudios Avanzados del Maule (CIEAM), Vicerrectoría de Investigación y Postgrado, Universidad Católica del Maule, Talca, Chile
e-mail: smanda@ucm.cl

S. Chirra
School of Sciences, Woxsen University, Kamkole, Telangana, India

B. V. B. Rao
Departamento de Física, Facultad de Ciencias Naturales Matemática y del Medio Ambiente, Universidad Tecnológica Metropolitana, Santiago, Chile

N. Singh
Department of Chemistry, Govt. Degree College Udhampur (A Constituent College of University of Jammu), Jammu, India



S. Jadoun, N. Kumari Jangid (eds.), *Tuning of Conducting Polymers*, Springer Series on Polymer and Composite Materials,
https://doi.org/10.1007/978-3-032-00461-1_10

sensitivity and selectivity with real-time detection abilities. The combination of nanotechnology with microfluidics and machine learning algorithms leads to substantial improvements in HCP performance. Advanced technologies combined with HCPs create opportunities for developing next-generation sensing platforms. Future research for HCPs must focus on enhancing stability and durability while achieving reproducibility and successful integration. The discussion examines how electronic device integration works with biocompatibility and toxicity issues.

**Keywords** Hybrid conducting polymers · Sensors · Biosensors · Nanoparticles · Sensing applications

## 10.1 Introduction

In 1977, Hideki Shirakawa innovated conductive polymers by synthesizing electrically conducting organic polymers from halogen derivatives of polyacetylene $(CH)_x$ and produced high-quality flexible copper-colored cis-isomer films alongside silvery trans-isomer films through acetylene polymerization using a Ziegler–Natta catalyst [1]. He improved reaction conditions to establish methods for managing the cis- and trans-isomer ratios in the final product [2, 3]. Both isomers exhibit crystalline structures made up of matted fibrils as demonstrated through X-ray diffraction and scanning electron microscopy. Trans-polyacetylene films showed a dramatic increase in conductivity when exposed to chlorine or bromine vapors and exhibited an increase of more than seven orders of magnitude specifically with iodine vapors. The "doping" process generated films with a silvery to silvery-black appearance that exhibited room temperature conductivity similar to metals [4, 5].

Conducting polymers, or intrinsically conducting polymers (ICPs), have several drawbacks, including: (a) *Poor mechanical strength*: Conductive polymers have weaker mechanical properties than other commercially available polymers. (b) *Poor solubility*: Conductive polymers are salt-like and less soluble in water and organic solvents, making them harder to process. (c) *Instability*: Conductive polymers are unstable in the environment and at high temperatures. They can also lose conductivity over time. (d) *Sensitivity to manufacturing modifications*: The properties of conductive polymers can be affected by manufacturing modifications, time, and temperature. (e) *Inability to degrade*: Conductive polymers cannot degrade, which limits their applications.

The field of designing high-performance conducting multifunctional nanocomposites has become an attractive research area with significant interest growing over recent years [6]. The category of advanced materials encompasses a broad spectrum of nanostructured materials that take on one-dimensional, two-dimensional, and three-dimensional nanoscale forms and demonstrate distinctive properties including electrical conductivity and mechanical strength as well as sensing capabilities. These emerging nanocomposites demonstrate flexibility and show potential for multiple uses across automotive industries, integrated circuits' development, sensor applications, medical device manufacturing, and biomedical applications [7–9]. The combination of conducting polymers with carbon-based materials in nano-composites stands out as particularly fascinating because this pairing generates a distinctive set of properties that attract significant research attention [10, 11]. Hybrid conducting polymers become attractive materials because of their simple production process combined with ductility and electrical conductivity and the mechanical strength, optical characteristics, and sensing properties of HCPs. Modern research efforts focus on enhancing the electrical properties of PANI/graphene-based nanocomposites at room temperature by studying numerous influential factors such as preparation methods, structural characteristics, component proportions, doping types, protonation levels, redox states, and thermal conditions [12].

## 10.2 Conduction Mechanism in Conductive Polymers and Hybrid Conductive Polymers

Conductive polymers and hybrid conductive polymers operate through a specific conduction mechanism. Researchers use electronic band structures to understand how materials exhibit their electrical properties. The classification of materials as insulators or conductors depends on the energy separation between their conduction band and valence band [13]. Intrinsically conducting materials feature overlapping conduction and valence bands due to their reduced bandgap. The electrical band theory successfully explained conducting polymers but other research revealed the transport characteristics of these materials beyond band theory. Conjugate bonds exist within the backbones of all conducting polymers and play a crucial role in electron transport. The single bond consists only of a localized $\sigma$ bond while the double bond includes both a $\sigma$ and a weaker $\pi$ bond. The $\pi$ bond that connects the first and second carbons migrates to establish a connection between the second and third carbons. Electrons move through the polymer chain as the $\pi$ bond between the third and fourth carbons shifts to an adjacent pair [14]. The conductivity of polymers changes drastically based on the type of dopant material used as well as the configuration and length of the polymer chain. The conductivity of materials increases with proper dopant concentration and pH settings such as polyaniline which achieves maximum conductivity at pH levels between 0 and 3. Virgin polymers display the properties of insulators and semiconductors before doping while their conductivity increases as more dopant is added. The dopants trigger redox reactions that enable the addition and removal of electrons from conducting polymers [15]. There are two types of doping: There are two doping methods known as p-type and n-type where dopants create both positive and negative polarons and bipolarons. The distribution of charge carriers across polymer chains enhances their electrical conductivity [16, 17]. P-type doping leads to electron movement from polymer backbone HOMO to dopant species which creates a hole and results in electron deficiency in the backbone. With n-type doping the electrons migrate to the

LUMO of the polymer backbone which increases electron density and improves conductivity. Scientists show interest in p-type doping because of the exceptional stability found in positively charged carriers [18–21]. Researchers investigated the band structure of polypyrrole (PPy) throughout different doping stages. Pristine undoped polypyrrole functions as a semiconductor with its bandgap measuring 3.16 eV. The introduction of p-type materials to the polymer backbone triggers oxidation which removes π electrons [21]. The loss of electrons alters the polymer structure from benzenoid to quinoid and generates one polaron in the backbone. A localized electronic level forms within the band structure when a polaron appears and bipolarons oxidize by shedding π electrons. Bipolarons undergo a more rapid transformation from benzenoid form to quinoid form. Oxidation enables bipolarons to overlap once more creating an additional minor bipolaronic band that integrates within the main band structure while decreasing the band energy from 3.16 to 1.4 eV [22]. Polymer backbone units form a valance band and a conduction band through their interaction with neighboring units. Ionization of conjugative polymers leads to high energy states which subsequently cause transformations in the band structure. The structural change can occur between benzenoid and quinoid forms [23]. The conducting polymer polyacetylene features a valance band filled with carbon atoms $sp^2$ orbitals and hydrogen atoms s orbitals while its conduction band remains empty. The bandgap energy level stands close to 10 eV demonstrating a high energy value [24, 25]. Trans-polyacetylene exhibits a degenerate energy state because it possesses two equivalent energy geometrical structures that vary through a single carbon-carbon interaction exchange. The exact process by which polyacetylene conducts electricity has received broad scientific confirmation. Polyacetylene behaves as an insulator before doping but becomes conductive after accepting electron-withdrawing dopants ($AsF_3$, $Br_2$, $I_2/CCl_4$) which initiate radical cations or polarons by extracting electrons from the π bond that is more readily removable than from a σ bond. After additional dopant treatment removes another electron from the polaron radical. Upon the creation of bipolarons the system moves π bonds between σ bonds to form solitons. The charge of solitons depends on which dopant is used [26]. The conduction mechanism in conducting polymers operates through two distinct aspects. The conduction mechanism in conducting polymers consists of a chemical aspect and a solid-state physics aspect. Chemists regard conducting polymers as systems formed through conjugative bonds while physicists describe them as systems with charge density waves. Double bonds exhibit higher electron density than σ bonds which leads to π electron movement being connected to bond order wave oscillations because electron density within the lattice remains unrestricted [27]. Conjugate polymers conduct electricity through the action of charge carriers. Polypyrrole achieves conductivity through bipolarons while solitons act as charge carriers for trans-polyacetylene. Conducting polymers enable charge carriers to hop through both defective regions and spaces between chains. Mott found that electrical conductivity changes according to both temperature levels and concentration measurements. Various theoretical models demonstrate the relationship between temperature and electrical conductivity [28].

**The Mott model** explained electronic conduction in amorphous materials and organic polymers while also examining temperature effects on conductivity. Mott's concept seeks to clarify the temperature response of distorted systems and reveals how phonon absorption enables charge carriers to move between localized states of differing energy [29]. The hopping (the movement of an electron from one localized energy state to another) mechanism is dependent on two key factors:

The tunneling factor operates as ( $e^{\frac{-L}{\upsilon}}$ ) (where the localization length is υ and L represents the hopping distance) and the Boltzmann factor: [ $Pij \propto e\left[\frac{2L}{\xi}-\frac{\Delta E}{KT}\right]$, where $L$ and $\Delta E$ have a connection with the density of states $N(E_F)$ and $L^d$ $\Delta E$ $N(E^F) \sim 1$.

$$\text{Now}, \sigma = \sigma_0 \exp\left[\left(\frac{-T_0}{T}\right)^{\frac{1}{n+1}}\right]$$

where the parameter n represents the dimensionality of charge transfer, $\sigma_0$ stands for conductivity at room temperature with $T_0$ being Mott's temperature constant. A graph of log s against T g permits exploration of the Mott model while the least squares-fitted curve reveals details about charge transport mechanisms and dimensionality [30].

**The Schaefer–Siebert–Roth model** provides insight into the electrical conduction process of conductive polymers. Charges called polarons and bipolarons form in polyacetylene when p and n dopants disrupt their pi bonds. Charge mobilizers incorporated into polymers cause multiple conjugation lengths to develop along their backbone [31]. The Schaefer–Siebert–Roth model demonstrates that $\sigma$ is equal to $\sigma_0$ multiplied by $\exp\left[\left(\frac{-T_0}{T}\right)^{-\gamma}\right]$; $\frac{1}{4} \le \gamma \le \frac{1}{2}$; with $\gamma$ confined to values between 1/4 and 1/2 while $\sigma_0$ and $T_0$ depend on localization length and $\gamma$ depends on the density of states at the Fermi level.

## 10.3 A Comparison of Multiple Polymerization Techniques Used to Create Hybrid Conducting Polymers (HCPs)

There are several techniques for synthesizing HCPs:

- Chemical polymerization.
- Sol-gel polymerization.
- Emulsion polymerization.
- Electrochemical polymerization.

## 10.4 Chemical Polymerization

The synthesis of HCPs benefits greatly from chemical polymerization which stands out as an effective method. It is further categorized into two types: Scientists categorize chemical polymerization into two processes: condensation polymerization which is step growth polymerization and addition polymerization which acts as chain growth polymerization. The popularity of chemical polymerization stems from its cost-effectiveness and simplicity alongside its ability to produce various metal/metalloid HCPs through multiple methods. The procedure does not require the use of an electrode [32]. Chemical polymerization typically requires a monomer to undergo oxidation when exposed to an oxidant (dopant ions) or catalyst. The primary drawback of this approach is its failure to produce thin films [33]. The primary limitation of this technique is its inability to produce thin films [34, 35]. Sindhu et al. Sindhu et al. [36] developed a Poly(p-phenylene)/$CaCO_3$ hybrid material through in situ chemical polymerization and found this material exhibits superior optical properties while being appropriate for optoelectronics and biological uses. Peng et al. [37] developed a PTh-g-PMA and zinc oxide hybrid by applying in situ chemical polymerization.

The research explored the structural characteristics as well as the morphological and optical properties of the hybrid materials. The newly synthesized material demonstrates an 85% reduction in photoluminescence which positions it as a valuable candidate for photovoltaic uses. Mostafaei et al. PANI and zinc oxide were combined into a hybrid nanocomposite through in situ chemical polymerization by Mostafaei et al. [38] Qui et al. created the Pt/PANI/graphene nanosheet (GNs) hybrid. The researchers created PANI/GNs through interfacial polymerization before coating Pt nanoparticles on their surface to establish the Pt/PANI/GNs hybrid system. Subsequent tests of the hybrid materials' electrocatalytic performance showed they displayed excellent sensing characteristics [39].

## 10.5 Sol-Gel Polymerization

The sol-gel process represents a versatile technology capable of producing hybrid conducting polymer composites and nanocomposite materials [40, 41]. The initial usage of the sol-gel process to create silica particles from silicon alkoxides took place in 1930. Silicon alkoxide plus tetraethyl orthosilicate (TEOS) and tetramethyl orthosilicate (TMOS) serve as common silica-based precursors during polymerization processes [42–46]. Silicon-carbon bonds in conductive polymers exhibit hydrolytic stability which allows researchers to effectively integrate organic components within inorganic frameworks. Stable chemical bonds connect organic and inorganic components together. The method offers benefits including low reaction temperature which simplifies the creation of hybrid composite materials and the integration of organic components into inorganic substances. Xiong et al. The research done by

Xiong et al. [47, 48] produced a PANI/$TiO_2$ nanocomposite through sol-gel synthesis and performed structural and morphological studies alongside electrochromic property assessments. The PANI/$TiO_2$ hybrid nanocomposite demonstrates reduced crystallinity and enhanced transparency which makes it an ideal candidate for electrochromic electrical devices. Kim et al. produced a PEDOT/silica hybrid film through the sol-gel process to demonstrate its use in polymer solar cells without ITO [49]. The hybrid film offers superior conductivity together with lower surface resistance which makes it the perfect anode material for solar cells. Zheng et al. [50] created organic–inorganic hybrid coatings through sol-gel processing before evaluating their ability to protect metals from corrosion. Clement et al. [51] created a PTh/silica composite and examined its optical and thermal properties. The exceptional optical properties of these hybrids enable their application as both photoactive and conductive materials. Wang et al. The PANI/(3-Glycidoxypropyl)trimethoxy silane (GPTMS) hybrid film was produced by Wang et al. [52] through the sol-gel process and its water resistance was studied. The hybrid composite film demonstrated superior water resistance and conductivity when compared to pure PANI with a conductivity level of 0.6 $\Omega$ $cm^{-1}$. Wang et al. [53] created the PANI/PU/Silica hybrid and analyzed its thermal stability along with glass transition and tensile strength.

## 10.6 Emulsion Polymerization

Emulsion polymerization stands as a viable oxidative polymerization technique for creating HCPs according to reference [54]. There are two forms of emulsions: Emulsion systems can exist as water-in-oil where water droplets are dispersed in oil or oil-in-water where oil droplets are dispersed in water. Amaike et al. The HPC/PPy hybrid was fabricated by Amaike et al. through emulsion polymerization [55]. The PPy nanoparticle size can be controlled by adjusting HPC concentration which achieves conductivity levels up to $10^{-1}$ S/cm. Ghaffar et al. [56] prepared polyaniline hybrid nanocomposites (PANI/Na+ -MMT) using ammonium peroxydisulfate (APS) as an initiator to incorporate Na+ into montmorillonite. The organophilization of clay was performed with CTAB to generate an MMT-CTAB complex which then helped to establish the PANI/Na+ -MMT nanocomposite. Dai et al. [57] created a hybrid PEDOT using PTEB as a surfactant and FeCl3 as an oxidant. As PTEB concentration rises thermal stability along with crystallinity and electrical conductivity shows improvement.

## 10.7 Electrochemical Polymerization

The process is executed using a three electrode system consisting of working, counter, and reference electrodes together with monomeric solution and appropriate solvent and electrolyte (dopant). Electrochemical polymerization uses three electrode systems consisting of working, counter, and reference electrodes along with monomeric solution, solvent and electrolyte which acts as dopant [58]. Diaz and Kanazaw [59] first introduced the mechanistic understanding of pyrrole electrochemical polymerization on platinum electrodes. Electro polymerization has three phases. During the first stage of electrochemical polymerization, the monomer undergoes oxidation at the electrode while soluble oligomers develop in the diffusion layers. During the second stage of electrochemical polymerization, oligomers deposit first before nucleation and growth occur. The final stage of electro polymerization transforms into solid-state polymerization which produces polymer molecules with extended chains and cross-links [60]. Wang et al. Wang et al. [61] created a flexible hybrid nanocomposite of graphene and PANI through electro polymerization and conducted an analysis of its properties. Electropolymerization of PANI on graphene paper required applying a steady potential of 0.75 V in relation to a saturated calomel electrode (SCE) during time periods of 60 s, 300 s, and 900 s. Research identified the synthesized material as a high-performance flexible electrode suitable for supercapacitors. Miyazaki et al. Miyazaki et al. [62] synthesized a polycarbazole (PCz)/Al hybrid using EP to evaluate its durability as a transparent organic electrode material. Liu et al. [63] described the process of synthesizing a PPy/$TiO_2$ hybrid material on a gold substrate. The researchers created the hybrid material through electropolymerization while maintaining a steady potential of 0.85 V and exposing the system to UV light. Zhang et al. [64] created a PCz/N-GE based hybrid electrochemical sensor to detect 4-nitrophenol (4-NP). Researchers used electrochemical polymerization to deposit PCz onto glassy carbon electrodes which had been modified with N-GE. Cai et al. Researchers generated $WO_3$/PANI core/shell nanowire array films using solvothermal and electro polymerization techniques. The electropolymerization of PANI on $WO_3$ nanowires was conducted using an Ag/AgCl reference electrode together with a Pt working electrode [65]. The $WO_3$/PANI nanowires demonstrate superior electrochromic performance because their donor–acceptor contacts along with a larger surface area enable better charge transfer. Crespo et al. Crespo and colleagues created a Ni/PPy hybrid bilayer coating specifically for corrosion protection in their study [66]. The research team tested this coating's electrochemical properties using a 3% NaCl solution.

**Comparison** After analysis of the various synthesis methods reveals both the vast potential for hybrid synthesis and the straightforward execution of synthesis using any of these methods. However, each option has drawbacks. Free-standing hybrid thin films remain unsynthesizable through chemical polymerization methods even though these methods provide homogenous mixing and low temperature synthesis alongside simple regulation and cost efficiency. Electropolymerizing enables straightforward production of hybrid thin films that exhibit enhanced conductivity and stability yet suffers from challenging reaction regulation. Sol-gel polymerization creates better homogeneity while preventing phase separation and requires lower reaction temperatures than emulsion polymerization. The viscosity reduction from emulsion polymerization enables faster reaction kinetics. The requirement for surfactants in bulk synthesis becomes problematic because breaking micelles after the process proves challenging. The preparation methods and applications of HCPs do not have a simple correlation because there are many polymerization procedures and numerous HCPs to choose from. Materials used in lithium-ion batteries and supercapacitors must demonstrate strong electrical conductivity combined with mechanical stability [67]. Material characteristics become accessible through both chemical methods [68–70] and electro polymerization procedures [71]. The morphology of the hybrid material plays a crucial role in achieving the desired traits. The way the hybrid's structure forms depends on the particular synthesis techniques used [72]. The flexible electrode with controlled shape produced through electro polymerization demonstrated improved electrochemical characteristics according to multiple studies [73, 74]. Sensors need materials that exhibit high surface areas along with excellent conductivity and redox capacity [75]. Hybrid properties can be achieved via chemical methods [76] and electropolymerization techniques [77]. The hybrid benefits from enhanced analyte interaction through its large surface area and improved sensitivity due to its excellent conductivity. Hybrid conducting polymers show potential in photo and electrocatalytic applications because they possess increased surface area. HCPs achieve better pollutant degradation efficiency when their surface area expands because it creates more active sites to engage in photocatalytic reactions. The material exhibits high surface area and good conductivity in addition to a core–shell structure [78, 79].

## 10.8 Hybrid Conducting Polymers as Sensors and Biosensors

### *10.8.1 Basic Requirements for Sensor Design*

Choosing materials forms a fundamental component of the sensor design process. The following features should be considered when selecting a material for sensor applications: The basic requirements for sensor material selection include selectivity, stability, and cyclability [80]. Sensor selectivity dictates whether it can detect multiple analytes or restricts detection to a single analyte. Sensor material stability is crucial for its consistent performance through time whereas cyclability defines the sensor's continuous operational duration with consistent reproducibility [81]. The sensor properties rely on factors such as size dimensions, extensive surface area, and electrical conductivity. When used by themselves conducting polymers failed to achieve the necessary selectivity and sensitivity levels to create effective sensors. Improved sensor properties result from the inorganic component's unique characteristics of high surface area and small size [82, 83].

### *10.8.2 Why Hybrid Conducting Polymers Are Ideal Candidates for Sensor Applications*

Filler particles such as CNT, $TiO_2$, $SiO_2$, $Al_2O_3$, and graphene enable the achievement of desirable material properties. These particles exhibit outstanding characteristics including flexibility together with sensitivity and robustness and stability [84]. Introducing nano-sized filler particles in suitable amounts allows for the creation of enhanced-performing gas, chemical, and bio sensors compared to pure conducting polymers. The combination of conducting polymers and inorganic components results in HCPs that offer synergistic properties suitable for chemical, gas, and biosensors [85]. HCPs benefit from a larger surface area which promotes more effective interactions between materials and analytes to achieve better sensitivity. HCPs demonstrate fast response times and high signal precision through their compact structure which enhances analyte adsorption and desorption interactions [86].

## 10.9 HCPs as Sensor and Biosensors Applications

Biosensors serve as diagnostic tools which find widespread application across several fields including clinical diagnostics and environmental monitoring as well as within the food industry and defense sector [87]. The unique attributes of intrinsically conducting polymers (ICPs) and their composite hybrid conducting polymers (HCPs), which include biological molecule compatibility and biomolecule immobilization capabilities along with efficient electron transfer in biochemical reactions and bioreagent loading ability, have led to their extensive investigation in the biosensing field. HCPs stand as potential candidates for biological detection applications due to their capability to detect different biological molecules and materials such as glucose with a detection limit of approximately 0.18 nm, DNA at 10 pm, cholesterol at 1 µm, aptamers at about 0.8 pm, and cancer cells at 5 pm $mL^{-1}$ [88].

Hybrid conductive polymers have become popular for biosensor development because they enable both biomolecule immobilization and fast electron transfer. The conductive polymer Polypyrrole (PPy) stands out as the most frequently utilized CP in biosensor technology because of its excellent biocompatibility and its ability to polymerize under neutral pH conditions. Enzymatic activities produce toxic hydrogen peroxide ($H_2O_2$) gas as a by-product which makes its detection essential. Researchers Aussawasathien et al. developed PANI/PS hybrid composite nanofibers [89]. using an electrospinning approach. The PANI-based hybrid material became a highly sought-after option for glucose and $H_2O_2$ sensors because of its immense surface area and exceptional electrochemical properties.

Fiorito et al. produced a cupper hexacyanoferrate/PPy hybrid composite with similar intentions [90]. in order to detect harmful $H_2O_2$. The synthesized material displayed high sensitivity of 726.00 m $mol^{-1}$ L $cm^{-2}$ along with strong stability for up to 20 cycles and outstanding electrocatalytic behavior toward sodium and potassium ions as opposed to the nickel hexacyanoferrate film which showed only 0.14 m $mol^{-1}$ L $cm^{-2}$ sensitivity. The combination of hexacyanoferrate and PPy led to increased conductivity and sensitivity in the hybrid material.

Baghayeri et al. created a poly(para-phenylenediamine) (PDPA)/$Fe_3O_4$ nanocomposite [91]. The $Fe_3O_4$ nanoparticle served as an enzyme mobilizing agent to improve its performance for sensing applications. The nanocomposite showed outstanding sensitivity and reliability in the reduction of $H_2O_2$ with a detection limit measured at 0.21 M.

Liu et al [92] developed polyaniline/carbon nanotube based hybrid multilayer films in neutral solution for stable low-potential detection of reduced β-Nicotinamide Adenine Dinucleotide (NADH). This developed multilayer hybrid films composed of poly (amino benzenesulfonic acid)-modified single-walled carbon nanotubes (PABS-SWNTs) combined with PANI. The as-synthesized multilayer film demonstrated enhanced electrocatalytic performance while detecting NADH at very low current levels. The synthesized film displayed remarkable efficiency after 500 cycles since its surface plasmon resonance (SPR) signal and current dropped by less than 5%. The durability and exceptional specificity of this material for NADH detection is demonstrated by this result. Qian et al. developed the Au/PPy/rGO hybrid sensor for detecting dopamine [93]. The synthetic electrode achieved an exceptional sensitivity of 16.40 mA mM1 while maintaining an extremely low detection limit of 18.29 M. The hybrid sensor demonstrated reproducibility with a relative standard deviation (RSD) of 3.12% following eight consecutive measurements in 5 M of DA. The electrode preserved 94.3% of its initial current after being stored for 2 weeks. The researchers proposed that the hybrid electrode could detect other biological substances including proteins, enzymes and nucleic acids. Feng et al. Feng et al. [94] developed a PANI/Au composite hollow sphere and tested its catalytic efficiency using a glassy carbon electrode (GCE). The combination of PANI's excellent catalytic and electrochromic traits with high biocompatibility from Au nanoparticles creates an ideal sensor material. Testing confirmed that the PANI/Au composite showed exceptional electrocatalytic activity and suitability for detecting dopamine. The synthesized hybrid electrode showed durability by maintaining approximately 99% of its catalytic peak current after 100 DA cycles at 2 mM concentration. Research findings reveal that this composite retains 99% of its catalytic peak current after 100 DA detection cycles. Liu et al. developed the $Fe_3O_4$/PANI/CNT hybrid nanocomposite [95]. using in situ chemical polymerization. The research revealed that the developed nanocomposite demonstrated exceptional performance as a glucose sensor. CNT helps immobilize more glucose oxidase by increasing both surface area and conductivity of the hybrid material. This process diminishes glucose oxidase's redox reaction ability while increasing sensor sensitivity. The researchers took seven measurements of five milligrams of glucose to evaluate reproducibility. The glucose sensor hybrid demonstrated acceptable repeatability with a relative standard deviation of 5.8%. The creation of a glucose biosensor based on a Pt/PANI hydrogel heterostructure has been established by Zhai et al. [96] The manufactured heterostructure displayed exceptional sensitivity at 96.1 A $M^{-1}$ $cm^{-2}$ and quick detection times of 3 s to measure extremely low gas concentrations of 0.7 M. Xu et al. developed the Au/PANI/GO hybrid electrode for glucose sensor applications [97]. The synthesized substance enabled direct electron transfer between the electrode and glucose oxidase and maintained stability. The nanocomposite structure enables glucose detection at minimal concentrations of 0.6 M. High blood triglyceride levels are the root cause of serious health conditions including diabetes and heart attacks. It is essential to manage elevated triglyceride levels for health reasons.

The glucose biosensor developed by Yan et al. [98] involved adsorption of glucose oxidase enzyme onto the Au/AgCl/PANI modified hybrid electrode.

The glucose biosensor developed by Yan and his team utilizes the Au/AgCl/PANI modified hybrid electrode technology [99]. The small size and extensive specific surface area of Au nanoparticles allow them to detect glucose down to 4 M while maintaining high stability and good reproducibility. Scientists have completed extensive research into attaching DNA onto HCPs to detect various DNA target sequences and microbial species. Xu et al. developed the PPy/MWCNT hybrid film according to their research [100, 101]. Xu et al. created the PPy/MWCNT hybrid film and subsequently electropolymerized it on a glassy carbon electrode (GCE). Due to its substantial surface area and positive electrochemical traits the produced electrode became a suitable choice for DNA sensors. Khan et al. developed the PANI/chitosan (CS) hybrid film. The PANI/chitosan (CS) hybrid film was developed by Khan et al. [102] to detect the hazardous chemical ochratoxin-A. The synthetic hybrid film demonstrated three times better performance than PANI and chitosan in the detection of ochratoxin-A. The PANI/CS matrix contains a higher concentration of carboxylic and hydroxyl groups allowing rabbit antibodies (IgGs) to bind more effectively which increases sensitivity. Radhapyari et al. reported the development of a PANI/GO-based biosensor capable of detecting the antimalarial drug artesunate [103]. The developed biosensor showed outstanding capabilities in detecting minimal drug quantities. The synthesized biosensor displayed user-friendly operation while remaining budget-friendly and showed outstanding repeatability and reproducibility. Lee et al. developed the PPy/iron oxyhydroxide (FeOOH) hybrid nanoparticle for their research. Lee et al. [104] developed the PPy/iron oxyhydroxide (FeOOH) hybrid nanoparticle specifically to detect nerve gas (DMMP). The hybrid nanoparticle revealed outstanding sensitivity at room temperature to detect extremely low levels of DMMP gas measured at 0.1 ppb. The interaction between the composite material with DMMP and FeOOH creates larger functional sites through both H-bonding and charge transfer mechanisms. The PPy's large surface area provided enhanced sensitivity through better DMMP adsorbent site performance [105]. The consequences of diabetes mellitus make it a major global health challenge that exceeds the risks of diabetes itself today. Developing a rapid and precise diagnostic test for diabetes detection remains an essential priority. Multiple published methods for diabetes monitoring use blood glucose measurements [106]. Researchers identify enzymatic glucose sensors alongside enzyme-free glucose sensors as two principal types [107–109]. NSCP-integrated electrochemical biosensors play a significant role in the electrochemical detection of glucose and $H_2O_2$. Basu et al. detail how PANI NSs fabrication with sodium dodecyl sulphate (NSPANI-SDS) functions as a glucose and $H_2O_2$ biosensor [110]. These NSs polymers provide an extensive specific surface area which helps to boost PANI conductivity and enable straightforward immobilization with high concentrations of glucose oxidase (GOx) and horseradish peroxidase (HRP). In addition to the other benefits, these NSs boost both current responsiveness and electron transfer rate. Modified PANI NSs showed excellent selectivity and stability while maintaining reproducibility together with a broad linear range and rapid response time when applied as optical and electrical biosensors. Abidian et al. demonstrated that PEDOT nanofibers can detect glucose electrochemically as they capture glucose oxidase enzyme (GOx) during galvanostatic polymerization at the Pt electrode [111]. Munteanu et al. created an opto-electrochemical dual electrochemical sensor to detect glucose and hydrogen peroxide together. Munteanu et al. developed a dual electrochemical sensor that used optical microscopy to detect both glucose and hydrogen peroxide [112]. The interaction between hydrogen peroxide and glucose with the enzyme leads to changes in the sensor's oxidation state due to its osmium complex-based redox polymer foundation. Aksoy et al. developed a selective electrochemical biosensor for dopamine detection using polyimide (PI) and polyimide-boron nitride (PI-BN) composite membranes. Aksoy and colleagues developed a selective electrochemical dopamine biosensor using polyimide (PI) and polyimide-boron nitride (PI-BN) composites as a selective membrane [113]. The addition of BN particles to the PI matrix leads to improved sensitivity and selectivity along with enhanced reversibility (meaning faster electron transfer) while achieving a LOD around $4 \times 10^{-8}$ M. The combination of PANI with nanomaterials results in increased electrical conductivity and the capacity to act as scaffolds for biological organisms which offers substantial promise for sensor technology development [114]. The emergence of Zika virus (flavivirus) infections in newborns has spurred increased efforts to develop rapid detection tests. Tancharoen et al. have developed a new type of electrochemical biosensor for ZIKV detection [115]. The created biosensor consisted of graphene oxide/surface imprinted polymers (SIPs) composites and successfully detected the Zika virus in blood samples with dengue virus interference while achieving a limit of detection equal to the RT-PCR method. A serious medical disorder-linked cellular metabolite includes lactate. According to Inal et al., researchers developed a micrometer scale polymer-based transistor platform for lactate detection which uses hydrophilic side chains in n-type electron-transporting organic polymers to improve ion transport/injection and enzyme conjugation while also functioning as redox centers [116] J. M. Pingarro'n et al. The electrode surface was coated with a hybrid material made from multiwalled carbon nanotubes (MWCNTs) and conducting polymer poly (3-methylthiophene) (P3MT) which was then used to bind lactate dehydrogenase (LDH) enzyme to a glassy carbon electrode forming a lactate biosensor. The hybrid material enabled creation of a reliable LDH electrochemical biosensor capable of operating without a redox mediator at working potentials of +300 mV by providing exceptional electrochemical oxidation properties for NADH to the electrode. Optimized conditions allowed researchers to establish a linear calibration graph for lactate across a concentration range of $1.0 \times 10^{-6}$ to $5.0 \times 10^{-4}$ M ($r = 0.9993$) and achieved measurements with a detection limit of $5.6 \times 10^{-7}$ M. Wing Cheung Mak and his team developed methods to create electrochemical biosensors by using a sophisticated enzyme-hybrid conductive composite of GOx-PtNPs-PEDOT-MSs. Researchers synthesized a biohybrid conducting polymer via template-assisted chemical polymerization which formed PEDOT microspheres (PEDOT-MSs) and then conducted in situ PtNP deposition along with electrostatic immobilization of glucose oxidase (GOx) to obtain water-processable GOx-PtNPs-PEDOT-MSs [117]. The biosensor demonstrated a linear glucose response in the range from 0.1 to 10 mM as indicated by a correlation coefficient of 0.9855 and exhibited a sensitivity of 116.25 $\mu A\ mM^{-1}\ cm^{-2}$ along with a detection limit of 1.55 μM which was calculated as three times the standard deviation divided by sensitivity. The sensitivity of the new PEDOT-MS-based biosensor exceeds that of the most effective PEDOT-based glucose biosensor published in scientific literature by a 2.7-fold margin. Researchers found that the apparent Michaelis-Menten constant (Km app) for the GOx PtNPs-PEDOT-MS biosensors measured 7.3 mM. The biosensor preserved 97% of its sensitivity after being stored for 12 days which indicates good storage stability [118]. Susan Daniel developed an uncomplicated method for producing conductive polymer that enabled the creation of solvent-assisted lipid bilayers (SLBs) supported on PEDOT:PSS conducting polymer surfaces through biomimetic processes. The research team created bacterial and mammalian membrane mimetics with this bioelectronic device to test polimyxin B's antibiotic effects and α-hemolysin's toxin activity [119].

Table 10.1 outlines the features of several biosensors based on graphene, carbon nanotubes, and gold nanoparticle ICP in addition to providing a summary of different biosensor types.

**Table 10.1** Biosensing application of hybrid conducting polymer (HCPs) composites

| ICP | Nanomaterial | Target analyte | Linear range | Detection limit | Ref |
|---|---|---|---|---|---|
| POT | CNT | Urea | 0.1–11 mm | 0.03 mm | [120] |
| Poly(o-aminophenol) | CNT | Glucose | Up to 5 mm | 0.01 mm | [121] |
| Polyaniline | Chitosan-CNT | Glucose | 1–20 mm | 0.1 mm | [122] |
| Polypyrrole | cMWCNT | Glucose | 1–10 mm | 0.1 mm | [123] |
| Poly-5,2′-5′,2″-terthiophene-3′-carboxylic acid | MWCNT | Glucose | 5–90 µm | 1 µm | [124] |
| Polyaniline | SWCNT | Triglyceride | 50–400 mg $dL^{-1}$ | – | [125] |
| PVP | MWCNT | Glucose | 10–700 µm | 2 µm | [126] |
| Polyaniline | cMWCNT | $H_2O_2$ | 2–600 µmol $L^{-1}$ | 0.2 µmol $L^{-1}$ | [127] |
| Poly (3-methylthiophene) | MWCNT | Lactate | $1 \times 10^{-6}$ to $5 \times 10^{-4}$ m | $5.6 \times 10^{-7}$ m | [128] |
| Polyaniline | CNT | Nisseria Gonnorhoeae | $1 \times 10^{-6}$ to $5 \times 10^{-17}$ m | $1.2 \times 10^{-17}$ m | [129] |
| Polypyrrole | cMWCNT | Cholesterol | $4 \times 10^{-4}$ to $6.5 \times 10^{-3}$ m $L^{-1}$ | 0.04 mm $L^{-1}$ | [130] |
| Poly-5,2′:5″,2″terthiophene-3′ carboxylic acid | MWCNT | IgG | 0.1–10 ng $mL^{-1}$ | $0.084 \pm 0.004$ ng $mL^{-1}$ | [131] |
| Polyaniline | MWCNT | Pesticide | 10–50 nmol $L^{-1}$ | 5 nmol $l^{-1}$ | [132] |
| Polypyrrole | cMWCNT | DNA | – | $5.0 \times 10^{-12}$ mol $L^{-1}$ | [133] |
| Poly (o-methoxyaniline) | AuNP | Dopamine | 10.0–300.0 µm | 0.062 µm | [134] |
| | | Folic acid | 0.5–900.0 µm | 0.090 µm | |
| Polyaniline | AuNP-peptide | Prostate specific antigen | 1–100 ng $mL^{-1}$ | 0.68 ng $mL^{-1}$ | [135] |
| Polyaniline | AuNP | *E. coli* | – | $4x10^6$ to 4 CFU | [136] |
| 4-([2,2′:5′,2″-terthiophen] 3′-yl) benzoic acid | AuNP | Non-small-cell lung cancer cells | 15 to 1 × 106 cells $mL^{-1}$ | 8 cells $mL^{-1}$ | [137] |
| [2,2:5,2-terthiophene-3-(p benzoic acid)] | AuNP | Hypoxia inducible factor 1 alpha (HIF1α) | 25–350 pm $mL^{-1}$ | $5.35 \pm 0.02$ pm $mL^{-1}$ | [138] |

(continued)

**Table 10.1** (continued)

| ICP | Nanomaterial | Target analyte | Linear range | Detection limit | Ref |
|---|---|---|---|---|---|
| Polypyrrole | AuNP+TB | MicroRNA-21 | 100 $am^{-1}$ nm | 78 am | [139] |
| Functionalized Polyterthiophene | AuNP | Brain-derived neutrophilic factor | 4.0 to 600.0 pg $mL^{-1}$ | 1.5 ± 0.012 pg $mL^{-1}$ | [140] |
| Polyaniline nanowire | AuNP | Glucose ssDNA strand Lamin A protein | 1–20 mm | 1 μm | [141] |
| Polypyrrole | Kappa-carrageen-AuNP | DNA of fish | 5 × 10–18 to 5 × 10–12 m | $5 \times 10^{-18}$ m | [142] |
| Polypyrrole | AuNP | DNA | 2 × 10–13 to 2 × 10–6 m | $8.4 \times 10^{-13}$ m | [143] |
| Polyaniline | AuNP | Uric acid | 4.0 × 10–7 to $6 \times 10^{-5}$ mol $L^{-1}$ | $2 \times 10^{-7}$ mol $L^{-1}$ | [144] |
| | | Epinephrine | 4.0 × 10–7 to 1 × 10–5 mol $L^{-1}$ | $8 \times 10^{-8}$ mol$L^{-1}$ | |
| Polypyrrole hydrogel | AuNP | Carcinoembryonic antigen | 1 fg mL − 1 to 200 ng mL − 1 | 0.16 fg $mL^{-1}$ | [145] |
| Polypyrrole | Luminol functional-AuNP | Carcinoembryonic antigen | 0.01 pg $mL^{-1}$ to 10 ng $mL^{-1}$ | 3 fg $mL^{-1}$ | [146] |
| PEDOT:PSS | GR + MIP | Salbutamol | 1 nm–1.2 μm | 100 pm | [147] |
| Poly p-phenylene diamine | GR | Neuron-specific enolase biomarker | 1.0–1000 ng $mL^{-1}$ | 0.3 ng $mL^{-1}$ | [148] |
| PEDOT | Laser scribbed GR | Dopamine | −150 μm | 0.33 μm | [149] |
| Polypyrrole | Sulfonated GR | Glucose | 0.02–12 mm | 0.02 mm | [150] |
| Polypyrrole | GR | Uric acid | 0.002–0.024 μmol $L^{-1}$ | 0.541 nmol $L^{-1}$ | [151] |
| Poly 2-amino-5-mercapto- 1,3,4-thiadiazole | rGO | Tyrosine | 0.1–400 μm | 0.046–0.0032 μm | [152] |
| | | Uric acid | 0.01–100 μm | | |
| PEDOT | rGO | Dopamine | 0.1–175 μm | 39 nm | [153] |
| Poly(sulfanilic acid) | rGO | Levofloxacin | 2.0–30.0 and 30.0–500.0 μm | 0.12 μm | [154] |

| | | | | | |
|---|---|---|---|---|---|
| Polyaniline | GR | Organophosphate pesticide | – | 20 ng $L^{-1}$ | [155] |
| PEDOT | GO | Hydrazine | 0.1–1.9 mm | 2.2 μm | [156] |
| | | H2O2 | 0.2–3.2 mm | 3.2 μm | |
| Poly (vinylferrocene) | GR | L-lysine | $9.9 \times 10^{-7}$—$3.1 \times 10^{-4}$ m | $2.3 \times 10^{-7}$ m | [157] |
| Polyaniline | GR | H2O2 | $1.0 \times 10^{-6}$ to $1.6 \times 10^{-4}$ m | $0.8 \times 10^{-7}$ m | [158] |
| Polyaniline nanowires | GO | DNA | – | $3.25 \times 10^{-13}$ Mol $L^{-1}$ | [159] |
| Poly(ionic liquids) functionalized Polypyrrole | GO | Dopamine | 4–18 μm | 73.3 nm | [160] |
| Poly (o-anisidine) | GR | NADH | – | 1.3 μm | [161] |
| Polyvinyl pyrrolidone/polyaniline | GR | Cholesterol | 50 μm–10 mm | 1.0 μm | [162] |

**Abbreviations:** The document uses these abbreviated terms: Carbon nanotubes (CNTs), single-walled CNT (SWCNT), multiwalled CNT (MWCNT), carboxy functionalized MWCNT (cMWCNT), immunoglobulin G (IgG), poly(o-toluidine) (POT), polyvinylpyridine (PVP)

## 10.10 HCPs as Gas Sensor

Gas sensors find application in multiple fields including environmental monitoring and medical treatment [163–167]. Interfacial polymerization created PANI nanofibers containing camphor sulfonic acid (CSA) show enhanced ammonia ($NH_3$) detection abilities compared to traditional PANI films according to Menegazzo et al. [168] The dopant used in PANI improves its sensitivity through increased permeability and conductivity. The performance of conducting polymer matrices as gas sensors is enhanced through the incorporation of metal salts according to research [169, 170]. Sadek et al. Sadek et al. researched the sensitivity of PANI–$In_2O_3$ hybrid composite nanofibers to detect $H_2$, CO, and $NO_2$ gases at room temperature following their creation through chemical polymerization. The newly synthesized material outperformed other tested materials like CO and $NO_2$ in terms of $H_2$ sensitivity. Two methods can achieve the detection of $H_2$ gas. The initial role of $In_2O_3$ involves generating indium–dihydrogen complexes that activate hydrogen gas. Chain alignment occurs as a result of PANI interacting with $In_2O_3$ which enables hydrogen to dissociate. Virji et al. In separate research, Virji and colleagues created a PANI/$CuCl_2$ composite to detect $H_2S$ gas. The synthetic hybrid material demonstrated enhanced detection effectiveness for $H_2S$ gas reaching up to four orders of magnitude based on the author's results. The film's rapid gas diffusion is enabled by PANI's small size together with its high surface area. Vijayakumar et al. reported that the combination of PANI and Poly vinyl pyrrolidone demonstrated higher sensitivity to $CHCl_3$ compared to $CH_2Cl_2$ [171–173]. PANI demonstrates lower sensitivity to $CHCl_3$ in comparison to its response to $C_6H_5Cl$ and $CH_2Cl_2$. The addition of PVP boosts the sensitivity of this material to $CHCl_3$. The $NH_3$ gas sensing properties of Reduced graphene oxide (RGO)/PANI hybrids were studied in another research work [174]. The research indicated that the hybrid sensor demonstrated a tenfold increase in response to NH3gas compared to bare graphene and virgin PANI sensors. Furthermore, Kaushik et al. Kaushik et al. [175] generated the hybrid PANI-$WO_3$ nanocomposite through vacuum thermal evaporation and tested its effectiveness at detecting $NO_x$ gas. The synthesis produced nanocomposites which exhibited an extremely fast response time of 30 seconds while demonstrating strong effectiveness for $NO_x$ gas detection. The ability to detect NOx gases improves when $NO_x$ interacts with PANI's p-electrons which results in higher conductivity and sensitivity. The research by Jian et. Al [176] found that PANI-$TiO_2$-Au ternary self-assembly nanocomposite thin films at room temperature provided superior $NH_3$-sensing performance compared to PANI-$TiO_2$ films. Researchers utilized an in situ self-assembly method to create a thin layer of PANI-$TiO_2$-Au ternary nanocomposite. PANI wrapping around $TiO_2$ nanoparticles and Au nanorods formed a traditional "core–shell" structure which led to the examination of probable physical-chemical interactions among the components. The PANI-$TiO_2$-Au ternary thin film sensor demonstrated superior performance with enhanced response levels and quicker response/recovery times compared to the PANI-$TiO_2$-based sensor. Zampetti et al. [177] developed a hybrid sensor made up of a nanofibrous layer of TiO2 coated with a thin film of PEDOT: Researchers examined polystyrene sulfonate (PEDOT:PSS) to evaluate its potential for NO gas detection. The conductometric sensor demonstrated detection of NO2 gas at concentrations down to 1 ppb through CrO3's stoichiometric oxidation of NO which makes it suitable for asthma monitoring applications. Shin et al. Shin et al. [178] created a sensor that functions at room temperature for detecting harmful $NO_2$ gas using multidimensional porous $Fe_2O_3$ nanorod-decorated carbon nanoparticles (MPFCNPs) which were produced through dual-nozzle electrospray followed by thermal stirring and heat treatment. The MPFCNP sensor demonstrated exceptional sensitivity when tested at 1 ppm concentration. The sensitivity increased when MPFCNP diameters shrank while $Fe_2O_3$ nanorod lengths grew. Huang et al. Huang et al. [179] synthesized PTh/$WO_3$ organic-inorganic hybrid materials through in situ chemical oxidative polymerization. Pure PTh exhibited less thermal stability compared to PTh/$WO_3$ hybrids which showed high response and strong selectivity for $NO_2$ detection at ppm levels with the strongest response observed for the 10 wt% PTh/$WO_3$ hybrid at 70 °C. Wei et al. Wei et al. [180] suggested a sulfonated poly(ether ether ketone) (SPEEK)/PPy core/shell nanofiber design to create an ultrasensitive and flexible NH3 sensor. SPEEK acted as a polymeric adsorbent which delivered stable and clear responses when exposed to extremely low $NH_3$ concentrations (20 ppb) up to a maximum value of 3.8. The reference material polyacrylonitrile (PAN)/PPy nanofibers showed limited response to NH3 gas (above 5 ppm). The PAN/PPy displayed reaction intensity below one twelfth compared to SPEEK/PPy when exposed to 10 ppm $NH_3$. An electrospinning process served as the base for creating the SPEEK which led to exceptional mechanical flexibility. Excellent anti-interference characteristics were achieved by introducing 100 ppm of each $H_2$, CO, $C_2H_2$, $CH_4$, methanol, ethanol, toluene, and $NH_3$ in sequence into the chamber at room temperature for real-world application (Fig. 10.1).

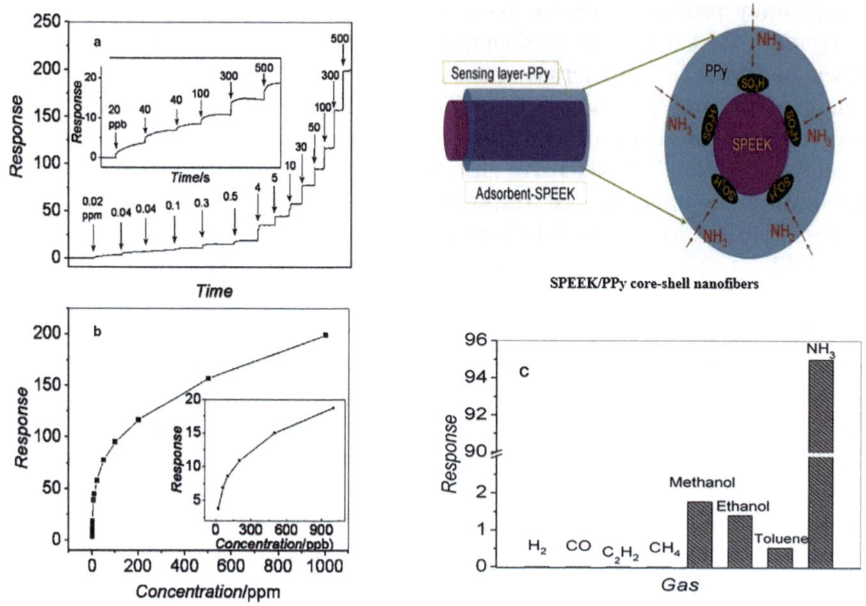


**Fig. 10.1** Illustration of the SPEEK/PPy core–shell nanofibers for $NH_3$ sensor; (**a**) real-time responses to $NH_3$ in the range of 0.02–500 ppm (inset: The sensor response to NH3 concentration is shown in (**b**) with an inset depicting the response curve between 20 and 1000 ppb). The SPEEK/PPy nanofibers demonstrated anti-interference capability when exposed to 100 ppm concentrations of $H_2$ $CH_4$ $C_2H_2$ CO methanol ethanol toluene and $NH_3$. (Copyright © 2012, American Chemical Society [180])

## 10.11 Engineering Functional Conductive Hydrogels

The formation of conductive hydrogels occurs through processes such as self-polymerizing conductive polymers/fillers, creating intertwined hydrogel networks by doping conductive polymers/fillers, employing free ions, and integrating conductive fillers/free ions into a non-conductive hydrogel matrix (Fig. 10.2). The classification of hydrogels depends on their conductive mechanism and includes electronic, ionic, and hybrid types [182–184]. Conductive polymer-based hydrogels function through electronic conduction because of the conjugated $\pi$ bond structure present in polymers such as PEDOT, PANI, polypyrrole (PPy), and polythiophene (PT) [185–187]. Conductive hydrogels with shape memory capabilities show great potential for applications in bioelectronic sensors and soft robotics as well as actuators [188–190]. Wang et al. Wang et al. [191] created advanced conductive hydrogels that exhibit robustness and fatigue resistance through interpenetrating networks of polyaniline with P(AAm-co-HEMA). These hydrogels demonstrate high sensitivity (GF = 11) when subjected to low strains together with outstanding linear behavior at high strains. These exceptional advantages

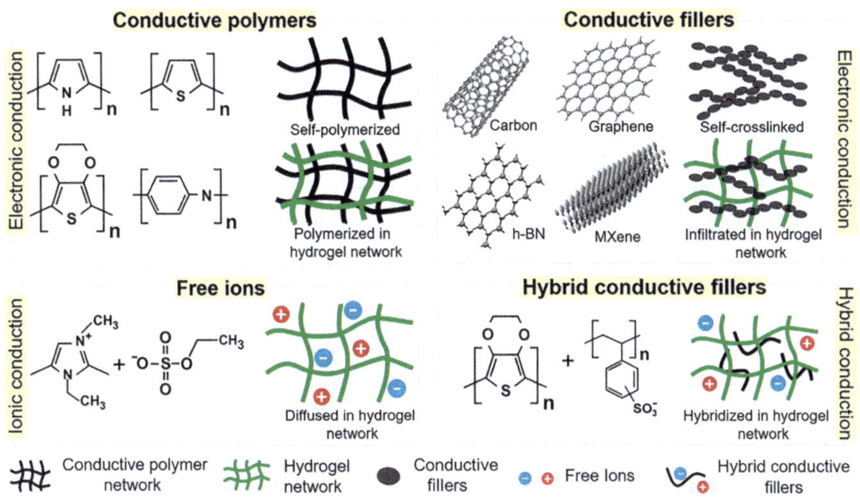


**Fig. 10.2** Multiple conductive hydrogel types display distinct structural properties. The creation of conductive hydrogels involves merging conductive polymers with conductive fillers and free ions along with their different combinations. Hydrogels exhibit electronic, ionic, or combined electronic-ionic conducting properties. (Copyright © 2020, American Chemical Society [181])

provide precise and reliable detection capabilities for small critical signals alongside human motion movements. Researchers constructed a sensor array using conductive hydrogels to detect small stresses in two-dimensional space. The strong conductive hydrogels could become essential materials for electronic skin applications as well as biomedical implants and movement monitoring technologies. Prof. Mingjie Liu and his team [192] created organohydrogels with anti-freezing properties as conductive materials using an $H_2O$/ethylene glycol binary fluid for dispersion. The organohydrogels demonstrate resilience to freezing temperatures which enables them to maintain flexible and strain-responsive properties from −55.0 to 44.6 °C. Solvent molecules formed hydrogen bonds with polyvinyl alcohol (PVA) chains which led to PVA crystallization and resulted in the organohydrogels gaining significant mechanical strength. The non-covalent cross-links provide conductive organohydrogels with excellent remoldability and self-healing abilities that prove valuable for real-world applications. The research by Meidan et al. [193] showed how soft conductive hydrogel comes from mixing silk fibroin with polyacrylamide along with graphene oxide and poly(3,4-ethylenedioxythiophene):poly(styrenesulfonate) in specific proportions. Because of its significant stretchability and compressibility, the [PAM-SF/GO-PEDOT: PSS hydrogel-based strain/pressure sensors demonstrate wide sensing capabilities with ranges from 2% to 600% strain and 0.5 to 119.4 kPa pressure while maintaining reliable stability.

## 10.12 Future Direction and Machine Learning Algorithms

Research on multifunctional and high-performance conducting nanocomposites stands as an engaging field that has recently gained substantial interes [6]. The combination of conducting polymers with carbon-based materials stands out as a highly promising nanocomposite system because of their unique property set. The insolubility of graphene in aqueous solutions and its propensity to clump together because of van der Waals interactions between its sheets restrict its potential applications. The design precision of PANI/Gr nanocomposite requires enhancement through advanced machine learning models and molecular input insights paired with outlier resilience to achieve superior electrical conductivity and gas sensing capabilities. Inas M. AlNashef built machine learning models to understand electrical conductivity and gas sensing responses of PANI/Gr nanocomposites using datasets from over 100 references [9–11, 194]. The research aims to detect and minimize structural and response-related outliers while validating the strength of suggested ANN models. The research emphasizes how application domain (AD) analysis plays a crucial role in establishing the dependability of models. The results in great accuracy for electrical conductivity (RMSE: The machine learning models showed high accuracy for electrical conductivity (RMSE: 0.408, R2: 0.984) and demonstrated precise gas sensing performance for ammonia (RMSE: 0.350, R2: 0.967), toluene (RMSE: 0.232, R2: 0.983), and benzene gases (RMSE: 0.081, R2: 0.976). Input contribution analysis determines vital parameters which affect performance outcomes. Chemical inputs become essential to machine learning models as the σ-profiles of additives emerge as major contributing factors. Research outcomes promote the creation of superior PANI/Gr nanocomposites which may be used in supercapacitors and energy storage devices as well as gas sensors (Fig. 10.3).

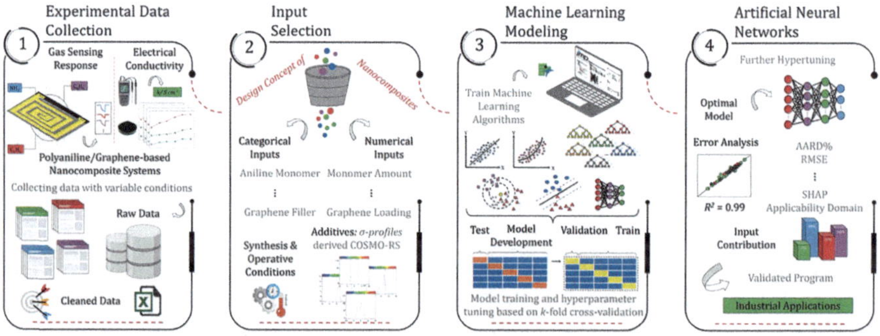


**Fig. 10.3** The process of building dependable models to predict electrical conductivity and gas sensing response through machine learning methods receives a complete description. (The Royal Society of Chemistry retains all copyright privileges for this work in 2024 [194])

## 10.13 Conclusion

Scientific research has extensively investigated HCPs for their application in sensor and biosensor technologies. HCPs created through the fabrication of conducting and intrinsic conducting materials with dopants like nanomaterials and metal oxides show improved sensing capabilities. The use of conducting polymers with nanomaterial dopants leads to enhanced sensing capabilities including rapid response times and low detection limits while maintaining high repeatability and selectivity alongside simple production methods. Now days sensors and biosensors are becoming more essential to track human health status, ecological conditions, food quality and safety standards the advancement of these systems into user-friendly formats remains crucial. The development of conducting polymers for flexible organic electronics makes them strong contenders for wearable platform components. New dopants synthesized with HCPs create unique properties that establish these materials as exceptional platforms for sensor and biosensor development.

The toxicity of nanomaterial doped HCPs becomes a critical concern when applied in sensing applications within living organisms. Research shows carbon-based nanomaterials as dopants in HCPs display little to no cytotoxicity in some studies but demonstrate potential toxicity and adverse effects both acutely and chronically in other studies. The inherent drawbacks of GO oxide-based HCPs such as poor solubility in water and their tendency to agglomerate combined with advanced machine learning models alongside outlier resilience and molecular input insights facilitate the enhancement of electrical conductivity and gas sensing performance to precisely design PANI/Gr nanocomposites.

**Conflicts of Interest** There are no conflicts to declare.

**Contribution of the Authors** Dr. Sahid Sk and Dr. Sathish Manda writing, editing, conceptualize the book chapter. Dr. Sathish Manda, Dr. B V Bhaskara Rao and Dr. Suman Chirra conceptualize the book chapter and helping in corrections during review of the book chapter. Dr. Nasseb Singh helps in writing the book chapter.

## References

1. Shirakawa H, Ikeda S (1971) Synthesis of electrically conducting organic polymers: halogen derivatives of polyacetylene (CH)x. Polym J 2:231
2. Shirakawa H, Ito T, Ikeda S (1973) Synthesis of electrically conducting organic polymers: polyacetylene, (CH)x. Polym J 4:460
3. Ito T, Shirakawa H, Ikeda S (1974) Synthesis of electrically conducting organic polymers: polyacetylene, (CH)x. J Polym Sci Part A-1: Polym Chem 12:11
4. Ito T, Shirakawa H, Ikeda S (1975) Synthesis of electrically conducting organic polymers: polyacetylene, (CH)x. J Polym Sci Part A-1: Polym Chem 13:1943
5. Shirakawa H, Louis EJ, MacDiarmid AG, Chiang CK, Heeger AJ (1977) Synthesis of electrically conducting organic polymers: halogen derivatives of polyacetylene (CH)x. J Chem Soc Chem Commun 1977:578–580

6. Raza S, Li X, Soyekwo F, Liao D, Xiang Y, Liu C (2021) Recent advances in the synthesis, properties and applications of conducting polymer nanocomposites. Eur Polym J 160:110773
7. Paul DR, Robeson LM (2008) Polymer nanotechnology: nanocomposites. Polymer 49:3187–3204
8. Ahuja T, Kumar D (2009) Recent progress in the development of nano-structured conducting polymers/nanocomposites for sensor applications. Sensors Actuators B Chem 136:275–286
9. Meer S, Kausar A, Iqbal T (2016) Conducting polymer nanocomposites: a comprehensive review. Polym-Plast Technol Eng 55:1416–1440
10. Hong X, Fu J, Liu Y, Li S, Wang X, Dong W, Yang S (2019) Recent advances in conducting polymer-based nanocomposites for electrochemical sensors. Materials 12:1451
11. Zhao H-Y, Yu M-Y, Liu J, Li X, Min P, Yu Z-Z (2022) Recent progress in flexible and stretchable conductive polymer composites. Nano-Micro Lett 14:129
12. Tian Z, Yu H, Wang L, Saleem M, Ren F, Ren P, Chen Y, Sun R, Sun Y, Huang L (2014) Conducting polymer nanocomposites: a review of recent advances. RSC Adv 4:28195–28208
13. Lee M-C, Simkovich G (1987) Metallurgical applications of conducting polymers. Metall Trans A 18:485–486
14. Le TH, Kim Y, Yoon H (2017) Electrical and electrochemical properties of conducting polymers. Polymers 9:150
15. MacDiarmid AG, Mammone RJ, Kaner RB, Porter SJ, Pethig R, Heeger AJ, Rosseinsky DR (1985) Conducting polymers and their applications. Phil Trans R Soc A 314:3–15
16. Brédas JL, Street GB (1985) Polarons, bipolarons, and solitons in conducting polymers. Acc Chem Res 18:309–316
17. Roth S, Bleier H (1987) Physics of conducting polymers. Adv Phys 36:385–462
18. Bajpai M, Srivastava R, Dhar R, Tiwari RS (2016) Review on optical and electrical properties of conducting polymers. Indian J Mater Scie 2016:5842763
19. Nollau M, Pfeiffer M, Fritz T, Leo K (2000) Electronic properties of organic semiconductors. J Appl Phys 87:4340–4343
20. Zhang Y, Blom PWM (2011) Charge transport in organic semiconductors. Appl Phys Lett 98:2011–2014
21. Patil AO, Heeger AJ, Wudl F (1988) Conducting polymers: fundamentals and applications. Chem Rev 88:183–200
22. Yussuf A, Al-Saleh M, Al-Enezi S, Abraham G (2018) Conducting polymer nanocomposites: synthesis and applications. Int J Polym Sci 2018:4191747
23. Baughman RH, Brédas JL, Chance RR, Elsenbaumer RL, Shacklette LW (1982) Conducting polymers: a review. Chem Rev 82:209–222
24. Grama A (2018) Laboring Along:27–78
25. Su WP, Schrieffer JR, Heeger AJ (1979) Solitons in polyacetylene. Phys Rev Lett 42:1698–1701
26. Borchert H (2014) Solar cells based on colloidal nanocrystals. Springer Ser Mater Sci 196
27. Wang CL, Wang WZ, Gu GL, Su ZB, Yu L (1994) Molecular crystals and liquid crystals. Mol Cryst Liq Cryst Sci Technol Sect A 256:903–908
28. Mzenda VM, Goodman SA, Auret FD (2002) Electrical properties of conducting polymers. Synth Met 127:285–289
29. Gosh M, Barman A, Meikap AK, De SK, Chatterjee S (1999) Electrical and magnetic properties of conducting polymers. Phys Lett A 260:138–148
30. Chandrasekhar P (2018) Conducting polymers: fundamentals and applications, pp 810–815
31. Chandrasekhar P, Chandrasekhar P (1999) Conducting polymers: fundamentals and applications. Conduct Polym Fundam Appl 1999:143–172
32. Guimard NK, Gomez N, Schmidt CE (2007) Conducting polymers in biomedical engineering. Prog Polym Sci 32:876
33. Beadle P, Armes SP, Gottesfeld S, Mombourquette C, Houlton R, Andrew WD (1992) Macromolecular structure of conducting polymers. Macromolecules 25:2526

34. Chen SA, Tsai CC (1993) Macromolecular structure of conducting polymers. Macromolecules 26:2234
35. Kumar D, Sharma RC (1998) Conducting polymers: synthesis and applications. Eur Polym J 34:1053
36. Sindhu S, Jegadesan S, Hairong L, Ajikumar PK, Vetrichelvan M, Valiyaveettil S (2007) Conducting polymers for advanced functional materials. Adv Funct Mater 17:1698
37. Peng X, Zhang L, Chen Y, Li F, Zhou W (2010) Surface applications of conducting polymers. Appl Surf Sci 256:2948
38. Mostafaei A, Zolriasatein A (2012) Progress in conducting polymer nanocomposites. Prog Nat Sci Mater Int 22:273
39. Qiu JD, Shi L, Liang RP, Wang GC, Xia XH (2012) Conducting polymers in analytical chemistry. Chem—A Eur J 18:7950
40. Tamaki R, Chujo Y (1998) Application of organometallic polymers. Appl Organomet Chem 12:755
41. Kickelbick G (2007) Hybrid materials. Hybrid Mater:1
42. Hsiue G, Kuo W, Huang Y, Jeng R (2000) Synthesis and properties of conducting polymers. Polymer 41:2813
43. Honma S, Hirakawa K, Yamada JMB (1999) Solid-state ionics in conducting polymers. Solid State Ionics 118:29
44. Honma Y, Takeda JMB (1999) Solid-state ionics in conducting polymers. Solid State Ionics 120:255
45. Wu C, Xu T, Yang W (2005) Synthesis and characterization of conducting polymers. Eur Polym J 41:1901
46. Schubert U, Huesing N, Lorenz A (1995) Sol-gel processing of hybrid materials. Chem Mater 7:2010
47. Schottner G (2001) Hybrid polymers—state of the art. Chem Mater 13:3422
48. Xiong S, Phua SL, Dunn BS, Ma J, Lu X (2010) Hybrid organic–inorganic materials. Chem Mater 22:255
49. Kim YS, Oh SB, Park JH, Cho MS, Lee Y (2010) Solar energy applications of conducting polymers. Sol Energy Mater Sol Cells 94:471
50. Shun JL, Zheng X (2010) Sol-gel science and technology for conducting polymers. J Sol-Gel Sci Technol 54:174
51. Clement S, Tizit A, Desbief S, Mehdi BA, De Winter J, Gerbaux P, Lazzaroni B, Boury R (2011) Synthesis and characterization of hybrid materials. J Mater Chem 21:2733
52. Wang Q, Liu N, Wang X, Li J, Zhao X, Wang F (2003) Macromolecular structure and properties of conducting polymers. Macromolecules 36:5760
53. Wang TL, Yang CH, Shieh YT, Yeh AC (2009) Synthesis and properties of conducting polymers. Eur Polym J 45:387
54. Mumtaz M, de Cuendias A, Putaux JL, Cloutet E, Cramail H (2006) Synthesis and characterization of conducting hydrogels. Macromol Rapid Commun 27:1446
55. Amaike M, Yamamoto H (2006) Synthesis of conducting polymers. Polym J 8:703
56. Abd El-Ghaffar MA, Youssef AM, Abd El-Hakim AA (2014) Conducting polymer nanocomposites. Arab J Chem 8:771
57. Dai C, Chang C, Chi H, Chien H, Su W, Chiu W (2008) Synthesis and characterization of polyaniline-based materials. J Polym Sci A Polym Chem 46:2536
58. Guimard NK, Gomez N, Schmidt CE (2007) Progress in polymer science: conducting polymers in biomedical engineering. Prog Polym Sci 32:876
59. Diaz AF, Kanazawa K (1979) Electrochemical properties of conducting polymers. J Chem Soc Chem Commun 14:635
60. Heinze J, Frontana-Uribe BA, Ludwigs S (2010) Electrochemistry of conducting polymers—persistent models and new concepts. Chem Rev 110:4724
61. Wang DW, Li F, Zhao J, Ren W, Chen ZG, Tan J, Wu ZS, Gentle I, Lu GQ, Cheng HM (2009) Carbon-based nanomaterials for energy storage. ACS Nano 3:1745

62. Miyazaki T, Kim SK, Hoshino K (2006) Synthesis and properties of conducting polymers. Chem Mater 18:5302
63. Liu YC, Huang JM, Tsai CE, Chuang TC, Wang CC (2004) Chemical physics of conducting polymers. Chem Phys Lett 387:155
64. Zhang Y, Wu L, Lei W, Xia X, Xia M, Hao Q (2014) Electrochemical properties of conducting polymers. Electrochim Acta 146:568
65. Cai GF, Tu JP, Zhou D, Zhang JH, Wang XL, Gu CD (2014) Solar energy materials and conducting polymers. Sol Energy Mater Sol Cells 122:51
66. Domínguez-Crespo MA, Torres-Huerta AM, Onofre-Bustamante E, Alanis-Valdelamar A, Escudero-Rincón ML, Brachetti-Sibaja SB (2014) Conducting polymers in solid-state electrochemistry. J Solid State Electrochem 19:1073
67. Zhang F, Cao H, Yue D, Zhang J, Qu M (2012) Inorganic chemistry of conducting polymers. Inorg Chem 51:9544
68. Lu X, Zhang F, Dou H, Yuan C, Yang S, Hao L, Shen L, Zhang L, Zhang X (2012) Electrochemical properties of conducting polymers. Electrochim Acta 69:160
69. Wang J, Lu L, Shi D, Tandiono R, Wang Z, Konstantinov K, Liu H (2013) Conducting polymers in chemical applications. ChemPlusChem 78:318
70. Wang B, Qiu J, Feng H, Sakai E (2015) Electrochemical properties of conducting polymers. Electrochim Acta 151:230
71. Zhou Y, Lachman N, Ghaffari M, Xu HDP, Bhattacharya D, Fattahi P, Abidian MRR, Wu S, Gleason KKK, Wardle BLL, Zhang QMM (2014) Materials chemistry of conducting polymers. J Mater Chem A 2:9964
72. Ramya R, Sivasubramanian R, Sangaranarayanan MV (2013) Electrochemical properties of conducting polymers. Electrochim Acta 101:109
73. Liu C, Cai Z, Zhao Y, Zhao H, Ge F (2016) Conducting polymers in cellulose composites. Cellulose 23:637
74. Yesi Y, Shown I, Ganguly A, Ngo TT, Chen LC, Chen KH (2016) Sustainable chemistry of conducting polymers. ChemSusChem 9:370
75. Sun J, Shu X, Tian Y, Tong Z, Bai S, Luo R, Li D, Chen A (2017) Sensors and actuators with conducting polymers. Sensors Actuators B Chem 238:510
76. Bai S, Zhao Y, Sun J, Tong Z, Luo R, Li D, Chen A (2017) Sensors and actuators with conducting polymers. Sensors Actuators B Chem 239:131
77. Barsan MM, Pifferi V, Falciola L, Brett CMA (2016) Analytical chemistry of conducting polymers. Anal Chim Acta 927:35
78. Swathy TS, Jose MA, Antony MJ (2016) Conducting polymers in advanced polymer science. Polymer 103:206
79. Gülce H, Eskizeybek V, Haspulat B, Sari F, Gülce A, Avci A (2013) Industrial chemistry of conducting polymers. Ind Eng Chem Res 52:10924
80. Thakur OP, Agrawal N (2014) Sensor applications of conducting polymers. IEEE Sensors J 14:2665
81. Ma J, Zhang J, Wang S, Wang Q, Jiao L, Yang J, Duan X, Liu Z, Lian J, Zheng W (2011) Crystal engineering of conducting polymers. CrystEngComm 13:6077
82. Park HW, Kim T, Huh J, Kang M, Lee JE, Yoon H (2012) Nanostructures in conducting polymers. ACS Nano 6:7624
83. Yoon J, Hong H, Jang JY (2007) Nanomaterials based on conducting polymers. Small 3:1774
84. Braik M, Barsan MM, Dridi C, Ben Ali M, Brett CMA (2016) Electrochemical sensors based on conducting polymers. Sensors Actuators B Chem 236:574
85. Eising M, Cava CE, Salvatierra RV, Zarbin AJG, Roman LS (2017) Sensors based on conducting polymer composites. Sensors Actuators B Chem 245:25
86. Rajesh T, Kumar AD (2009) Sensor applications of conducting polymers. Sensors Actuators B Chem 136:275
87. Nambiar S, Yeow JTW (2011) Biosensors based on conducting polymers. Biosens Bioelectron 26:1825

88. Timmer B, Olthuis W, Van Den Berg A (2005) Chemical sensors using conducting polymers. Sensors Actuators B Chem 107:666
89. Aussawasathien D, Dong JH, Dai L (2005) Synthesis of conducting polymer nanofibers. Synth Met 154:37
90. Fiorito PA, De Torresi SIC (2005) Electroanalytical chemistry of conducting polymers. J Electroanal Chem 581:31
91. Baghayeri M, Zare EN, Lakouraj MM (2014) Biosensors based on polyaniline nanostructures. Biosens Bioelectron 55:259
92. Liu J, Tian S, Knoll W (2005) Surface chemistry of conducting polymers. Langmuir 21:5596
93. Qian T, Yu C, Zhou X, Wu S, Shen J (2014) Conducting polymers for chemical sensors. Sensors Actuators B Chem 193:759
94. Feng X, Mao C, Yang G, Hou W, Zhu J-J (2006) Conducting polymers in biosensing. Langmuir 22:4384
95. Liu Z, Wang J, Xie D, Chen G (2008) Nanostructured conducting polymers. Small 4:462
96. RZ, Zhai GYD, Liu B, Shi Y, Pan L, Wang Y, Li W (2013) Conducting polymer nanocomposites. ACS Nano 7:3540
97. Xu Q, Gu S-X, Jin L, Zhou Y, Yang Z, Wang W, Hu X (2014) Conducting polymers in chemical sensors. Sensors Actuators B Chem 190:562
98. Yan W, Feng X, Chen X, Hou W, Zhu JJ (2008) Conducting polymers in biosensors. Biosens Bioelectron 23:925
99. Singh R, Prasad R, Sumana G, Arora K, Sood S, Gupta RK, Malhotra BD (2009) Conducting polymers for bio-recognition. Biosens Bioelectron 24:2232
100. Booth MA, Harbison S, Travas-Sejdic J (2011) Electrochemical DNA sensors based on conducting polymers. Biosens Bioelectron 28:362
101. Xu Y, Ye X, Yang L, He P, Fang Y (2006) Electrochemical behavior of DNA at conducting polymer modified electrodes. Electroanalysis 18:1471
102. Khan R, Dhayal M (2009) Conducting polymer-based biosensors for detection of DNA hybridization. Biosens Bioelectron 24:1700
103. Radhapyari K, Kotoky P, Das MR, Khan R (2013) Conducting polymer nanocomposites for biosensing applications. Talanta 111:47
104. Lee JS, Shin DH, Jun J, Jang J (2013) Conducting polymer nanostructures for biosensor applications. ACS Nano 7:10139
105. Yang S, Li G, Wang D, Qiao Z, Qu L (2017) Conducting polymer-based sensors for environmental monitoring. Sens Actuator B 238:588
106. Zhu H, Li L, Zhou W, Shao Z, Chen X (2016) Conducting polymer nanocomposites for biomedical applications. J Mater Chem B 4:7333
107. Elkholy AE, Heakal FE-T, El-Said WA (2019) Application of conducting polymers in catalysis. Appl Catal A Gen 580:28
108. Lee J-H, El-Said WA, Oh B-K, Choi J-W (2009) Conducting polymers for biosensor applications. Electroanalysis 21:386
109. Salazar P, Rico V, González-Elipe AR (2016) Conducting polymers in chemical sensors. Sens. Actuator B 226:436
110. Saini D, Chauhan R, Basu T (2011) Int Res J Biotechnol 2:145
111. Yang G, Kampstra KL, Abidian MR (2014) Conducting polymers for neural interfaces. Adv Mater 26:4954
112. Munteanu R-E, Ye R, Polonschii C, Ru A, Gheorghiu M, Gheorghiu E, Boukherroub R, Schuhmann W, Melinte S, Gáspár S (2019) Conducting polymers in biosensing. Sci Rep 9:15196
113. Aksoy B, Pasahan A, Güngör Ö, Köytepe S, Seçkin T (2017) Conducting polymer composites for biomedical applications. Int J Polym Mater Polym Biomater 66:203
114. El-Said WA, Yea C-H, Kim H, Choi J-W (2009) Conducting polymers for cell-based biosensors. Curr Appl Phys 9:76

115. Tancharoen C, Sukjee W, Thepparit C, Jaimipuk T, Auewarakul P, Thitithanyanont A, Sangma C (2019) Conducting polymer-based sensors for virus detection. ACS Sens 4:69
116. Pappa AM, Ohayon D, Giovannitti A, Maria IP, Savva A, Uguz I, Rivnay J, McCulloch I, Owens RM, Inal S (2018) Organic electrochemical transistors based on conducting polymers. Sci Adv 4:eaat0911
117. Aguí L, Eguilaz M, Peña-Farfal C, Yáñez-Sedeño P, Pingarrón JM (2009) Conducting polymer-based electrochemical sensors. Electroanalysis 21:386
118. Liu Y, Turner APF, Zhao M, Mak WC (2018) Conducting polymers for biosensors. Biosens Bioelectron 100:374
119. Su H, Liu H-Y, Pappa A-M, Hidalgo TC, Cavassin P, Inal S, Owens RM, Daniel S (2019) Conducting polymers in bioelectronics. ACS Appl Mater Interfaces 11:43799
120. Hassan RYA, Kamel AM, Hashem MS, Hassan HNA, Abd El-Ghaffar MA (2018) Recent advances in biosensors for detection of hydrogen peroxide: a review. J Solid State Electrochem 22:1817–1839
121. Pan D, Chen J, Yao S, Tao W, Nie L (2005) A DNA biosensor based on polyaniline-modified electrode for sequence-specific detection. Anal Sci 21:367–371
122. Wan D, Yuan S, Li GL, Neoh KG, Kang ET (2010) Electrochemical biosensors based on conducting polymer nanocomposites. ACS Appl Mater Interfaces 2:3083–3091
123. Singh K, Singh BP, Chauhan R, Basu T (2012) Cholesterol biosensor based on polyaniline–gold nanocomposite. J Appl Polym Sci 125:E235–E243
124. Rahman MM, Shiddiky MJA, Rahman MA, Shim YB (2009) Electrochemical detection of dopamine using polypyrrole modified electrode. Anal Biochem 384:159–165
125. Dhand C, Solanki PR, Datta M, Malhotra BD (2010) Polyaniline nanostructures and their applications in biosensors. Electroanalysis 22:2683–2700
126. Li J, Qiu JD, Xu JJ, Chen HY, Xia XH (2007) Electrochemical DNA biosensor based on polyaniline nanowire-modified glassy carbon electrode. Adv Funct Mater 17:1574–1580
127. Aghamiri ZS, Mohsennia M, Rafiee-Pour HA (2018) Preparation and characterization of polyaniline nanocomposite films for sensor applications. Thin Solid Films 660:484–490
128. Agüí L, Eguílaz M, Peña-Farfal C, Yáñez-Sedeño P, Pingarrón JM (2009) Electrochemical sensors based on conducting polymers. Electroanalysis 21:386–401
129. Singh R, Dhand C, Sumana G, Verma R, Sood S, Gupta RK, Malhotra BD (2010) Cholesterol biosensor based on polyaniline–gold nanocomposite. J Mol Recognit 23:472–478
130. Singh K, Solanki PR, Basu T, Malhotra BD (2012) Polyaniline–gold nanocomposite based cholesterol biosensor. Polym Adv Technol 23:1084–1091
131. Zhu Y, Koh WCA, Shim YB (2010) A polyaniline film modified electrode for the simultaneous determination of ascorbic acid, dopamine, and uric acid. Electroanalysis 22:2908–2916
132. Cesarino FC, Moraes SAS (2011) Machado, electrochemical sensors based on polyaniline and its derivatives: a review. Electroanalysis 23:2586–2595
133. Xu Y, Ye X, Yang L, He P, Fang Y (2006) Electrochemical behavior of DNA at polyaniline modified electrode. Electroanalysis 18:1471–1477
134. Sangamithirai D, Munusamy S, Narayanan V, Stephen A (2018) A highly sensitive amperometric glucose biosensor based on polyaniline nanofibers. Mater Sci Eng C 91:512–520
135. Vural T, Yaman YT, Ozturk S, Abaci S, Denkbas EB (2018) Development of a polyaniline-based biosensor for glucose detection. J Colloid Interface Sci 510:318–325
136. Shoaie N, Forouzandeh M, Omidfar K (2018) Recent advances in biosensors for detection of glucose and hydrogen peroxide: a review. Microchim Acta 185:1–19
137. Mir TA, Yoon JH, Gurudatt NG, Won MS, Shim YB (2015) A polyaniline nanofiber-based biosensor for detection of glucose. Biosens Bioelectron 74:594–601
138. Hussain KK, Gurudatt NG, Mir TA, Shim YB (2016) Electrochemical biosensors based on conducting polymers for detection of glucose and hydrogen peroxide. Biosens Bioelectron 83:312–319
139. Tian L, Qian K, Qi J, Liu Q, Yao C, Song W, Wang Y (2018) A polyaniline-based biosensor for detection of glucose. Biosens Bioelectron 99:564–570

140. Akhtar MH, Hussain KK, Gurudatt NG, Chandra P, Shim YB (2018) A polyaniline nanofiber-based biosensor for detection of glucose. Biosens Bioelectron 116:108–115
141. Chowdhury AD, Gangopadhyay R, De A (2014) Conducting polymer nanocomposite-based biosensors for detection of glucose and hydrogen peroxide. Sensors Actuators B Chem 190:348–356
142. Esmaeili C, Heng LY, Chiang CP, Rashid ZA, Safitri E, Malon Marugan RSP (2017) A polyaniline-based biosensor for detection of glucose. Sensors Actuators B Chem 242:616–622
143. Nowicka AM, Fau M, Rapecki T, Donten M (2014) Electrochemical biosensors based on conducting polymers for detection of glucose and hydrogen peroxide. Electrochim Acta 140:65–72
144. Zou L, Li Y, Cao S, Ye B (2013) A polyaniline-based biosensor for detection of glucose. Talanta 117:333–338
145. Rong Q, Han H, Feng F, Ma Z (2015) A polyaniline-based biosensor for detection of glucose. Sci Rep 5:11440
146. Zhu W, Wang Q, Ma H, Lv X, Wu D, Sun X, Du B, Wei Q (2016) A polyaniline-based biosensor for detection of glucose. Sci Rep 6:24599
147. Dechtrirat D, Sookcharoenpinyo B, Prajongtat P, Sriprachuabwong C, Sanguankiat A, Tuantranont A, Hannongbua S (2018) A polyaniline-based biosensor for detection of glucose. RSC Adv 8:206–212
148. Amani J, Maleki M, Khoshroo A, Sobhani-Nasab A, Rahimi Nasrabadi M (2018) A polyaniline-based biosensor for detection of glucose. Anal Biochem 548:53–60
149. Xu G, Jarjes ZA, Desprez V, Kilmartin PA, Travas-Sejdic J (2018) A polyaniline-based biosensor for detection of glucose. Biosens Bioelectron 107:184–191
150. Ott C, Raicopol MD, Andronescu C, Vasile E, Hanganu A, Pruna A, Pilan L (2018) A polyaniline-based biosensor for detection of glucose. Synth Met 235:20–27
151. de Fátima Giarola J, Mano V, Pereira AC (2018) A polyaniline-based biosensor for detection of glucose. Electroanalysis 30:119–126
152. Zheng W, Zhao M, Liu W, Yu S, Niu L, Li G, Li H, Liu W (2018) A polyaniline-based biosensor for detection of glucose. J Electroanal Chem 813:75–82
153. Wang W, Xu G, Cui XT, Sheng G, Luo X (2014) A polyaniline-based biosensor for detection of glucose. Biosens Bioelectron 58:153–160
154. Han L, Zhao Y, Chang C, Li F (2018) A polyaniline-based biosensor for detection of glucose. J Electroanal Chem 817:141–148
155. Li Y, Zhang Y, Han G, Xiao Y, Li M, Zhou W (2016) A polyaniline-based biosensor for detection of glucose. Chin J Chem 34:82–89
156. Huang TY, Kung CW, Wang JY, Lee MH, Chen LC, Chu CW, Ho KC (2015) A polyaniline-based biosensor for detection of glucose. Electrochim Acta 172:61–68
157. Kaçar C, Erden PE, Kılıç E (2017) A polyaniline-based biosensor for detection of glucose. Electroanalysis 29:2114–2121
158. Feng XM, Li RM, Ma YW, Chen RF, Shi NE, Fan QL, Huang W (2011) A polyaniline-based biosensor for detection of glucose. Adv Funct Mater 21:2989–2996
159. Bo Y, Yang H, Hu Y, Yao T, Huang S (2011) A polyaniline-based biosensor for detection of glucose. Electrochim Acta 56:2676–2683
160. Mao H, Liang J, Zhang H, Pei Q, Liu D, Wu S, Zhang Y, Song XM (2015) A polyaniline-based biosensor for detection of glucose. Biosens Bioelectron 70:289–296
161. Sangamithirai D, Narayanan V, Muthuraaman B, Stephen A (2015) A polyaniline-based biosensor for detection of glucose. Mater Sci Eng C 55:579–586
162. Ruecha N, Rangkupan R, Rodthongkum N, Chailapakul O (2014) A polyaniline-based biosensor for detection of glucose. Biosens Bioelectron 52:13–19
163. Kodogiannis V, Wadge E (2005) A polyaniline-based biosensor for detection of glucose. Int J Neural Syst 15:363–370

164. Pavlou AK, Magan N, Jones JM, Brown J, Klatser P, Turner APF (2004) Detection of Mycobacterium tuberculosis using conducting polymer-based sensors. Biosens Bioelectron 20:538–544
165. Ameer Q, Adeloju SB (2005) Polypyrrole-based electronic noses for detection of volatile organic compounds. Sens. Actuators B Chem. 106:541–552
166. Doleman BJ, Lewis NS (2001) Comparison of odor detection thresholds and odor discrimination using conducting polymer sensor arrays. Sens Actuators B Chem 72:41–51
167. Jin G, Norrish J, Too C, Wallace G (2004) Conducting polymer gas sensors: a review. Curr Appl Phys 4:366–368
168. Menegazzo N, Herbert B, Banerji S, Booksh KS (2011) Conducting polymer-based sensors for environmental monitoring. Talanta 85:1369–1375
169. Chen Y, Li Y, Wang H, Yang M (2007) Conducting polymer composites for gas sensing applications. Carbon N Y 45:357–364
170. Bai H, Shi G (2007) Conducting polymer-based sensors for detection of volatile organic compounds. Sensors 7:267–307
171. Sadek AZ, Wlodarski W, Shin K, Kaner RB, Kalantar-Zadeh K (2006) A polyaniline nanofiber-based sensor for detection of ammonia. Nanotechnology 17:4488–4492
172. Virji S, Fowler JD, Baker CO, Huang J, Kaner RB, Weiller BH (2005) Polyaniline nanofiber-based sensors for detection of ammonia. Small 1:624–627
173. Vijayakumar N, Subramanian E, Padiyan DP (2012) Conducting polymer-based sensors for detection of ammonia. Int J Polym Mater 61:847–853
174. Huang X, Hu N, Gao R, Yu Y, Wang Y, Yang Z, Kong ES-W, Wei H, Zhang Y (2012) Polyaniline nanofiber-based sensors for detection of ammonia. J Mater Chem 22:22488–22495
175. S.A., S.P.S., Kaushik A, Khan R, Gupta V, Malhotra BD (2009) Polyaniline nanofiber-based sensors for detection of ammonia. J Nanosci Nanotechnol 9:1792–1797
176. Liu C, Tai H, Zhang P, Ye Z, Su Y, Jiang Y (2017) A polyaniline nanofiber-based sensor for detection of ammonia. Sens Actuators B: Chem 246:85–95
177. Zampetti E, Pantalei S, Muzyczuk A, Bearzotti A, De Cesare F, Spinella C, Macagnano A (2013) Polyaniline nanofiber-based sensors for detection of ammonia. Sensors Actuators B Chem 176:390–398
178. Shin DH, Lee JS, Jun J, Kim SG, Jang J (2015) Highly sensitive and selective ammonia sensor based on polyaniline nanofibers. ACS Appl Mater Interfaces 7:1746–1751
179. Huang J, Kang Y, Yang T, Wang Y, Wang S (2011) A polyaniline nanofiber-based sensor for detection of ammonia. J Nat Gas Chem 20:403–407
180. Wang W, Li Z, Jiang T, Zhao Z, Li Y, Wang Z, Wang C (2012) A polyaniline nanofiber-based sensor for detection of ammonia. ACS Appl Mater Interfaces 4:6080–6084
181. Fu F, Wang J, Zeng H, Yu J (2020) Recent advances in conductive hydrogels: classifications, properties and applications. ACS Mater Lett 2:1287–1301
182. D. Mawad, A. Lauto, G.G. Wallace, Conductive polymer hydrogels, in: Polymeric Hydrogels as Smart Biomaterials, (2016) 19–45
183. Wang Z, Cong Y, Fu J (2020) Stretchable and tough conductive hydrogels for flexible pressure and strain sensors. J Mater Chem B 8:3437–3459
184. Chung J, Khot A, Savoie BM, Boudouris BW (2020) 100th anniversary of macromolecular science viewpoint: recent advances and opportunities for mixed ion and charge conducting polymers. ACS Macro Lett 9:646–655
185. Chalmers E, Lee H, Zhu C, Liu X (2020) Increasing the conductivity and adhesion of polypyrrole hydrogels with electro-polymerized polydopamine. Chem Mater 32:234–244
186. Chen J, Liu J, Thundat T, Zeng H (2019) Polypyrrole-doped conductive supramolecular elastomer with stretchability, rapid self-healing, and adhesive property for flexible electronic sensors. ACS Appl Mater Interfaces 11:18720–18729
187. Sekine S, Ido Y, Miyake T, Nagamine K, Nishizawa M (2010) Conducting polymer electrodes printed on hydrogel. J Am Chem Soc 132:13174–13175
188. Gao H, Li J, Zhang F, Liu Y (2019) Recent advances in stretchable and self-healing electronic materials. Mater Horiz 6:931–944

189. Willner I (2017) Stimuli-controlled hydrogels and their applications. Acc Chem Res 50:657–658
190. Teymouri M, Kokabi M, Alamdarnejad G (2020) Preparation and characterization of polyaniline-based hydrogels for sensor applications. J Appl Polym Sci 137:48602
191. Wang Z, Chen J, Cong Y, Zhang H, Xu T, Nie L, Fu J (2018) A polyaniline-based hydrogel for flexible sensors. Chem Mater 30:8062–8069
192. Rong Q, Lei W, Chen L, Yin Y, Zhou J, Liu M (2017) A polyaniline-based hydrogel for flexible sensors. Angew Chem Int Ed 56:14159–14163
193. He F, You X, Gong H, Yang Y, Bai T, Wang W, Guo W, Liu X, Ye M (2020) A polyaniline-based hydrogel for flexible sensors. ACS Appl Mater Interfaces 12:6442–6450
194. Boublia Z, Guezzout N, Haddaoui M, Badawi AS, Darwish T, Lemaoui F, Banat KK, Yadav B-H, Jeon IM, AlNashef (2024) Recent progress in the development of conducting polymer nanocomposites for energy applications. J Mater Chem A 12:2209–2236

# Chapter 11
# Tuning of Conducting Polymers for Applications in Material Science

**Chetna Kumari, Anshul Yadav, Astha Kamboj, Kavita Poonia, and Nirmala Kumari Jangid**

**Abstract** Conducting polymers has garnered much interest due to their efficient mechanical, optical, and electronic properties, electrical conductivity, and economic significance. Conducting polymers are used in diodes, transistors, artificial nerves, conducting adhesives, and electromagnetic interference shielding. Some possible uses of nanotubes and nanofibers are discussed in this chapter, including drug delivery, protein purification, neural interfaces, supercapacitors and energy storage, and actuators. This chapter also focuses on environmental remediation (adsorption of pollutants, electrocatalytic oxidation/reduction of contaminants on conducting polymer-based electrodes, and cation and anion trapping), use of 3D material, gas removal, photocatalysis, and dioxygen removal.

**Keywords** Conducting polymer · Applications · Materials · Energy storage · Environmental remediation

## 11.1 Introduction

Intrinsically conducting polymers are an interesting class of organic materials that may be readily synthesized with an extensive range of micro- and nano-morphologies and chemical structures to achieve specific macroscopic chemical and physical properties. The first conducting polymer, polyacetylene, was discovered in 1977 by Shirakawa, Heeger, and Mac Diarmid. They were jointly awarded the Nobel Prize in Chemistry in 2000 [1]. In terms of chemistry, CPs are π-conjugated organic polymers. This results in a semiconductor electronic framework in neutral CPs, with a vacant π*-band (conduction band) and a filled π-band (valence band), separated by an energy difference of around 1 eV. Doping, or the withdrawal or introduction of

C. Kumari · A. Yadav · A. Kamboj · K. Poonia · N. K. Jangid (✉)
Department of Chemistry, Banasthali Vidyapith, Banasthali, Rajasthan, India
e-mail: nirmalakumari@banasthali.in


S. Jadoun, N. Kumari Jangid (eds.), *Tuning of Conducting Polymers*, Springer Series on Polymer and Composite Materials,
https://doi.org/10.1007/978-3-032-00461-1_11

electrons onto the interlinked polymeric chain, produces intrinsic conductivity, but adding counter ions, or so-called dopants, maintains overall electroneutrality. Simple reduction (n-doping) or oxidation (p-doping) processes produce the doped state, which results in the development of delocalized charged structural defects (solitons, polarons, and bipolarons) that function as charge carriers and are energetically situated inside the energy gap. Therefore, the electrical conductivity can be changed from insulating to metallic using straightforward and reversible electrochemical doping/dedoping. The electronic structure of CPs can be engineered by creating new chemical structures using synthetic electrochemical or chemical techniques, in addition to the straightforward control of the doping level. This allows for adjustable electronic, optical, and electrochemical properties [2]. This idea has developed into a wide range of chemically varied, flexible, and lightweight tuned polymer structures that exhibit appealing qualities for a wide range of applications with huge technological effects, many of which make it to market. They are essential components of solar cells, thin-film transistors, biosensors, electrochromic displays, light-emitting diodes, actuators, supercapacitors, artificial muscles, and secondary batteries [1]. Other novel applications in tissue engineering and medicinal science, adsorbent materials for environmental problems have been attractive areas in recent decades [3]. Even though CPs have clear advantages over their metallic and inorganic semiconducting counterparts in terms of cost, flexibility, low density, and adjustable conductivity, much work remains to be done to overcome their limitations in terms of long-term stability, conductivity, and solubility/processability. Because of this, the creation of hybrid composites based on CPs that have improved qualities and regulated morphology and shape [4]. Examples include metal oxide nanoparticles, hydrogel inorganic matrices, and CP nanocomposites containing various carbon nanomaterials. Much research has been done recently on using polymer-based materials in biomedical science and healthcare.

## 11.2 Properties of Conducting Polymers

CP has remarkable magnetic, electrical, wetting, optical, microwave-absorbing, and mechanical qualities in addition to their conducting properties as shown in Fig. 11.1.

### *11.2.1 Magnetic Properties*

Conducting polymer magnetic characteristics has been researched because they reveal important information about single spins and molecules that carry charges [5]. Lu et al. examined the magnetic activity of PANI/$Fe_3O_4$ material nanotubes generated using ultrasonic irradiation [6]. The magnetic activity of PANI/$Fe_3O_4$ material nanorods made using a self-assembly process was investigated by Long et al. [7] $Fe_3O_4$ particle dispersion is facilitated by the ultrasonic irradiation-synthesized samples instead of the self-assembly method. The ultrasonic

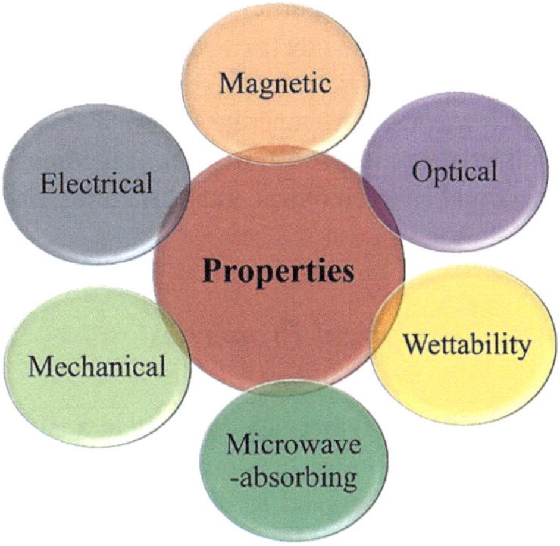


**Fig. 11.1** Different properties of conducting polymers

irradiation method produced composite nanotubes that exhibited superparamagnetic behavior.

### *11.2.2 Electrical-Conducting Properties*

Doping allows conducting polymers to achieve the metallic conducting state in electrical conductivity. Martin demonstrated that a single nanofiber's electrical conductivity is a couple of orders of magnitude superior to that of the material found in nanowires or pellet nanotubes [8]. When 1D-conducting polymer nanoparticles are combined with an insulating component, electrical conductivity is often reduced because the insulating component partially blocks the conductive channel. According to Long et al., beta-naphthalene sulfonic acid (NSA)-doped PANI/$Fe_3O_4$ material nanowire pellets exhibited a characteristic semiconducting behavior in which their resistance increased as the temperature decreased [7]. The enhanced charge transport distribution between NSA-doped PANI/$Fe_3O_4$ nanoparticles is responsible for the reduced composite conductivity. Several types of 1D-CP nanocomposite systems also demonstrated a comparable reduction in electrical conductivity [9]. However, the conductivity of nanocomposites may be improved by adding a different high-electric conductivity nanocomponent to conducting polymers. The electrical conductivities of CNT/PANI material nanocables were investigated by Long et al., who discovered that as CNT loading increases, increasing the conductivity of pure polyaniline [10].

The insertion of carbon nanotubes (CNTs) into CP may enhance electrical conductivities by acting as a "conducting bridge" between conducting domains [11]. Moreover, as the temperature dropped, the conductivity of well-aligned CNT/PANI nanocomposites reduced as well, demonstrating typical semiconductor behavior [12]. For CNT/PPY composite nanocables, comparable outcomes have also been attained [13]. Conducting polymers electrical conductivity can also be increased by adding metal nanoparticles to them [14]. Single Au/PANI nanocables have a substantially greater electrical conductivity than a pure PANI nanotube doped with camphor sulfonic acid (CSA) [15, 16]. A template free electro-polymerization technique was used to generate different functionalized PEDOT nanostructures on an indium-tin-oxide substrate. The outcome demonstrated the connection between electrical characteristics, nanostructures, and functional groups [17].

### *11.2.3 Optical Properties*

Conducting polymers special optical qualities have been thoroughly studied due to their potential usage in nanophotonic devices. The semiconductors with a 1D nanostructure can be used to create photonic wire lasers, photodetectors, and photochemical sensors [18, 19]. The optical characteristics of CdS/PANI material nanocables were examined by Xi et al., who discovered that while signal intensities were higher, the photoluminescence spectrum shared characteristics with CdS nanowires [20]. This rise was caused by photogenerated transporters moving from the polyaniline layer into CdS nanowires [21].

### *11.2.4 Mechanical Properties*

Recently, there has been a lot of interest in the mechanical characteristics of a single nanotube [22]. The polypyrrole (PPY) nanotubes elastic tensile modulus was published by Cuenot et al. after they investigated force-curve or resonance-frequency measurements. It was discovered that when the outer diameter or wall thickness of the polypyrrole nanotube was reduced, the elastic modulus significantly increased [23]. Other single nanofibers have also been found to exhibit comparable size-dependent mechanical behavior [24]. Inorganic nanowires like CuO [25], silicon [26], silver, and lead [27] have also been shown to exhibit comparable size-dependent mechanical behavior.

### *11.2.5 Wettability*

One of a solid surface's most crucial characteristics is wettability, which opens up possible uses in bio-separation, controlled drug administration, microfluidics, and self-cleaning surfaces [28]. Hydrophilia is a characteristic of conducting polymers [29]. By doping hydrophobic acids, a coating of conducting polymers with superhydrophobic qualities can be created [30]. The chemical composition of CP can be changed to provide a reversibly switchable superhydrophilic and superhydrophobic surface [31]. Chemical dual-responsive properties have also been shown to be present in the wettability of polyacrylonitrile/polyaniline (PANI) coaxial nanofiber films

[32]. On substrates with decreasing hydrophilicity, a single-layered photopolymerized nanocomposite film comprising $TiO_2$ and polystyrene nanorods exhibits a modification in wetting characteristics from hydrophobic to hydrophilic [33].

### *11.2.6 Microwave-Absorbing Properties*

CP has been investigated as a novel microwave-absorbing composite because of its ease of processing and lower density. Wan et al., discovered that PANI-naphthalene sulfonic acid and PANI-NSA/glucose micro-nanotubes made without a template exhibit superior electromagnetic loss [34]. Doped PANI with fiber-like design has a superior ability to absorb electromagnetic waves than polyaniline with particle-like morphology, according to Liu et al. [35] According to the findings, CP nanotubes can be employed as high-absorption, lightweight, and wide-frequency microwave absorbents.

## 11.3 Applications

Conducting polymers have found extensive applications in materials science due to their unique combination of electrical conductivity, flexibility, and chemical stability. The versatility of conducting polymers arises from their tunable conductivity, ease of synthesis, and functionalization, making them vital in advancing various fields within materials science. Some applications are discussed below as shown in Fig. 11.2.

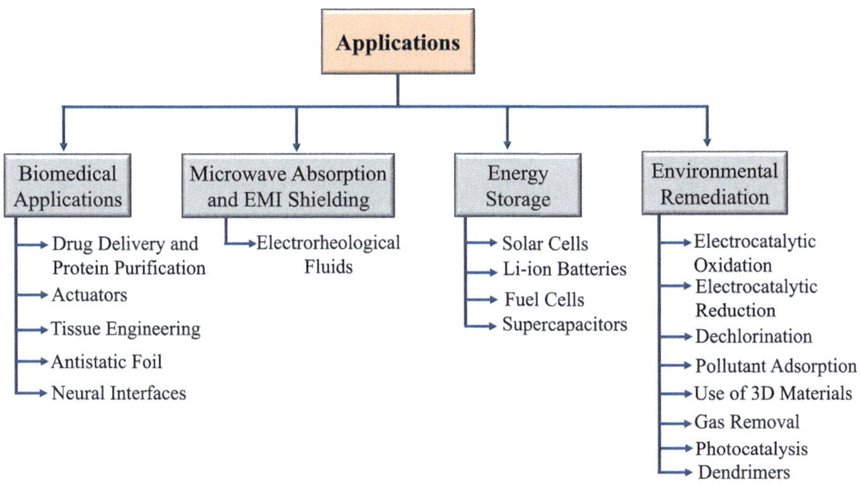


**Fig. 11.2** Different applications of conducting polymers

## *11.3.1 Biomedical Applications*

### 11.3.1.1 Drug Delivery and Protein Purification

Over the past few decades, numerous drug delivery system gadgets have been designed to improve drug-targeting specificity and minimize overall drug toxicity. These devices have been utilized to treat a variety of diseases. The various delivery techniques include micro-nanogels, polymeric microspheres, polymeric micelles, and polymeric nanofibers. Maintaining rigorous control over the ON/OFF state is one of the delivery system's principal disadvantages. Conducting polymers have been utilized to work around this problem because of their reversible electrochemical response, which causes them to expand upon oxidation and compress upon reduction. Therefore, the controlled release of different types of medicines is anticipated to be favored by this induced volume change [36, 37]. For this, a variety of conducting polymers have been employed, including PEDOT nanotubes, polyacrylic acid (PAA) microspheres, and PPy film [38, 39]. Using polymer nanocomposites to create controlled medication release is a practical way to create electronically active devices using living tissues. Easy loading, minimal impact on drug action, and regulated release rate are a few benefits of conducting polymer nanocomposites. For protein purification, chemically altered PPy microtubes have been reported to function as an affinity matrix; upon protein elution, the desorption ratio was high [30].

### 11.3.1.2 Actuators

By interacting with polymer films, actuators can be either tiny or macroscopic devices or chains of polymers that convert energy from many sources (such as electric fields, light, current, pressure, temperature, and many more) into macroscopic mechanical energy [40]. Artificial muscles are another name for polymeric actuators in the literature. There are two families among them: (a) electrochemomechanical actuators and (b) electromechanical actuators, where the presence of reactive or nonreactive electrodes is the primary distinction between them [41]. Given that their reaction is predicated on the faradaic processes and the modifications to the polymeric framework, we are particularly interested in the electrochemomechanical actuators in terms of electrochemical applications [42]. This type of electrochemical actuator usually consists of laminated electrodes with a simple bilayer (CP/plastic), asymmetrical bilayer (CP1/CP2), and triple layer (CP/plastic/CP), where the doping/dedoping process causes the electrodes to bend in response to an electrical stimulus. CPs can be utilized as composite or pure materials. Two examples of the latter are the silk-polypyrrole, and silk-polyaniline actuators, both of which are said to be biocompatible [43].

#### 11.3.1.3 Tissue Engineering

Several characteristics of CP, including their electrical characteristics, high surface area, and electroactivities, make them viable options for tissue engineering purposes. Electrical stimulation can alter cell-to-cell adhesion, migration, differentiation, and proliferation in living cells. A conductive material nanofiber with good biocompatibility, high porosity, biodegradability, and high surface area is produced by electrospinning a blended polymer solution that contains CP and other biocompatible polymers. These nanofibers have been used in tissue engineering. Polygalacturonic acid (PGA) and the natural protein gelatine are also utilized as scaffolds in tissue engineering. A variety of conducting polymer types, such as PANI nanofibers, poly (L- lactide-co-ε-caprolactone) (PLCL) nanofibers, polylactic acid/polypyrrole nanofibers, poly-ε-caprolactone (PCL) nanofibers, and poly (L-lactide-co-glycolide) (PLGA) nanofibers, have been used in tissue engineering [38]. Khan and his colleagues have used bio cellulose-nanofiber polyurethane nanomaterial as scaffolds in bone tissue transplantation [44].

#### 11.3.1.4 Antistatic Foils

Antistatic matter prevents triboelectric charging, which is the accumulation of electric charge caused by contact between various composites. In certain industries, static charges pose a significant problem due to the potential for flammable vapor ignition, damage to electronic systems, and dust attraction, in these cases charge dissipation using appropriate conducting materials may be essential. Antistatic foils, frequently used for protection and packing, were developed due to the potential to create aqueous dispersions of CP employing polyanions to balance the positive charges inherent in the charged structure [43]. When doped, CPs benefit from their conductivities and the corresponding flexibility of films made from water-soluble dispersions. For this application, the most widely utilized water-soluble polymer is the conducting poly-(3,4-ethylenedioxithiophene) polymer with poly-(styrene sulfonate) polyanions (PEDOT: PSS). In the 1980s, it was patented by the Bayer and AFGA firms as an antistatic coating for photographic films, which is still in use today. PEDOT: PSS exhibits a high conductivity (300–500 $Scm^{-1}$), stability at ambient settings, and the ability to withstand up to 1000 h at 100 °C without losing conductivity [45]. This idea has developed into the foundation of the buffer layer in both perovskite solar panels and organic. Other CP families, such as polythiophenes (PTh), polypyrroles (PPy), and polyanilines (PANI), have been investigated for this purpose. Other textiles and packing materials have been suggested in place of films to satisfy market demands [43].

#### 11.3.1.5 Neural Interfaces

Abidian et al. [46] and Kim et al. [47] have constructed conducting polymer nanofibers with a high charge density, low impedance, and the ability to release drugs under regulated conditions from a hydrogel scaffold placed on neural microelectrodes. Drug release is regulated by the biodegradable electrospun nanofiber, and it is maintained by the hydrogel layer. PEDOT nanotubes are a promising biosensing technique for detecting the change from acute to chronic responses in brain tissues based on spectroscopic studies [46]. Furthermore, polypyrrole and poly (3,4-ethylenedioxithiophene) nanotubes have been shown to stick to the electrode surface more effectively than their film equivalents [38].

### *11.3.2 Microwave Absorption and EMI Shielding*

An important problem caused by the quick development of navigation, electronics, wireless systems, space technology, and many more is electromagnetic frequency interference. Electric devices and other living things including humans are impacted by electromagnetic interference (EMI). For this reason, conducting polymers, metals, and carbon compounds have been used as shielding materials to stop electromagnetic noise. Because conducting polymers are easily processed and have strong electrical conductivity, their applications as shielding materials have gained more attention. EMI shielding performance for conducting polymers is improved by PANI when paired with other nanocomponents such as CNT. PANI-multiwalled CNT nanocomposites and polyaniline microtube nanofibers can be employed as materials to shield against electromagnetic interference and absorb microwave radiation [38].

#### 11.3.2.1 Electrorheological Fluids

Electrorheological (ER) fluid's tunable vibration control characteristics under external stimuli make them useful in hydraulic valves, dampers, clutch systems, and brakes. ER fluids typically consist of low-conductivity, high-dielectric constant particles scattered throughout a non-conductive fluid medium. This characteristic can be rapidly and reversibly altered in an electric field. The enhanced physical features of conducting polymers, such as their high polarizability and improved environmental durability, have led to their recent widespread application as polarizable particles [38].

## 11.3.3 Energy Storage

Fossil fuels will eventually run out, making energy the most pressing global issue. An alternative for the creation of next-generation devices that store energy is CP nanostructures, which generally have higher capacitance values [38].

### 11.3.3.1 Solar Cells

Energy-conversion devices called solar cells transform sunlight into electrical energy. Combining CP with metal oxide can enhance solar efficiency by utilizing their special light absorption and hole transportation capabilities. $TiO_2$ nanotube arrays have favorable electron migration paths, making them attractive options for solar cell fabrication [38].

### 11.3.3.2 Lithium-Ion Batteries

Traditional rechargeable Ni-metal hydride or Ni-cadmium batteries have limitations in terms of durability and capacity. However, Li-ion batteries, which are lighter and have a significantly higher capacity, have been regarded as the most useful rechargeable batteries [48]. Therefore, due to their significant specific capacity and excellent cycle efficiency, 1D-nanostructured materials have shown promise as Li-ion battery electrodes [49].

### 11.3.3.3 Fuel Cells

Fuel cells use electrochemical reactions to transform chemical energy into electrical energy. The use of fuel cells in electric cars has garnered interest in recent decades. Energy applications research has concentrated on direct methanol fuel cells (DMFCs) because of their high energy-conversion efficiency, fuel mobility, and environmental friendliness. CPs with 1D nanostructures have emerged as a promising option for electrocatalyst support in the well-researched impacts of an electrocatalyst on DMFC performance [38].

### 11.3.3.4 Supercapacitors

Supercapacitors are among the most promising energy storage technologies for a variety of uses, such as uninterruptible power supplies and electric cars. The specific power of supercapacitors is higher than that of lithium-ion batteries. CPs, carbon, and metal oxides are the primary electrode materials used in supercapacitors. The cyclic stability of CP is low, despite their high specific capacitance. Conducting polymer nanocomposites has been created to overcome this limitation [38].

### *11.3.4 Environmental Remediation*

Conducting polymers (CPs) have several uses in detecting, measuring, and eliminating pollutants. In terms of removing pollutants, polymers can function as filters, membranes, sorbents, ion exchangers, and chemical reactors that use redox reactions. Doping/dedoping, electrocatalytic reduction/oxidation, and pollutant absorption (or uptake) and release are the primary processes aided by CP and their composites with an applied energy [43]. CPs are frequently appropriate for the transfer of electrons, which is beneficial for removing pollutants. They need less energy than demineralization, photolytic, or adsorption methods to remove contaminants from samples and exhibit more hydrophobicity and stability than monolayers of adsorbed catalysts [50].

#### 11.3.4.1 Electrocatalytic Oxidation

In this case, a substance with electrocatalytic qualities for oxidizing target pollutants is added to the polymeric framework; the polymer changes its oxidation state throughout the process, but the oxidizer remains unchanged.

A general reaction is as follows:

$$\mathrm{xCP_{monomer}}\left(\mathrm{aq}\right)+n\mathrm{A}^{-}\left(\mathrm{aq}\right)+\mathrm{Z}^{2y+}\left(\mathrm{aq}\right)+y\mathrm{H_2O}\left(\mathrm{l}\right)\rightarrow\mathrm{CP}^{n+}.n\mathrm{A}^{-}.\mathrm{ZO}y\left(\mathrm{s}\right)$$
$$+2y\mathrm{H}^{+}\left(\mathrm{aq}\right)+ne^{-} \quad (11.1)$$

$$\mathrm{CP}^{n+}.n\mathrm{A}^{-}.\mathrm{ZO}y\left(\mathrm{s}\right)+\mathrm{Red}\left(\mathrm{aq}\right)\rightarrow\mathrm{CP}^{0}.\mathrm{ZO}y\left(\mathrm{s}\right)+\mathrm{Ox}^{n+}\left(\mathrm{aq}\right)+n\mathrm{A}^{-}\left(\mathrm{aq}\right) \quad (11.2)$$

where ZOy is the electrocatalytic metal oxide (for example, lead oxide using PPy to form a hydrophobic material on the surface of a tin oxide/Ti substrate), $Z^{2y+}$ is its metal precursor (Pb (II)), a pollutant in its reduced state represented by red (example, 2-chlorophenol), and Ox is a less dangerous oxidized state [50]. The interpretation of the other symbols is identical to that of the latter equations.

The electrode regeneration process is then:

$$\mathrm{CP}^{0}.\mathrm{ZO}y\left(\mathrm{s}\right)+n\mathrm{A}^{-}\left(\mathrm{aq}\right)\rightarrow\mathrm{CP}^{n+}.n\mathrm{A}^{-}.\mathrm{ZO}y\left(\mathrm{s}\right)+ne^{-} \quad (11.3)$$

The increase of kinetic and thermodynamic features of desired oxidation reactions makes catalysts useful for energy production and storage. In several applications (such as 2-chlorophenol oxidation) [43], $PbO_2$ is chosen because of its low cost, large dioxygen overpotential, relative chemical inertness, and high electrical conductivity [50]. As an alternative, NiOx films on graphite-grown PPy are capable of effectively oxidizing ethanol, most likely due to the creation of oxidizing Ni (III) active sites within the polymer framework [43]. An additional use is for dangerous As (III) species, both man-made and natural. Their oxidation to the far less toxic and readily soluble As (V) is a common treatment. As (III) to As(V) oxidation is effectively catalyzed by nanocomposites made by incorporating metal nanoparticles scattered throughout a CP matrix [43].

#### 11.3.4.2 Electrocatalytic Reduction

The reduction of hazardous and highly mobile $Cr_2O_7^{2-}/CrO_4^{2-}$ ions with (a) polypyrrole, (b) polypyrrole composites with $Fe_3O_4$, cellulose fiber, CNTs or 2,5-diamino benzene sulfonic acid, (c) polyaniline, (d) polypyrrole-polyaniline material nanofibers, (e) polyaniline-titanotungstate materials, and (f) palladium-decorated polyaniline, are possibly the most studied system. This reduction is primarily based on two processes: (i) an anion exchange process triggered by the electroneutrality demand resulting from the polymer oxidation step; and (ii) a spontaneous chemical reaction between $CrO_4^{2-}$ and polypyrrole caused by their varying reduction potentials [43].

The reaction steps are as follows:

$$\mathrm{xCP}_{\mathrm{monomer}}\left(\mathrm{aq}\right)+n\mathrm{A}^{-}\left(\mathrm{aq}\right)\rightarrow\mathrm{CP}^{n+}.n\mathrm{A}^{-}\left(\mathrm{s}\right)+ne^{-}\tag{11.4}$$

$$\mathrm{CP}^{n+}.n\mathrm{A}^{-}\left(\mathrm{s}\right)+ne^{-}\rightarrow\mathrm{CP}^{0}\left(\mathrm{s}\right)+n\mathrm{A}^{-}\left(\mathrm{aq}\right)\tag{11.5}$$

$$\mathrm{CP}^{0}\left(\mathrm{s}\right)+\mathrm{Ox}^{n+}\left(\mathrm{aq}\right)+nA^{-}\left(\mathrm{aq}\right)\rightarrow\mathrm{CP}^{n+}.n\mathrm{A}^{-}\left(\mathrm{s}\right)+\mathrm{Red}\left(\mathrm{aq}\right)\tag{11.6}$$

The conducting polymers monomer is represented by CP, the pollutant's oxidized form is represented by Ox ($CrO_4^{2-}$), its reduced form is represented by red (Cr (III)), and all other symbols are identical to those in the preceding equations. Because the pollutant itself regenerates the active form of the conductive polymer, a regeneration phase is not required in this system (Eq. 11.6). However, there is a problem with the conversion dropping with repeated cycling [43]. An alternative elimination method that can further enhance the process is the pollutant's adsorption onto the polymer [51]. According to X-ray photoelectron spectroscopy (XPS), the product, Cr(III), is also capable of being adsorbed. Furthermore, the electrode material itself, such as reticulated vitreous carbon (RVC), may aid in the decrease of Cr(VI). To calculate the mass transfer parameters from concentration-time profiles, the simple reduction route on a polyaniline/vitreous carbon electrode was modeled taking into account both a continuously stirred tank reactor (CSTR), and plug-flow reactor (PFR). For the reduction of Cr(VI) to Cr(III), the employment of Pd zerovalent nanoparticles in nanofibrous polymeric materials with a reducing agent offers superior catalytic efficiency and reusability. Another remediation method for hazardous halogenated substances is the electrocatalytic dehalogenation of organics employing zerovalent palladium in a CP matrix (polypyrrole) [52]. $NO_3^-$ ion electroreduction is of interest due to its synthetic and environmental uses. Indirect $NO_3^-$ toxicity, such as the blue-baby syndrome, has been extensively studied [43]. However, with metal electrodes, $NO_3^-$ reduction yields valuable $NH_3$, $NH_2OH$, and $N_2H_4$; for instance, $NH_3$ is the primary result when utilizing polypyrrole-copper electrodes [53]. The electron transfer from polypyrrole via the nitrogen sites in its ring causes $NO_3^-$ reduction, whereas the ion exchange mechanism previously mentioned between $SO4^{2-}$ and $Cl^-$ counterions at PANI causes $NO_3^-$ removal from water, as determined by FTIR and XPS [54]. From a synthetic and environmental standpoint, $CO_2$ fixation/reduction is also significant. The method utilizing its heterogeneous methanation at elevated temperatures is arguably the most evident. Unfortunately, this procedure is not very practical [43]. Electrochemical pathways are worthwhile to investigate and don't require so high temperatures. By combining the electrophilicity of the carbon atom with the amino groups in Fe(II)-immobilized polyaniline/Prussian blue (PB) laminated electrodes, for instance, it has been possible to produce formic, acetic, lactic, and formic acids, along with ethanol and methanol, from carbon dioxide reduction through effective chemical activation of $CO_2$ [43]. In addition to the interaction between the complex's Fe(II) and $CO_2$ basic oxygen atom, this also facilitates the hydrogenation of carbon dioxide to produce the aforementioned alcohols and organic acids by the catalytic production of H(ads) atoms at PB's zeolitic structure.

#### 11.3.4.3 Dechlorination

A long-standing option for the cleanup of hazardous halogenated organics is reductive dechlorination. It is commonly recognized that an aromatic structure becomes more poisonous the more halogens it contains [52]. Thus, eliminating halogen atoms is a viable approach. Remarkably, progressive electrochemical halogen elimination is possible [55]. As previously mentioned, Pd (0) nanoparticles immobilized in a polypyrrole matrix are used in a conducting polymer dechlorination example to reductively dechlorinate hazardous 3,3′,4,4′-tetrachlorobiphenyl in the presence of hydrogen. Eliminating Cl atoms eventually produces biphenyl, which lowers contaminants [52].

#### 11.3.4.4 Pollutant Adsorption

CP can act as an adsorber for pollutants. A sustainable polymer/biopolymer composite (PANI/chitosan), for instance, may efficiently absorb a lot of dyes from aqueous solutions. It was confirmed by FTIR that hydroxyl and amino groups are involved in this type of adsorption. SEM shows that polymer roughness increases the potential for dye adsorption [56]. As an example, a PANI/clinoptilolite composite adsorbed methyl orange [57]. By attaching to the $\pi$-electrons at the polymers backbone, these ions create a strong contact between the As and S atoms, making polythiophene an effective adsorbent material for the elimination of hazardous As (III) [58]. Finally, PANI-Zr (IV) nanocomposites facilitate the simultaneous sorption photocatalytic remediation of dyes [59].

#### 11.3.4.5 Use of 3D Materials

As contamination is often present at low levels, their mass transfer rates toward the electrodes are highly restricted. Therefore, the utilization of CPs is deposited on 3-D materials (graphite felts or porous glassy carbon). Moreover, the high electroactive regions that are thus produced typically raise the desired response rates [43]. PANI hollow nanospheres with controlled discontinuous nanocavities are a potential structure because of their low effective density, and large specific surface area [60].

#### 11.3.4.6 Use of Conducting Polymer Dendrimers

Water purification and the detection and elimination of pollutants are two environmental applications that can benefit from the exciting features of these highly branched macromolecules [61]. According to the current review, certain dendrimers of the new generation may be altered to become electrically conductive. To increase a dendrimer's conductivity, for instance, a high density of aromatic anion radicals affixed to its outside edges may cause intramolecular delocalization [43]. Numerous water-soluble ($Cu^{2+}$, $NO_3^-$) and insoluble (like polyaromatic hydrocarbons, and humic acids) cationic, anionic, aromatic, and complex contaminants can be selectively eliminated by dendrimers. They may also aid in the dispersion of oil and the elimination of certain nanopollutants. In addition to the size, geometry, and charge of the pollutant species to be eliminated (polyvalent nature of $SO_4^{2-}$, $AsO_4^{3-}$, $CrO_4^{2-}$, or $PO_4^{3-}$ species, or the trigonal geometry of monovalent $NO_3^-$ ions versus the tetrahedral monovalent nature of $TcO_4^-$), high selectivity is frequently controlled by the structure of the dendrimer [61]. In this case, the pH may also be important because Lewis's acid-base/complex-forming interactions or the addition or removal of protons may control or alter the chemistries of certain groups within the dendrimers [61].

#### 11.3.4.7 Gas Removal

In general, physical adsorption is the preferred technique for eliminating small-molecule undesirable gases such as NOx, $H_2S$, $NH_3$, $SO_2$, CO, and $CO_2$. However, the intensity of the adsorption forces toward the receptive substrates significantly decreases when these gases are to be removed while moisture is present. Reactive adsorption is recommended in these situations [62]. There is a highly effective reduction of $NO_2$ to NO at graphite oxide-polyaniline composites. This is significant because NO is far less hazardous than $NO_2$. S-doped microporous carbon materials, which are produced by chemically activating a reduced graphene oxide/polythiophene substrate, may also be used to remove $CO_2$ from a combination of $N_2$, $H_2$, $CH_4$, and $CO_2$, in a very selective manner. This is a viable strategy because of its huge specific surface area and consistent recycling adsorption capability [63]. This technique might theoretically collect or convert the majority of analytes detected by CP sensors. Dioxygen, ethanol, methanol, acetone, pentanone, butanone, ammonia, acetonitrile, triethylamine, hexane, toluene, benzene, xylene, methanol, and ethylbenzene are among the other species [64]. When employing a PANI-$TiO_2$ composite, for instance, the photocatalytic method works well in particular gases, such as ammonia [65].

#### 11.3.4.8 Photocatalysis

The oxidation processes that are triggered by the creation of holes in their valence bands are the primary mechanism by which photocatalysts are known to destroy hazardous contaminants. Additionally reduction mechanisms are appropriate mediated by photogenerated electrons in the conduction bands [66]. A polymer (PANI) can have specific physical properties (conductivity) tuned at will by adding inorganic spherical particles ($TiO_2$) during polymerization [67]. Additionally, this has the positive consequence of preventing the discharge of toxic $TiO_2$ nanoparticles into the environment [68]. This type of CP alteration offers different routes for the breakdown of pollutants. In a $TiO_2$ photocatalyst structure made by a straightforward impregnation-hydrothermal technique, for instance, PANI serves as a sensitizer and support to degrade specific dyes (such as reactive brilliant blue, and anthraquinone or KN-R) when exposed to visible light [69]. Furthermore, $TiO_2$ has been linked to conductive poly(3-hexylthiophene) (P3HT) to effectively photocatalyze the decomposition of methylene blue utilizing visible light [70]. Effective charge separation triggered by the heterojunction created between the semiconductor and the conducting polymer facilitates the process. A photocatalytic process's efficiency can be increased by microwave irradiation, such as in the breakdown of Orange G dye using a poly(1-naphthylamine), peptide nucleic acid (PNA)/bentonite clay nanohybrid [68]. Because of their increased porosity, improved charge separation, and photo response range expansion produced by the synergistic interaction of polyaniline/polypyrrole, $TiO_2$, and $WO_3$ carbonaceous $TiO_2/WO_3$ composites with polyaniline and polypyrrole also effectively degrade [71]. Apart from $TiO_2$, various inorganic oxides are also employed as photocatalysts. For instance, the poly(aniline-co-o-aminophenol) [poly (ANIco-oAP)] CP in a ZnO composite is an effective photocatalyst for the degradation of MB [72]. Recognizing the necessity of quick electron transfer kinetics to stop photogenerated electron and hole recombination is a logical step in designing an effective system. The polypyrrole/$Bi_2WO_6$ material was created to facilitate effective charge separation throughout the photodegradation of phenol, a model pollutant, under simulated solar irradiation because conducting polymers have significant charge carrier mobility. Rapid photoinduced charge separation resulted from the transfer of the photogenerated holes in the valence band of $Bi_2WO_6$ to the highest occupied molecular orbital (HOMO) of polypyrrole [73]. Finally, highly ordered $TiO_2$ nanotube arrays that are CP-sensitized can be used to deactivate microorganisms and degrade organic pollutants [74].

## 11.4 Conclusion and Future Prospects

Significant advancements have been made in the synthesis, structural and morphological characteristics, and chemical and physical properties of conducting polymer nanotubes and nanofibers over the past 20 years. The unique characteristics of CP include high conductivities, structural alterations, catalytic behavior, and specific electrochemical and optical properties as covered in this chapter. Conducting polymer nanotubes and the nanofiber's application potential in fields ranging from drug delivery, protein purification, neural interfaces, supercapacitors, energy storage, actuators, environmental remediation (adsorption of pollutants, oxidation/reduction of pollutants, use of 3D material, gas removal, photocatalysis, dioxygen removal), electronic devices to biochemical sensing and energy harvesting also covered in this chapter. These issues include the reproducibility and controllability of polymer nanotubes or wires, the stability of the doping level, and enhancing the processability of CP nanostructures. Because of their high specific capacitance, supercapacitors made of CP are still gaining popularity in the field of energy applications. Nevertheless, they are not very stable. It is anticipated that using an additional nanocomponent as the conducting polymer scaffold could improve the supercapacitor device's stability. Research is still needed, to develop innovative techniques for producing these materials, identify their intriguing and improved qualities, and broaden their range of uses.

## References


1. Le T-H, Kim Y, Yoon H (2017) Electrical and electrochemical properties of conducting polymers. Polymers 9(4):150
2. Jangid NK, Chauhan NPS, Meghwal K, Ameta R, Punjabi PB (2015) Synthesis of dye-substituted polyanilines and study of their conducting and antimicrobial behavior. Cogent Chem 1(1):1084666
3. Jangid NK, Chauhan NPS, Punjabi PB (2014) Novel dye-substituted polyanilines: conducting and antimicrobial properties. Polym Bull 71:2611–2630
4. Chaubisa P, Dharmendra D, Vyas Y, Chundawat P, Jangid NK, Ameta C (2024) Synthesis and characterization of PANI and PANI-indole copolymer and study of their antimalarial and anti-tuberculosis activity. Polym Bull 81(4):3333–3353
5. Long Y, Chen Z, Shen J, Zhang Z, Zhang L, Xiao H, Wan M, Duvail JL (2006) Magnetic properties of conducting polymer nanostructures. J Phys Chem B 110(46):23228–23233
6. Lu X, Mao H, Chao D, Zhang W, Wei Y (2006) Ultrasonic synthesis of polyaniline nanotubes containing $Fe_3O_4$ nanoparticles. J Solid-State Chem 179(8):2609–2615

7. Long Y, Chen Z, Duvail JL, Zhang Z, Wan M (2005) Electrical and magnetic properties of polyaniline/$Fe_3O_4$ nanostructures. Phys B Condens Matter 370(1–4):121–130
8. Martin CR (1994) Nanomaterials: a membrane-based synthetic approach. Science 266(5193):1961–1966
9. Zhang Z, Deng J, Shen J, Wan M, Chen Z (2007) Chemical one-step method to prepare polyaniline nanofibers with electromagnetic function. Macromol Rapid Commun 28(5):585–590
10. Long Y, Chen Z, Zhang X, Zhang J, Liu Z (2004) Synthesis and electrical properties of carbon nanotube polyaniline composites. Appl Phys Lett 85(10):1796–1798
11. Su C, Wang G, Huang F (2007) Preparation and characterization of composites of polyaniline nanorods and multiwalled carbon nanotubes coated with polyaniline. J Appl Polym Sci 106(6):4241–4247
12. Feng W, Bai XD, Lian YQ, Liang J, Wang XG, Yoshino K (2003) Well-aligned polyaniline/carbon-nanotube composite films grown by in-situ aniline polymerization. Carbon 41(8):1551–1557
13. Long Y, Chen Z, Zhang X, Zhang J, Liu Z (2004) Electrical properties of multi-walled carbon nanotube/polypyrrole nano cables: percolation-dominated conductivity. J Phys D Appl Phys 37(14):1965
14. Pillalamarri SK, Blum FD, Tokuhiro AT, Bertino MF (2005) One-pot synthesis of polyaniline– metal nanocomposites. Chem Mater 17(24):5941–5944
15. Huang K, Zhang Y, Long Y, Yuan J, Han D, Wang Z, Niu L, Chen Z (2006) Preparation of highly conductive, self-assembled gold/polyaniline nanocables and polyaniline nanotubes. Chem Eur J 12(20):5314–5319
16. Long Y, Huang K, Yuan J, Han D, Niu L, Chen Z, Changzhi G, Jin A, Duvail JL (2006) Electrical conductivity of a single Au/polyaniline microfiber. Appl Phys Lett 88(16)
17. Luo S-C, Sekine J, Zhu B, Zhao H, Nakao A, Hsiao-hua Y (2012) Polydioxythiophene nanodots, nonowires, nano-networks, and tubular structures: the effect of functional groups and temperature in template-free electropolymerization. ACS Nano 6(4):3018–3026
18. Cui Y, Lieber CM (2001) Functional nanoscale electronic devices assembled using silicon nanowire building blocks. Science 291(5505):851–853
19. Zhang JP, Chu DY, Wu SL, Seng-Tiong Ho WG, Bi CWT, Tiberio RC (1995) Photonic-wire laser. Phys Rev Lett 75(14):2678
20. Xi Y, Zhou J, Guo H, Cai C, Lin Z (2005) Enhanced photoluminescence in core-sheath CdS-PANI coaxial nano cables: a charge transfer mechanism. Chem Phys Lett 412(1–3):60–64
21. Turac E, Sahmetlioglu E, Toppare L, Yuruk H (2010) Synthesis, characterization and opto-electrochemical properties of poly (2, 5-di (thiophen-2-yl-) 1-(4-(thiophen-3-yl) phenyl)-1H-pyrrole-co-EDOT). Des Monomers Polym 13(3):261–275
22. Tan EPS, Lim CT (2006) Mechanical characterization of nanofibers–a review. Compos Sci Technol 66(9):1102–1111
23. Cuenot S, Demoustier-Champagne S, Nysten B (2000) Elastic modulus of polypyrrole nanotubes. Phys Rev Lett 85(8):1690
24. Sun L, Han RPS, Wang J, Lim CT (2008) Modeling the size-dependent elastic properties of polymeric nanofibers. Nanotechnology 19(45):455706
25. Muraoka M, Tobe R (2009) Mechanical characterization of nanowires based on optical diffraction images of the bent shape. J Nanosci Nanotechnol 9(8):4566–4574
26. Gordon MJ, Baron T, Dhalluin F, Gentile P, Ferret P (2009) Size effects in mechanical deformation and fracture of cantilevered silicon nanowires. Nano Lett 9(2):525–529
27. Guo J-G, Zhao Y-P (2007) The size-dependent bending elastic properties of nanobeams with surface effects. Nanotechnology 18(29):295701
28. Xia F, Jiang L (2008) Bio-inspired, smart, multiscale interfacial materials. Adv Mater 20(15):2842–2858

29. Zhong W, Chen X, Liu S, Wang Y, Yang W (2006) Synthesis of highly hydrophilic polyaniline nanowires and sub-micro/nanostructured dendrites on poly (propylene) film surfaces. Macromol Rapid Commun 27(7):563–569
30. Zhu Y, Li J, Wan M, Jiang L (2008) Superhydrophobic 3D microstructures assembled from 1D nanofibers of polyaniline. Macromol Rapid Commun 29(3):239–243
31. Xu L, Chen Z, Chen W, Mulchandani A, Yan Y (2008) Electrochemical synthesis of perfluorinated ion doped conducting polyaniline films consisting of helical fibers and their reversible switching between superhydrophobicity and superhydrophilicity. Macromol Rapid Commun 29(10):832–838
32. Zhu Y, Feng L, Xia F, Zhai J, Wan M, Jiang L (2007) Chemical dual-responsive wettability of superhydrophobic PANI-PAN coaxial nanofibers. Macromol Rapid Commun 28(10):1135–1141
33. Villafiorita-Monteleone F, Canale C, Gianvito Caputo P, Cozzoli D, Cingolani R, Fragouli D, Athanassiou A (2011) Controlled swapping of nanocomposite surface wettability by multilayer photopolymerization. Langmuir 27(13):8522–8529
34. Wan M, Li J, Li S (2001) Microtubules of polyaniline as new microwave absorbent materials. Polym Adv Technol 12(11–12):651–657
35. Liu C, Jiao Y, Zhang L, Xue M, Zhang F (2007) Electromagnetic wave absorbing property of polyaniline/polystyrene composites. Acta Metallurgica Sinica-Chinese Edition 43(4):409
36. Kim S, Kim J-H, Jeon O, Kwon IC, Park K (2009) Engineered polymers for advanced drug delivery. Eur J Pharm Biopharm 71(3):420–430
37. Oh JK, Drumright R, Siegwart DJ, Matyjaszewski K (2008) The development of microgels/nanogels for drug delivery applications. Prog Polym Sci 33(4):448–477
38. Das TK, Prusty S (2012) Review on conducting polymers and their applications. Polym-Plast Technol Eng 51(14):1487–1500
39. Dai Y, Zhang C, Cheng Z, Li C, Kang X, Yang D, Lin J (2012) pH-responsive drug delivery system based on luminescent $CaF_2$: $Ce^{3+}/Tb^{3+}$-poly (acrylic acid) hybrid microspheres. Biomaterials 33(8):2583–2592
40. Otero TF, Martinez JG, Arias-Pardilla J (2012) Biomimetic electrochemistry from conducting polymers. A review: artificial muscles, smart membranes, smart drug delivery, and computer/neuron interfaces. Electrochim Acta 84:112–128
41. Conzuelo LV, Arias-Pardilla J, Cauich-Rodríguez JV, Smit MA, Otero TF (2010) Sensing and tactile artificial muscles from reactive materials. Sensors 10(4):2638–2674
42. Kiefer R, Martinez JG, Kesküla A, Anbarjafari G, Aabloo A, Otero TF (2016) Polymeric actuators: solvents tune reaction-driven cation to reaction-driven anion actuation. Sensors Actuators B Chem 233:328–336
43. Ibanez JG, Rincón ME, Gutierrez-Granados S, Chahma M'h, Jaramillo-Quintero OA, Frontana-Uribe BA (2018) Conducting polymers in the fields of energy, environmental remediation, and chemical–chiral sensors. Chem Rev 118(9):4731–4816
44. Khan F, Dahman Y (2012) A novel approach for the utilization of biocellulose nanofibres in polyurethane nanocomposites for potential applications in bone tissue implants. Des Monomers Polym 15(1):1–29
45. Lövenich W (2014) PEDOT-properties and applications. Polym Sci Ser C 56(1):135–143
46. Abidian MR, Martin DC (2009) Multifunctional nanobiomaterials for neural interfaces. Adv Funct Mater 19(4):573–585
47. Kim D-H, Richardson-Burns S, Povlich L, Abidian MR, Spanninga S, Hendricks JL, Martin DC (2008) Soft, fuzzy, and bioactive conducting polymers for improving the chronic performance of neural prosthetic devices. In: Indwelling neural implants: strategies for contending with the in vivo environment, pp 165–207
48. Kang K, Meng YS, Breger J, Grey CP, Ceder G (2006) Electrodes with high power and high capacity for rechargeable lithium batteries. Science 311(5763):977–980
49. Chen J, Cheng F (2009) Combination of lightweight elements and nanostructured materials for batteries. Acc Chem Res 42(6):713–723

50. Zhao X, Lv L, Pan B, Zhang W, Zhang S, Zhang Q (2011) Polymer-supported nanocomposites for environmental application: a review. Chem Eng J 170(2–3):381–394
51. Krishnani KK, Sira Srinives BC, Mohapatra VM, Boddu JH, Meng X, Mulchandani A (2013) Hexavalent chromium removal mechanism using conducting polymers. J Hazard Mater 252:99–106
52. Venkatachalam K, Arzuaga X, Chopra N, Gavalas VG, Jian X, Bhattacharyya D, Hennig B, Bachas LG (2008) Reductive dechlorination of 3, 3′, 4, 4′-tetrachlorobiphenyl (PCB77) using palladium or palladium/iron nanoparticles and assessment of the reduction in toxic potency in vascular endothelial cells. J Hazard Mater 159(2–3):483–491
53. Çirmi D, Aydın R, Köleli F (2015) The electrochemical reduction of nitrate ion on polypyrrole coated copper electrode. J Electroanal Chem 736:101–106
54. García-Fernández MJ, Sara Sancho-Querol M, Pastor-Blas M, Sepúlveda-Escribano A (2017) Surfactant-assisted synthesis of conducting polymers. Application to the removal of nitrates from water. J Colloid Interface Sci 494:98–106
55. Gallardo-Rivas NV, Guzman J, Gutierrez-Granados S, Garcia-Jimenez MG, Ibanez JG, Paramo-Garcia U (2015) Kinetics and product analysis of the electrochemical dehalogenation of hexachlorobenzene. J Braz Chem Soc 26(4):790–796
56. Janaki V, Byung-Taek O, Shanthi K, Lee K-J, Ramasamy AK, Kamala-Kannan S (2012) Polyaniline/chitosan composite: an eco-friendly polymer for enhanced removal of dyes from aqueous solution. Synth Met 162(11–12):974–980
57. Zaremotlagh S, Hezarkhani A (2014) Removal of textile dyes from aqueous solution by conducting polymer modified clinoptilolite. Environ Earth Sci 71:2999–3006
58. Din MI, Ata S, Mohsin IU, Rasool A, Aziz AA (2014) Evaluation of conductive polymers as an adsorbent for eradication of As (III) from aqueous solution using inductively coupled plasma optical emission spectroscopy (ICP-OES). Int J Sci Eng 6(2):154–162
59. Pathania D, Sharma G, Amit Kumar M, Naushad SK, Sharma A, Zeid Abdullah AL, Othman. (2015) Combined sorptional–photocatalytic remediation of dyes by polyaniline Zr (IV) selenotungstophosphate nanocomposite. Toxicol Environ Chem 97(5):526–537
60. Han J, Fang P, Dai J, Guo R (2012) One-pot surfactant less route to polyaniline hollow nanospheres with in continuous multi cavities and application for the removal of lead ions from water. Langmuir 28(15):6468–6475
61. Bhattacharya P, Geitner NK, Sarupria S, Ke PC (2013) Exploiting the physicochemical properties of dendritic polymers for environmental and biological applications. Phys Chem Chem Phys 15(13):4477–4490
62. Seredych M, Pietrzak R, Bandosz TJ (2007) Role of graphite oxide (GO) and polyaniline (PANI) in $NO_2$ reduction on GO-PANI composites. Ind Eng Chem Res 46(21):6925–6935
63. Seema H, Christian Kemp K, Le NH, Park S-W, Chandra V, Lee JW, Kim KS (2014) Highly selective $CO_2$ capture by S-doped microporous carbon materials. Carbon 66:320–326
64. Lakard B, Carquigny S, Segut O, Patois T, Lakard S (2015) Gas sensors based on electrodeposited polymers. Metals 5(3):1371–1386
65. Yang S-B, Jo W-K, Cho S-B, Mee-Seon Y (2015) Removal of ammonia gas via conducting polymer-assisted titania under visible-light or UV exposure. Asian J Chem 27(11)
66. Chen H, Nanayakkara CE, Grassian VH (2012) Titanium dioxide photocatalysis in atmospheric chemistry. Chem Rev 112(11):5919–5948
67. Arora R, Mandal UK, Sharma P, Srivastav A (2015) $TiO_2$ and PVA based polyaniline composite materials-a review. Mater Today Proc 2(4–5):2767–2775
68. Riaz U, Ashraf SM, Ruhela A (2015) Catalytic degradation of orange G under microwave irradiation with a novel nanohybrid catalyst. J Environ Chem Eng 3(1):20–29
69. Singh S, Mahalingam H, Singh PK (2013) Polymer-supported titanium dioxide photocatalysts for environmental remediation: a review. Appl Catal A Gen 462:178–195
70. Liao G, Shuo Chen XIE, Quan HC, Zhang Y (2010) Photonic crystal coupled $TiO_2$/polymer hybrid for efficient photocatalysis under visible light irradiation. Environ Sci Technol 44(9):3481–3485

71. Ghoreishi KB, Asim N, Ramli ZAC, Emdadi Z, Ambar Yarmo M (2016) Highly efficient photocatalytic degradation of methylene blue using carbonaceous $WO_3/TiO_2$ composites. J Porous Mater 23:629–637
72. Gharebaghi A, Magham AHJ, Hokmabadi L (2024) The methylene blue and phenol red photocatalytic degradation study by cyclodextrin-ZnO nanostructures. Struct Chem 35(3):801–812
73. Zhang Z, Wang W, Gao E (2014) Polypyrrole/$Bi_2$ $WO_6$ composite with high charge separation efficiency and enhanced photocatalytic activity. J Mater Sci 49:7325–7332
74. Zhang Q, Hao X, Yan W (2012) Highly ordered $TiO_2$ nanotube arrays: recent advances in fabrication and environmental applications review. Nanosci Nanotechnol Lett 4(5):505–519

# Chapter 12
# Biodegradable Conducting Polymers for Smart Drug Delivery

**Anupama Rajput, Sadiqa Abbas, Prachika Rajput, and Shagufta Jabin**

**Abstract** Although polymers like poly (ethylene-vinyl acetate) (PEVA) which is a nonbiodegradable in nature have been proved useful to a greater extent for drug delivery applications, they have limitations like accumulation in the body, surgical removal after drug delivery, and many more adverse biological responses. Therefore, biodegradable polymers like chitosan and PLA can be implied for more sustainable and act as an efficient biocompatible alternative in such cases. Furthermore, to enhance the effect of biodegradability, polymers are engineered in such a way that they can produce a synergistic response to electrical stimuli to promote one of the major requirements in drug delivery, that is, controlled drug release. On the one hand where the property of the polymer enables effective elimination after drug release, the same offers the stimuli responsive system for variable drug dosing. Overall, this chapter discusses the overview of various biodegradable polymers, their advantages over the conventional polymers and the additional impact of their conductive behavior to generate enhanced efficacy, reduce toxicity, better decomposition, and positive environmental impact.

**Keywords** Biodegradable polymer · Drug delivery · Conducting polymer · Sustainable

A. Rajput · S. Abbas · S. Jabin
Manav Rachna International Institute of Research and Studies, Faridabad, India
e-mail: anupama.set@mriu.edu.in; sadiqa.set@mriu.edu.in; shagufta.set@mriu.edu.in

P. Rajput (✉)
Netaji Subhas University of Technology, Delhi, India
e-mail: prachika.rajput.phd23@nsut.ac.in


S. Jadoun, N. Kumari Jangid (eds.), *Tuning of Conducting Polymers*, Springer Series on Polymer and Composite Materials,
https://doi.org/10.1007/978-3-032-00461-1_12

## 12.1 Introduction

The evolution of biomedical applications, especially in drug delivery, has been marked by significant advancements in materials and technologies [1].

Drug delivery systems (DDSs) are engineered to transport a drug to a targeted location within the body at a regulated pace, enhancing therapeutic effectiveness while reducing adverse effects [2].

The objective of these systems is to enhance the therapeutic outcomes of medications and boost patient adherence. Drug delivery systems exhibit a range of complexity, spanning from basic formulations to advanced systems that react to external stimuli or the physiological conditions of the body [3].

The primary objective is to improve the drug's bioavailability, efficacy, and targeted delivery, ultimately leading to better treatment outcomes. Nonbiodegradable drug delivery systems (DDS) have found extensive application in the realm of drug delivery; however, they are accompanied by a range of considerable limitations that impact their overall efficacy [4].

Although certain systems like liposomes and nanoparticles offer sustained release, many nonbiodegradable carriers fall short due to the absence of an inherent stimuli-responsive mechanism, which restricts their ability for dynamic control, such as responsiveness to pH, temperature, or electric fields. Nonbiodegradable carriers may occasionally facilitate the early release of the encapsulated drug prior to its arrival at the intended site. This may lead to diminished effectiveness of the drug and increased toxicity to unintended tissues. Biomedical drug delivery systems have seen significant progress, enhancing the accuracy, effectiveness, and safety of therapeutic agents.

The progression of biodegradable polymers, nanoparticles, intelligent drug delivery systems, and conducting polymers has produced novel solutions to issues such as targeted delivery, sustained release, and the navigation of biological barriers [5]. Such developments present substantial potential for addressing chronic illnesses, infections, cancers, and genetic conditions, with the capacity to markedly enhance patient outcomes and overall quality of life [6]. In drug delivery systems, the synergistic benefits of biodegradable polymers and conducting polymers (BCPs) for targeted drug delivery, controlled release, and site-specific action are very appealing [7].

This holds true for a wide range of therapeutic applications, including anticancer, antimicrobial, and other biomedical treatments. The unique characteristics of biodegradable polymers, including their biocompatibility, biodegradability, and controlled-release capabilities, alongside those of conducting polymers, which exhibit electrical conductivity, responsiveness to external stimuli, and antimicrobial properties, can collaboratively enhance drug delivery outcomes [8]. The amalgamation of these technologies holds the promise to revolutionize the domain of drug delivery, providing enhanced efficiency, tailored solutions, and reduced invasiveness in treatment methodologies [9].

In light of the wide range of research on BCPs, this chapter investigates into the successful relationship between the biodegradability and conductive properties of polymers, along with their classification. The biomedical applications of these polymers are made apparent by their extensive reach across various domains, such as tissue engineering, where their synergistic properties prove especially advantageous in the field of smart drug delivery systems. Within the discipline of antimicrobial therapies, conducting polymers possess the capability to augment the release of antimicrobial agents, simultaneously delivering localized electrical stimuli to improve efficacy [10]. In the arena of anticancer therapies, these polymers facilitate precise delivery to tumor locations, minimizing adverse effects and enhancing treatment efficacy [11]. Furthermore, their application in ocular and dermal drug delivery facilitates enhanced drug penetration, prolonged release, and minimized systemic exposure, presenting innovative pathways for addressing eye diseases and skin conditions.

This chapter emphasizes the significance of biodegradable and conducting polymers in the development of smart drug delivery systems. It concentrates on their use in antimicrobial, anticancer, ocular, and dermal drug delivery, underscoring their potential to transform modern medicine by enabling more efficient, targeted, and controlled therapeutic strategies [12].

## 12.2 Biodegradable Polymers

Conventional petroleum-derived plastics endure in the environment for extended periods owing to their resistance to degradation. On the other hand, polymers—like biodegradable polymers—are a family of materials that can decompose spontaneously over time [13]. The breakdown of these polymers may occur through the influence of microorganisms, enzymes, or various environmental factors [14]. They decompose into non-toxic constituents, ultimately contributing positively to the environment [15]. A wide variety of products, including packaging, medical equipment, and agricultural goods, rely on these polymers, which are often made from renewable resources like starch, plant oils, or microbes. A plethora of biodegradable alternatives have been developed in response to the growing concern over plastic pollution [16]. One potential application of these polymers is the development of drug delivery systems that allow for the controlled, progressive release of medicinal substances. Ideal for illnesses requiring long-term medicine, these controlled-release formulations decrease the need for frequent dosages, enhance patient compliance, and assure prolonged therapeutic effects (Fig. 12.1). Some of the examples in the field include chitosan (deacylated form of chitin), polyethylene glycol (PEG), polylactic acid (PLA), and alginate [17]. Properties on which the inclusivity of biodegradable polymers in the area of smart drug delivery can be justified are illustrated in Fig. 12.1. Apart from many advantages, these polymers may possess limitations like inconsistent drug release, poor mechanical strength, limited eletrical

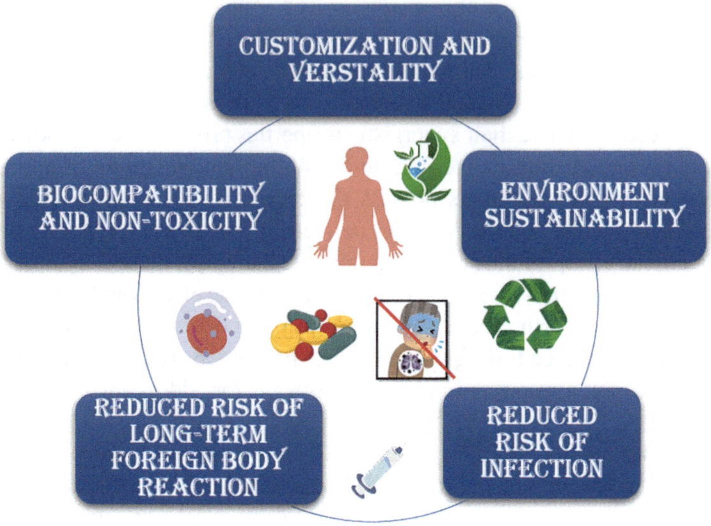


**Fig. 12.1** The advantageous characteristics of biodegradable polymers in the area of biomedical applications

control, and uncontrolled rapid degradation [18]. Therefore, to overcome these, another class of polymers known biodegradable conducting polymers are introduced for effective therapeutic actions.

## 12.3 Biodegradable Conducting Polymers

Biodegradable conducting polymers constitute a distinctive class of polymers that amalgamate the electrical conductivity inherent in synthetic polymers with the environmentally sustainable qualities of biodegradability [19]. The materials in question are attracting significant attention in fields such as biomedical applications, electronics, and sensors, where the necessity for environmental sustainability harmonizes with functional performance [20]. The underlying concept of biodegradable conducting polymers implies the integration of the properties of traditional conducting polymers with elements capable of degradation, enabling them to disintegrate over time under the influence of environmental or biological factors, thus reducing long-term waste accumulation [21]. The utilization of conducting polymers facilitates the modulation of release rates through the establishment of electrostatic interactions with drug molecules. Through the meticulous adjustment of both the composition and the quantity of conducting polymer within the composite, one can achieve a more precise control over the degradation rate, thereby facilitating the attainment of more consistent drug release profiles [22]. The utilization of an electrical current or potential allows the conducting polymer to participate in redox reactions or electrochemical doping, thereby enabling the release of the encapsulated drug or modulating the release rate. This responsiveness enables on-demand drug delivery, demonstrating particular advantages for targeted therapies and neural stimulation [23]. BCPs enable the advancement of electro-responsive drug delivery and stimuli-responsive drug delivery systems, thereby improving the efficacy of intelligent drug delivery mechanisms. The $\pi - \pi$ conjugation in chemical structure of polymers makes them electrically conductive. They lack biodegradability on their own thus are further combined with other materials to enhance the said property [24]. Following we have exemplified some types of conducting polymers and their respective monomers (Table 12.1).

I. **Poly(3,4-ethylenedioxythiophene) (PEDOT)**
   It is a polymer of EDOT (3,4-ethylenedioxythiophene) monomers. These monomers are joined at 2,5-positions of each five-membered thiophene ring to create linear polymer chains (Table 12.1a). It is significant to note that it is not naturally degraded, thereby it is combined with biodegradable polymers like Chitosan to promote environment sustainability [25].

**Table 12.1** Fabrication of conducting polymers including (a) PEDOT, (b) PANI, (c) polypyrrole

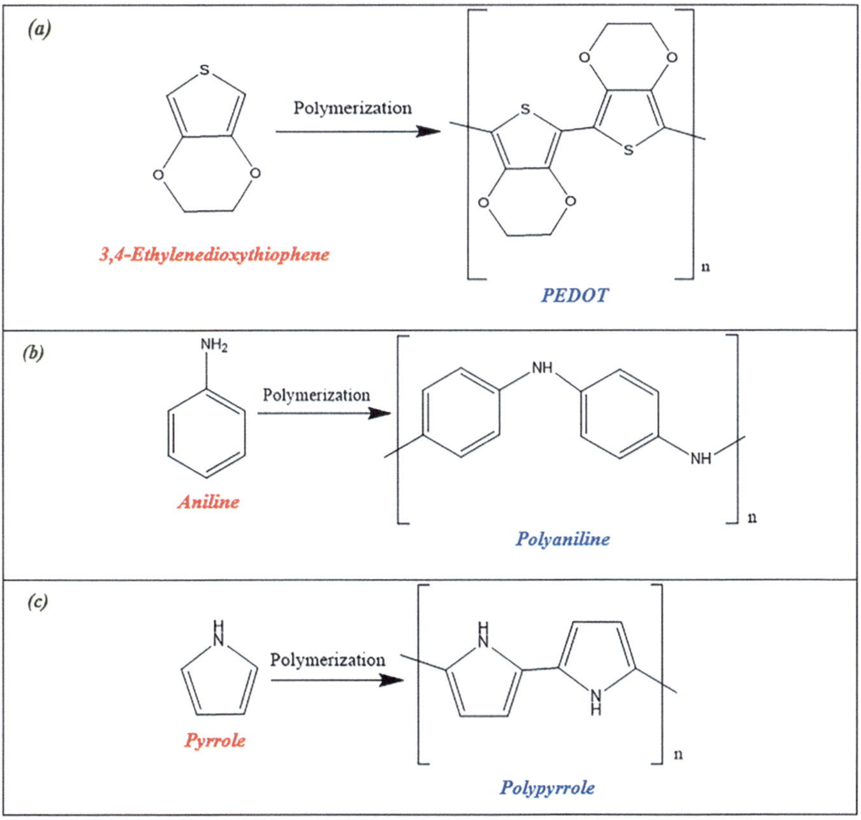

II. **Polyaniline (PANI)**
Polymerization of aniline results in the fabrication of polyaniline which is highly useful polymers in terms of electrically modified applications (Table 12.1b). Incorporating polyaniline with polysaccharides can produce more environmental-friendly materials. The synthesis of polyaniline (PANI) merges the electrical conductivity characteristic of metals with the fascinating attributes of plastics, such as ease of processing and the ability to control both physical and chemical properties [26].

III. **Polypyrrole (PPy)**
Briefly, when five-membered ring bearing nitrogen, i.e., pyrrole is subjected to polymerization, it produces polypyrrole which is highly conductive in nature (Table 12.1c). It is further combined with biodegradable materials like polylactic acid which is obtained from 100% renewable source [27].

## 12.4 Classification of Biodegradable Conducting Polymers

### *12.4.1 Based on Source (Natural vs. Synthetic and Hybrid)*

#### 12.4.1.1 Natural Polymer-Based Biodegradable Conducting Polymers

These polymers come from natural sources like polysaccharides (such starch and alginate) or chitosan (found in crustaceans). For use in biomedicine, they can be combined with conductive polymers similar to alginate + polyaniline and chitosan + polypyrrole. Although synthetic biodegradable polymers are also frequently used in medical contexts, both natural and manmade sources are capable of withstanding breakdown. This is probably due to the fact that they are more chemically adjustable and enable the production of personalized designs [28]. Combining synthetic and natural polymers, as well as various synthetic blends, has resulted in polymeric materials that exhibit improved mechanical properties, processing capabilities, and compatibility with cells. By promoting the quick exchange of protons in water and the incorporation of bioactive components in ceramics, it can help in enhancing the hydrophilicity and water absorption properties of the polymer matrix. Consequently, this impacts the polymer degradation process.

These polymers come from natural sources like polysaccharides (such starch and alginate) or chitosan (found in crustaceans). For use in biomedical applications, they can be combined with conductive polymers.

#### 12.4.1.2 Synthetic Polymer-Based Biodegradable Conducting Polymers

These are made from monomers or oligomers and usually have more stable and controlled characteristics, such as PCL + polypyrrole and PLA + PEDOT.

Similar to their related CPs, conducting oligomers often have well-defined structures, electroactivity, and redox behavior [29]. Nevertheless, the use of oligomers presents numerous benefits compared to CPs, including a well-defined structure, enhanced flexibility in production and treating, improved solubility, and the ability to degrade over time and be eliminated by the kidneys [30]. Actually, studies on the pyrrole, aniline, and thiopene oligomers have shown that they have better properties of biodegradability allow for easier processing and copolymerization with biodegradable polymers. Their electro activity is comparable to that of their conductive polymer counterparts [31].

Multiple studies have shown that $C_6H_5NH_2$(Aniline) oligomers, particularly $C_6H_5NH_2$ (aniline) trimer, $C_6H_5NH_2$ (aniline) tetramer, and $C_6H_5NH_2$ (aniline) pentamer, exhibit additional straightforward also accessible production and treating stages than pyrrole and thiophene oligomers [29]. Most importantly, the macrophages have a far easier time absorbing and internalizing these oligomers, which the kidneys can then eliminate. As a result, conducting oligomers have become a promising choice for creating conducting and fully biodegradable polymeric structures. Oligomer-based biodegradable conducting polymers are generally assembled in a linear macromolecular structure by linking hetero aromatic oligomers to biodegradable polymer segments through biodegradable ester bonds. A significant amount of investigation has been conducted on this approach for developing decomposable transitory supports intended used for nerve manufacturing solicitations [19].

#### 12.4.1.3 Hybrid Polymer-Based Biodegradable Conducting Polymers

The decomposable portion of the recyclable CPs has been integrated into polyesters for example PLA, PCL, and PGA, along with mixture inorganic-organic polymers, especially derivative of poly phosphazene [32]. Because of their adaptable cross chains, which may be situated altered through a variety of efficient sets, these polymers are special. The number of oligomers in the recyclable CPs may be easily and precisely adjusted thanks to this side chain alteration, controlling both their total conductivity and biodegradability.

The adaptability of performing oligomers combined by the large variety of biodegradable polymers and macromolecular structure made it possible to functionalize $C_6H_5NH_2$ (aniline) pentamer and glycine ethyl ester on PGAP (polyphosphazene) copolymer to create a biodegradable electroactive CP that may find use in nerve tissue engineering [33].

### *12.4.2 Polymers Based on Biodegradation Mechanism*

Biodegradable conducting polymers are often formed by integrating conducting polymers with either natural or synthetic biodegradable polymers. These mixtures enhance the conducting polymer's biodegradability without compromising its electrical characteristics. Polypyrrole and Chitosan, Polyaniline and Polycaprolactone (PCL), PEDOT and polylactic acid (PLA), and so on are examples of such combination of biodegradable conducting polymers.

Depending on the type of biodegradable polymer employed in the blend, these polymers break down by microbial activity, enzymatic action, or hydrolysis. These mixtures are frequently utilized in wound healing, medication delivery, and biomedical equipment.

These include crosslinked biodegradable electroactive elastomers (illustrated in Fig. 12.2), Linear, star-shaped branched biodegradable electroactive copolymers derived from $C_6H_5NH_2$ (aniline) trimer and PLA, hyper-branched biodegradable electroactive copolymers formulated from $C_6H_5NH_2$ (aniline) tetramer and PLA68, as well as from $C_6H_5NH_2$(aniline) pentamer and PCL, alongside hydrogels created from $C_6H_5NH_2$(aniline) trimer, PLLA, and PU-urea, and those based on aniline tetramer, chitosan, and glutaraldehyde, also utilizing $C_6H_5NH_2$ (aniline) tetramer [34].

PU, poly (glycerol sebacate), $C_6H_5NH_2$ (aniline) pentamer, PLA, glycidyl methacrylate, and ethylene glycol dimethacrylate are all significant materials in the field of study. These biodegradable CPs based on $C_6H_5NH_2$ (aniline) oligomers usually exhibit improved electrical [35].

These biodegradable CPs based on aniline oligomers usually exhibit improved electrical conductivity between 10–6 and 10–4 S/cm [36]. Biodegradable polymers have been synthesized through the integration of the chemical degradability inherent in aliphatic polyesters and the electroactive properties of conducting polymers derived from PCL, PLA, and $C_6H_5NH_2$(aniline) oligomers.

### 12.4.3 Polymers Based on Chemical Composition

A conducting polymer that has been extensively researched, polyaniline can be made more biodegradable by combining it with biodegradable substances like polysaccharides (like chitosan). Another typical conducting polymer with strong conductivity and good environmental resilience is polypyrrole. Like PANI, PPy is

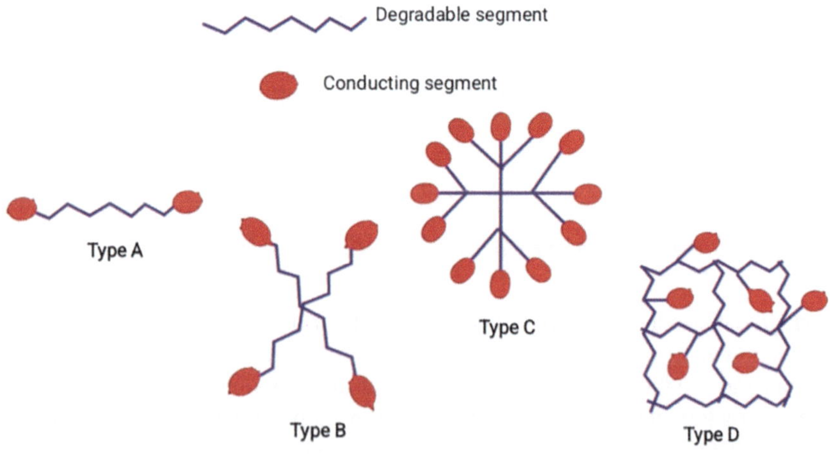


**Fig. 12.2** Various macromolecular structures of the BCPS built on oligomers: **a**. line copolymer, **b**. star-shaped copolymer. **c**. hyper-branched copolymer, and d. crosslinked network architectures

not biodegradable on its own; however, it can be made more biodegradable by blending it with biodegradable substances like PCL or PLA. PEDOT is a highly stable and conductive polymer that is frequently seen in energy storage and electronic devices. Although PEDOT alone is not biodegradable, it can be mixed with biodegradable substances like polyesters, chitosan, or starch to create biodegradable composites. Design and fabrication of an innovative electroactive macro monomer utilizing PDLLA and an end group functionalized with 3,4-ethylenedioxythiophene (EDOT). With a focus on the proposed synthesis for biomedical applications was executed through either an enzymatic process that uses either PS-IM or CAL-B, two distinct lipases, to promote the ring-opening polymerization of PLA, or through a conventional method employing an organometallic catalyst, specifically stannous octoate [37].

The incorporation of biodegradable imidazole units into the conjugated backbone facilitated the development of biodegradable fluorescent conjugated polymer nanoparticles (CPNs) for biomedical imaging applications, as evidenced by a recent study [38].

In this case, completely conjugated biodegradable CPNs were generated by copolymerizing the thiophene monomer unit combined through the imidazole monomer. In order to make the degraded products more soluble and to make the resultant nanoparticles more stable in water, The thiophene monomer underwent modification with either oligo ethyleneoxy (OEG) or methoxy (OMe) side groups. An imidazole degradative oxidation mechanism was noted to break down the constructed CPNs upon exposure to reactive oxygen species (ROS) [38].

### 12.4.4 BCPs Based on Modified Monomers

Conducting polymers (CPs) represent a category of materials that combine electrical conductivity with the lightweight, flexible, and moldable qualities of ordinary polymers. Their prospective uses in a variability of industries, as well as bioelectronics, energy loading devices, flexible electronics, and sensors, have garnered a lot of interest. However, the lack of biodegradability of many conventional conducting polymers (such polypyrrole, polyaniline, and polythiophene) raises questions regarding their long-term sustainability and potential effects on the environment.

Researchers are looking on biodegradable conducting polymers, which combine the benefits of electrical conductivity with biodegradability, to address these problems. The creation of biodegradable conducting polymers from modified monomers is a promising strategy. Biodegradability and electrical conductivity can be achieved by adding functional groups or changing the chemical structure of the monomers used to create these polymers.

Toppare and Yagci have conducted significant work utilizing the "grafting through" method for electroactive macro-monomers, primarily focusing on pyrrole and thiophene. The methodology employed led to the acquisition of ICPs that exhibit enhanced processability and mechanical characteristics, with a particular emphasis on biosensing applications [39].

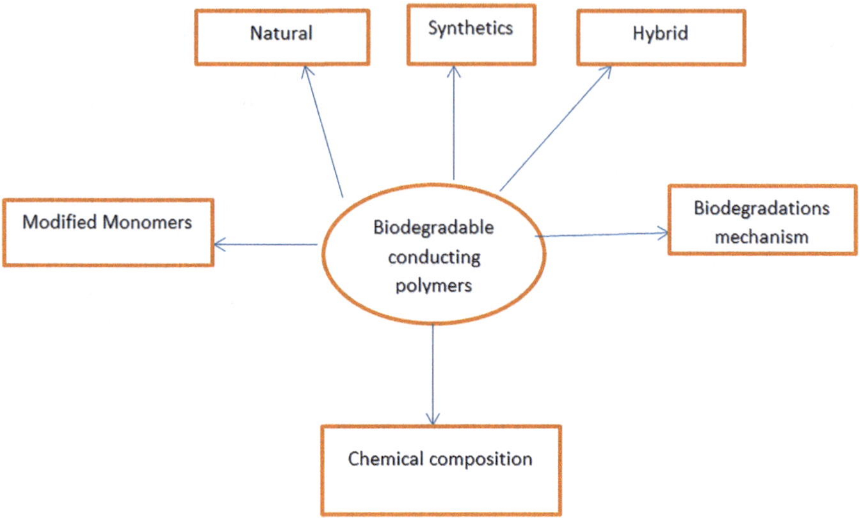


**Fig. 12.3** Classification of BCPs

By altering the monomers to include biodegradable portions, conducting polymers can become more biodegradable. Usually, these biodegradable segments come from bio-based materials or natural sources, like:Based on lactic acid, monomers Polymers with a reputation for biodegradability include polylactic acid (PLA). It is possible to create a biodegradable conductive polymer by mixing PLA with conductive groups [40].

It is possible to design monomers with ester bonds to hydrolyze in the presence of environmental conditions. For instance, biodegradable conducting polymers can be created by copolymerizing ester-based monomers, such as glycolide or lactide derivatives, with conductive units [41]. It is also possible to integrate monomers made from natural polysaccharides such as starch, chitosan, and cellulose. These monomers can offer extra biocompatibility in addition to improving biodegradability, which is crucial for biomedical applications [42]. Fig. 12.3 illustrates the flowchart including classification of BCPs.

## 12.5 Biomedical Applications

### *12.5.1 Tissue Engineering for Skin*

Since the skin is in direct contact with the outside world, there is a great possibility of microbial infection and breakage. For wound dressings, numerous biomaterials having antibacterial properties, such as polyaniline (PANI), have been produced. It has also been demonstrated that conductive polymers can activate biological processes such as fibroblasts and keratinocyte [43].

### *12.5.2 Tissue Engineering for Neural Applications*

Neurons are specialized cells that utilize electrical signals to interact with other nerve cells. Researchers created the conducting polymers PPY and PANI as conductive nanofiber structures to support courage (axon) muscle. Neuritis and neuronal cells grow longer thanks to electroactive PANI/PSS/(TMV) nanofibers [44, 45].

### *12.5.3 Tissue Engineering for Bone*

Bones provide crucial roles in movement, support, and organ protection. However, a number of issues, including fractures, infections, trauma injuries, aging, and illnesses, can weaken bones. Currently, a variety of cutting-edge polymers, such as the development of injectable nanocomposites with biodegradable polymers, enable bone mending. Because of their mechanical qualities, biodegradable polymers are used for bone mending instead of tempered steel [46].

### *12.5.4 Tissue Engineering for Skeletal Muscles*

Since incredibly tiny muscles give the bones strength and voluntary control, it is crucial for a high quality of life that they heal quickly after an injury. Tissue engineering produces difference and influence stem cell development on the scaffold in a number of ways, such as the advancement of innovative electroactive and biodegradable polyurethane-urea (PUU) copolymers and the development of conductive PEG-co-PGES combined with aniline pentamer copolymer facilitates the arrangement of myotubes through the differentiation of myoblast cells [47].

The most researched materials used for film manufacturing and use as wound dressings include probable polymers for example collagen, chitosan, carboxy methyl cellulose, and alginate. There have been reports of the potential usage of man-made polymers, such as PVA, PEG, and PCL, as materials for wound dressings [48, 49].

### *12.5.5 Applications of Electrochromic*

Electrochromic biodegradable polymers like polythiophene (PTh), polypyrrole (PPy), and polyaniline (PANI) are being utilized as electrode materials. These create electrochromic devices and light-emitting diodes (LEDs) based on polymers. In bio-electrochemical systems, PANI-based electrodes have demonstrated high power density. A composite of polyaniline (PANI) and multi-walled carbon nanotubes (MWCNT) has been utilized as a cathode in the microbial electrochemical cell, demonstrating an improvement in hydrogen production. Early breast cancer detection has also been accomplished using the chitosan-PANI electrode [50].

Polyaniline has a wide range of uses, including but not limited to biosensors, antioxidants, drug delivery vehicles, bio-actuators, the food manufacturing, conductive grit, and material production. Poly(ε-caprolactone) is utilized for a variety of reasons, including but not limited to structural integrity, biological qualities (including adhesion, migration, and differentiation of Schwann cells, along with their cosmetic applications. (Different types of application of BCPs have been illustrated in Fig. 12.4.)

Drug delivery, neural prosthetics, and antioxidants all use poly(3,4-ethylenedioxythiophene) (PEDO). Three-dimensional printing allows for the creation of electrochemical sensors that can recognize a wide variety of chemicals. These sensors are built with PLA nanocomposites. By combining PLA with carbon nanotubes at a weight ratio of 4%, a melt spinning technique is utilized to create a textile sensor that is capable of detecting the amount of moisture present [51].

On the other hand, a sensor utilizing a PLA/graphene hybrid filament has the capability to detect 2,4,6-trinitrotoluene (TNT) as well as other nitro-explosives. An electrochemical sensor that uses a graphite and PLA electrode can detect the presence of lead and cadmium ions in jewelry. Dichloromethane, chloroform, tetra-hydro furan, acetone, ethyl acetate, and ethanol are some of the solvents that can be identified by other sensors that are based on PLA and carbon black compounds.

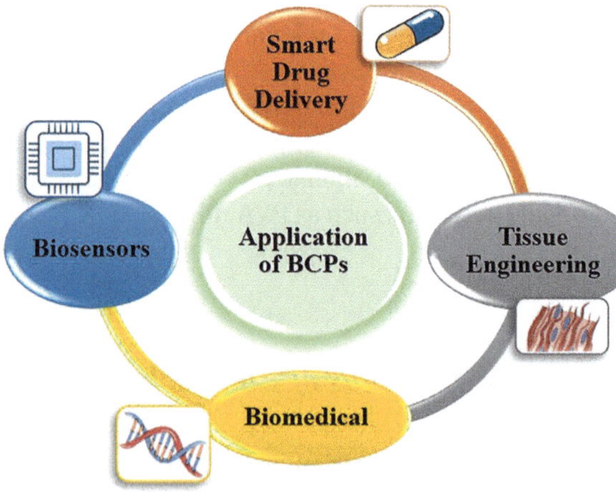


**Fig. 12.4** Application of biodegradable conducting polymers

## 12.6 Biological Applications

The increasing significance of biodegradable conducting polymers (BCPs) has driven quick advancements in this exclusive category of polymeric ingredients. These polymers combine electrical conductivity comparable to metallic conductors with the biodegradability and biocompatibility of natural polymers, making them highly desirable. They have become standout biomaterials with vast potential for enhancing various biomedical applications [52]. They have gained significant attention for their non-toxic properties, environmental friendliness, and excellent automatic power. The detrimental effects of nonbiodegradable plastics have spurred the creation of sustainable alternatives, specifically biodegradable polymers [53].

The demand for biodegradable polymers is increasing significantly and contributing to a global market valued at billions of dollars annually [54]. These polymers have diverse applications, including food packaging, computer peripherals, automotive interior components, and medical uses such as implantable devices, drug delivery systems, and tissue engineering [55, 56]. The following section discusses the applications of biodegradable polymers in some areas.

### *12.6.1 Antibacterial Applications*

Extensive research has explored the role of polymers in antibacterial treatments. While most polymers do not possess intrinsic antibacterial properties, they are frequently utilized as carriers for delivering antibacterial agents to targeted locations. Among them, chitosan is notable for its inherent antibacterial activity. Polymeric nanostructures demonstrate bactericidal effects due to their ability to improve medicine adsorption, provide measured issue, enhance cytocompatibility, and increase solubility. These materials remain highly adaptable, enabling the formation of diverse structures such as vesicles and nanoparticles of varying sizes [57].

Masim et al. (2019) demonstrated that incorporating $ZrO_2$ into a conducting polymer composite significantly enhanced its antibacterial efficacy against *E. coli* and *S. aureus* in contrast to the unadulterated conducting polymers (CPs). Abou Hammad et al. (2019) synthesized a biodegradable nanocomposite comprising polyaniline, cellulose, and cobalt ferrite using an in situ polymerization method. The resulting nanocomposites exhibited notable antibacterial action against *Escherichia coli* and *Bacillus subtilis* as well as antifungal properties against *Candida albicans*. These nanocomposites have a $3.5 \times 10^{-3}$ S/cm natural conductivity and were biodegradable. However, as the cobalt ferrite concentration increased, the biodegradability of the nanocomposites decreased, while their antibacterial effectiveness improved.

Using a chemical polymerization process, a biodegradable film made of polyaniline and chitosan exhibited antibacterial, antifungal, and conductive qualities. It was found that incorporating different amounts of aniline improved the antibacterial performance along with the automated and electrical belongings of the chitosan-based flick. The resulting nanocomposite displayed notable antibacterial activity, highlighting its strong antimicrobial potential [58].

Membranes incorporating PCL and CuO nanoparticles exhibited potent antibacterial activity resistant to *S. aureus*, *E. coli*, and pathogenic bacteria. These membranes significantly reduced the risk of postoperative infections, demonstrating their potential in medical applications [59]. A conductive, biodegradable, and biocompatible hydrogel developed using a bio nanocomposite made of polypyrrole, chitosan, and polyacrylic acid enhanced with varying concentrations of silver nanoparticles (CS/PAA/PPy/Ag-NPs). This composition notably improved the hydrogel's antibacterial action against a range of microorganisms [60].

A conductive composite film developed by mixing chitosan, polycaprolactone, and polypyrrole (PCL/chitosan/PPy) using the casting process. When compared to the PCL/chitosan film, the polypyrrole addition greatly enhanced the antibacterial characteristics of the composite film against *S. aureus* and *E. coli* [61].

A PANI-CS-modified TiO2 ternary nanocomposite fabricated using a simple solution polymerization method enhanced conductivity of nanocomposite effectively and inhibited the growth of various bacterial strains, making it a promising candidate as both an antistatic and antibacterial agent [62].

β-chitin-g-PANI composites synthesized via in situ graft copolymerization of β-chitin with conducting polyaniline (PANI) revealed that resulting composites exhibited improved antibacterial properties, biodegradability, and conductivity. These unique characteristics highlight the potential of these composites for use in wound care applications [63].

Fabrication of PGS-U/PPy/ZnO scaffolds using the salty leaking method enhanced the electrical conductivity due to addition of polypyrrole (PPy), further the inclusion of ZnO particles refined the scaffolds' antibacterial properties and surface hydrophilicity. Additionally, research might look at other adjustments to make these nanocomposites more suitable for usage as efficient electrically conductive scaffolds in tissue engineering applications.

For upcoming wound dressing applications, the composite with a higher particle concentration is the best option. Electroconductivity, antibacterial activity, hemostatic ability, and biocompatibility are among the important qualities that this composite exhibit [64].

### *12.6.2 Antimicrobial Applications*

Microbial infection or contamination presents a significant challenge in various settings, including clinical and hospital environments, clean yields, aquatic refinement schemes, fabric ingredients, medical devices, and food packaging. Antibiotics are formulated to interfere with essential bacterial functions, including the production of cell walls, proteins, DNA, and various other cellular processes to prevent bacterial growth and proliferation (Kapoor, G. et al., 2017). Bacteria have evolved mechanisms to counteract antibiotics by producing specific enzymes. For example, β-lactamase enzymes can break open the β-lactam ring of β-lactam antibiotics, making them ineffective or significantly reducing their potency. This enzymatic defense allows bacteria to resist the effects of these antibiotics and continue to proliferate [65].

Polypyrrole (PPy) is a conjugated polymer having a positively charged backbone that facilitates interaction with the negatively charged bacterial cell surface via electrostatic adherence. Gram-positive bacteria, characterized by polysaccharides, and Gram-negative bacteria, with teichoic acids linked to their peptidoglycan layer, typically exhibit negative zeta potentials. For example, the mean zeta potential of *E. coli* is −44.2 mV, while that of Gram-positive *S. aureus* is −35.6 mV [66].

Conducting polymers like PPy often need extended contact times, sometimes many hours, to effectively bring down viable pathogens, limiting their practicality for many real-world applications. In practical settings, effective antimicrobial activity is typically needed within seconds to a few minutes. Metal and metal oxide nanoparticles offer a promising alternative by disrupting microbial cells through mechanisms such as oxidative stress, membrane damage, and the denaturation of proteins and DNA. Nevertheless, nanoparticles are prone to aggregation and agglomeration due to interactions like van der Waals forces and chemical bonding. This tendency leads to the formation of micron-scale aggregates, reducing their total surface area and consequently diminishing their antimicrobial effectiveness. Nanoparticles can be stabilized in solutions by interacting with long-chain organic molecules, which provide steric stabilization, or through electrosteric stabilization when these molecules carry an electrostatic charge. Conducting polymers with positively charged surfaces play a crucial role in stabilizing nanoparticle suspensions, enabling the maintenance of high nanoparticle concentrations and enhancing their antimicrobial efficacy [67].

Polyaniline (PANI) exhibits germ destroying activity against specific pathogens due to several factors, including a reduced concentration of residual low molecular weight byproducts, the polymer chain length, electrostatic interactions, and the presence of amino groups in its structure. The antimicrobial properties of PANI are expected, as phenol—a component with structural similarity—acts as a potent antimicrobial agent capable of disrupting bacterial membranes. Similar to polypyrrole (PPy), polyaniline (PANI) has been widely utilized to create composites with metals and metal oxides to enhance antibacterial performance. For instance, the synthesis of an Ag-doped ZnO/PANI composite has demonstrated significant germ destroying activities, highlighting the synergistic effects of the components in combating microbial pathogens [68].

## 12.7 Smart Drug Delivery

### *12.7.1 BCPs-Based Dermal Drug Delivery System*

For instance, hyper-branched polyglycerols with oligo caprolactone segments in their backbone (hPGOCs) serve as effective intradermal drug delivery systems. The ability of these polymers to penetrate the skin enhances the transport of hydrophobic guest molecules through the stratum corneum into deeper layers [69]. Notably, hPGOCs have the capability to break down into smaller fragments when exposed to enzymes found in skin lysates [70]. This property ensures efficient drug release while minimizing tissue accumulation and avoiding significant effects on viable dermal layers [69].

### *12.7.2 BCPs Based Drug Delivery for Ocular Therapy*

Delivering drugs to the eye presents significant challenges due to its complex anatomical and physiological barriers. Biodegradable polymers have shown great potential as effective tools for achieving controlled and targeted drug delivery in the treatment of ocular conditions [71]. This recognition in ocular drug delivery systems (DDSs) is due to their capacity to degrade over time, allowing for controlled and sustained drug release at the selected site. Their customizable degradation profiles allow precise therapeutic delivery, aligning well with the eye's structural and functional requirements. Additionally, the intrinsic characteristics of biodegradable polymers—such as flexibility, biocompatibility, and formulation adaptability—make them ideal for addressing ocular challenges. These polymer-based DDSs offer numerous advantages, including enhanced drug stability, prolonged retention in ocular tissues, refined bioavailability, reduced dosing frequency, and targeted delivery to specific ocular regions, effectively overcoming the eye's natural barriers to meet therapeutic needs [72].

### *12.7.3 BCPs for Anticancer Drug Delivery*

Cancer is a leading global health concern, accounting for an estimated 9.6 million deaths in 2018 [73]. In the recent decades, biodegradable polymeric nanoparticles (BPNPs) have garnered significant attention as carriers for therapeutic agents in anticancer drug delivery systems. These nanoparticles enhance the efficacy of active compounds by enabling targeted delivery to specific sites (cancerous cells) within the body. Biodegradable conducting polymers are particularly well-suited as nanocarriers for drugs with poor solubility, instability, rapid metabolism, or swift systemic elimination. Such therapeutic agents can be efficiently encapsulated and released in a controlled manner, improving their stability and bioavailability. Biodegradable conducting polymers are ideal candidates for use in various delivery platforms due to their biocompatibility and biodegradability [74]. Additionally, the release of drugs from BCP-based carriers can be controlled through electrical stimulation. For example, applying a voltage can modulate the drug release rate, enhancing precision and reducing side effects. The *combination of biodegradable and conducting polymers* can create *electro-responsive drug delivery systems* that can enhance drug penetration into tumors, *reduce off-target effects*, and *provide controlled, on-demand release* of chemotherapy agents. PPy is one of the most studied conducting polymers for biomedical applications. Its ability to encapsulate drugs and respond to electrical fields makes it a good candidate for controlled release and targeted delivery. It has been used to deliver drugs like doxorubicin, with promising results in vitro and in vivo [75]. Addressing cancer effectively requires a multidisciplinary approach that involves working of multidisciplinary teams of biomedical engineers, biologists, and clinicians [76]. Present treatment methods include chemotherapy, radiation, and surgery, but these approaches can harm healthy cells along with cancerous ones. This many a times lead to systemic toxicity and unfavorable side effects, limiting the upper limit of safe dose of anticancer drugs and reducing their effectiveness. Surgery and radiotherapy are typically preferred for treating localized, non-metastatic cancers, whereas chemotherapy, hormonal, and biological therapies are commonly used for metastatic cancers and as part of adjuvant treatments [77].

## 12.8 Conclusion

This chapter has explored the biodegradable conducting polymers which have come up as a transformative class of materials with vast potential across various areas of healthcare management. The unique characteristics of electrical conductivity, biodegradability, and biocompatibility make them an ideal candidate for advanced applications in areas like biosensors, antimicrobial agents and smart drug delivery systems.

Conducting polymers not only enhance the functionality of drug delivery platforms through controlled-release mechanisms but also address critical challenges such as improving bioavailability and minimizing systemic side effects. Additionally, their inherent antimicrobial and antibacterial properties make them invaluable in combating infections and maintaining sterile environments, particularly in clinical and hospital settings.

The integration of biodegradable conducting polymers in smart drug delivery systems signifies a step forward in personalized medicine, enabling targeted and responsive drug release. This approach holds promise for treating complex diseases such as cancer and chronic infections, where precision and minimal invasiveness are of paramount importance.

Continued research and innovation in this field will likely uncover new possibilities for these polymers, including their incorporation into multifunctional systems that combine diagnostics, treatment, and monitoring in a single platform. The development of eco-friendly and cost-effective production methods will further facilitate their widespread adoption, making biodegradable conducting polymers a cornerstone of future biomedical advancements.

## References


1. Tiwari G, Tiwari R, Bannerjee S, Bhati L, Pandey S, Pandey P, Sriwastawa B (2012) Int J Pharm Investig 2:2
2. Chandrakala V, Aruna V, Angajala G (2022) Emerg Mater 5:1593–1615
3. Adepu S, Ramakrishna S (2021) Molecules 26:5905
4. Sultana A, Zare M, Thomas V, Kumar TSS, Ramakrishna S (2022) Med Drug Discov 15:100134
5. Liechty WB, Kryscio DR, Slaughter BV, Peppas NA (2010) Annu Rev Chem Biomol Eng 1:149–173
6. Aoki K, Saito N (2020) Pharmaceutics 12:95
7. Paramshetti S, Angolkar M, Al Fatease A, Alshahrani SM, Hani U, Garg A, Ravi G, Osmani RAM (2023) Pharmaceutics 15:1204
8. Kumari A, Yadav SK, Yadav SC (2010) Colloids Surf B Biointerfaces 75:1–18
9. Han WB, Yang SM, Rajaram K, Hwang S Adv Sustain Syst. https://doi.org/10.1002/adsu.202100075
10. Amer MS, Ibrahim HAH (2019) Egypt J Aquat Res 45:359–365
11. Gagliardi A, Giuliano E, Venkateswararao E, Fresta M, Bulotta S, Awasthi V, Cosco D Front Pharmacol. https://doi.org/10.3389/fphar.2021.601626
12. Anju S, Prajitha N, Sukanya VS, Mohanan PV (2020) Mater Today Chem 16:100236
13. Samir A, Ashour FH, Hakim AAA, Bassyouni M (2022) NPJ Mater Degrad 6:68
14. Jha K, Kataria R, Verma J, Pradhan S (2019) AIMS Mater Sci 6:119–138
15. Parthasarathy M, John AA (2023) Tribology of polymers, polymer composites, and polymer nanocomposites. Elsevier, pp 235–263
16. Ghanbarzadeh B, Almasi H (2013) Biodegradation—life of science. InTech
17. Deeraj BDS, Jayan JS, Saritha A, Joseph K (2022) Biodegradable polymers, blends and composites. Elsevier, pp 237–281
18. Balaji AB, Pakalapati H, Khalid M, Walvekar R, Siddiqui H (2018) Biodegradable and biocompatible polymer composites. Elsevier, pp 3–32
19. Guo B, Glavas L, Albertsson A-C (2013) Prog Polym Sci 38:1263–1286
20. Das TK, Prusty S (2012) Polym Plast Technol Eng 51:1487–1500
21. George A, Sanjay MR, Srisuk R, Parameswaranpillai J, Siengchin S (2020) Int J Biol Macromol 154:329–338
22. Puiggalí-Jou A, del Valle LJ, Alemán C (2019) J Control Release 309:244–264
23. Mongkolkitikul S, Paradee N, Sirivat A (2018) Eur J Pharm Sci 112:20–27
24. Chang HC, Sun T, Sultana N, Lim MM, Khan TH, Ismail AF (2016) Mater Sci Eng C 61:396–410
25. Bhat MA, Rather RA, Shalla AH (2021) Synth Met 273:116709
26. Kazemi F, Naghib SM, Zare Y, Rhee KY (2021) Polym Rev 61:553–597
27. Street GB, Lindsey SE, Nazzal AI, Wynne KJ (1985) Mol Cryst Liq Cryst 118:137–148
28. Gomes ME, Reis RL (2004) Int Mater Rev 49:261–273
29. Wei Z, Faul CFJ (2008) Macromol Rapid Commun 29:280–292
30. Wang Y, Tran HD, Kaner RB (2011) Macromol Rapid Commun 32:35–49
31. Stejskal J, Trchová M (2012) Polym Int 61:240–251

32. Potin P, De Jaeger R (1991) Eur Polym J 27:341–348
33. Zhang Q, Yan Y, Li S, Feng T (2010) Mater Sci Eng C 30:160–166
34. Chen J, Dong R, Ge J, Guo B, Ma PX (2015) ACS Appl Mater Interfaces 7:28273–28285
35. Guo B, Finne-Wistrand A, Albertsson A-C (2011) Biomacromolecules 12:2601–2609
36. Guo B, Finne-Wistrand A, Albertsson A-C (2011) Chem Mater 23:1254–1262
37. da Silva AC, Higgins MJ, Córdoba de Torresi SI (2019) Mater Sci Eng C 99:468–478
38. Repenko T, Rix A, Ludwanowski S, Go D, Kiessling F, Lederle W, Kuehne AJC (2017) Nat Commun 8:470
39. Yagci Y, Toppare L (2003) Polym Int 52:1573–1578
40. Guo B, Finne-Wistrand A, Albertsson A-C (2010) Macromolecules 43:4472–4480
41. Wang S, Sun Q, Gröning O, Widmer R, Pignedoli CA, Cai L, Yu X, Yuan B, Li C, Ju H, Zhu J, Ruffieux P, Fasel R, Xu W (2019) Nat Chem 11:924–930
42. Maquet V, Boccaccini AR, Pravata L, Notingher I, Jérôme R (2004) Biomaterials 25:4185–4194
43. GhavamiNejad A, Park CH, Kim CS (2016) Biomacromolecules 17:1213–1223
44. Zhang Z, Rouabhia M, Wang Z, Roberge C, Shi G, Roche P, Li J, Dao LH (2007) Artif Organs 31:13–22
45. Wu Y, Wang L, Guo B, Shao Y, Ma PX (2016) Biomaterials 87:18–31
46. Hu D, Qiang T, Wang L (2017) Wound Med 16:15–21
47. Colobatiu L, Gavan A, Potarniche A-V, Rus V, Diaconeasa Z, Mocan A, Tomuta I, Mirel S, Mihaiu M (2019) React Funct Polym 145:104369
48. Ge L, Xu Y, Li X, Yuan L, Tan H, Li D, Mu C (2018) ACS Sustain Chem Eng 6:9153–9166
49. Türe H (2019) Int J Biol Macromol 123:878–888
50. Faraji S, Danesh E, Tate DJ, Turner ML, Majewski LA (2016) J Phys D Appl Phys 49:185102
51. Guo B, Finne-Wistrand A, Albertsson A (2011) J Polym Sci A Polym Chem 49:2097–2105
52. Kenry, Liu B (2018) Biomacromolecules 19:1783–1803
53. Meghana MC, Nandhini C, Benny L, George L, Varghese A (2023) Polym Bull 80:11507–11556
54. Manavitehrani I, Fathi A, Badr H, Daly S, Negahi Shirazi A, Dehghani F (2016) Polymers (Basel) 8:20
55. Yin G-Z, Yang X-M (2020) J Polym Res 27:38
56. Farachi F, Ardisson GB, Innocenti FD (2020) Handbook of biodegradable polymers. De Gruyter, pp 45–74
57. Mao Z, Li Y, Yang Y, Fang Z, Chen X, Wang Y, Kang J, Qu X, Yuan W, Dai K, Yue B Front Pharmacol. https://doi.org/10.3389/fphar.2018.00368
58. Mohammadi B, Pirsa S, Alizadeh M (2019) Polym Polym Compos 27:507–517
59. Muñoz-Escobar A, Ruíz-Baltazar Á d J, Reyes-López SY (2019) Dose-Response 17:155932581986950
60. Youssef AM, El-Aziz MEA, Abd El-Sayed ES, Moussa MA, Turky G, Kamel S (2019) Int J Biol Macromol 140:886–894
61. Talebi A, Labbaf S, Karimzadeh F (2019) Polym Test 75:254–261
62. Mirmohseni A, Rastgar M, Olad A J Appl Polym Sci. https://doi.org/10.1002/app.47629
63. Balitaan JNI, Martin GAV, Santiago KS (2021) J Bioact Compat Polym 36:13–28
64. Unnikrishnan G, Joy A, Megha M, Thomas J, Haris M, Kolanthai E, Muthuswamy S (2024) Polym Compos 45:267–285
65. Bush K, Bradford PA (2016) Cold Spring Harb Perspect Med 6:a025247
66. Halder S, Yadav KK, Sarkar R, Mukherjee S, Saha P, Haldar S, Karmakar S, Sen T (2015) Springerplus 4:672
67. Maruthapandi M, Saravanan A, Gupta A, Luong JHT, Gedanken A (2022) Macromolecules 2:78–99
68. Hou Y, Feng J, Wang Y, Li L (2016) J Mater Sci Mater Electron 27:6615–6622
69. Mohammadifar E, Zabihi F, Tu Z, Hedtrich S, Nemati Kharat A, Adeli M, Haag R (2017) Polym Chem 8:7375–7383
70. Neupane R, Boddu SHS, Renukuntla J, Babu RJ, Tiwari AK (2020) Pharmaceutics 12:152
71. Tsung T-H, Tsai Y-C, Lee H-P, Chen Y-H, Lu D-W (2023) Int J Mol Sci 24:12976

72. Allyn MM, Luo RH, Hellwarth EB, Swindle-Reilly KE Front Med (Lausanne). https://doi.org/10.3389/fmed.2021.787644
73. Bray F, Ferlay J, Soerjomataram I, Siegel RL, Torre LA, Jemal A (2018) CA Cancer J Clin 68:394–424
74. Song R, Murphy M, Li C, Ting K, Soo C, Zheng Z (2018) Drug Des Devel Ther 12:3117–3145
75. Siegel RL, Miller KD, Fedewa SA, Ahnen DJ, Meester RGS, Barzi A, Jemal A (2017) CA Cancer J Clin 67:177–193
76. Danhier F, Ansorena E, Silva JM, Coco R, Le Breton A, Préat V (2012) J Control Release 161:505–522
77. Tran S, DeGiovanni P, Piel B, Rai P Clin Transl Med. https://doi.org/10.1186/s40169-017-0175-0

# Chapter 13
# Tailored Conducting Polymers for Biomedical Engineering

**S. Vimal Kumar, S. Ram Prasanth, S. Neetika Singh, and Suryasarathi Bose**

**Abstract** Conductive polymers (CPs) are organic materials known for their distinct optical and electrical properties resembling metals and inorganic semiconductors. CPs such as polyaniline (PANI), polypyrrole(PPy), polythiophene(PTh), and derivatives like aniline oligomer and poly(3,4-ethylenedioxythiophene) (PEDOT) can be synthesized using simple, adaptable, and cost-efficient approaches. These materials can be established into multifunctional supramolecular structures via basic electropolymerization techniques. Various methods have been developed to modify and optimize CPs, making them suitable for integration into biomedical applications, including biomaterials and biosensors. Advances in CP technology are particularly significant in bioengineering, regenerative medicine, and biosensing, offering the potential for future innovation. CPs have demonstrated the ability to activate different cellular mechanisms, broadening their applicability in the biomedical sector. This review includes information on desired CP qualities pertinent to biomedical applications and specifics on how CPs have been modified to create these traits.

**Keywords** Conducting polymers · Nerve tissue engineering · Sensors · Drug delivery · Polypyrrole

## 13.1 Introduction

Conductive polymers (CPs) are a unique class of organic materials exhibiting electrical and optical properties comparable to inorganic semiconductors and metals. They can be synthesized through simple, versatile, and cost-effective methods and easily assembled into multifunctional supramolecular structures using

S. V. Kumar · S. R. Prasanth · S. N. Singh (✉) · S. Bose (✉)
Department of Materials Engineering, Indian Institute of Science, Bangalore, India
e-mail: vimalkumars@iisc.ac.in; sram@iisc.ac.in; sbose@iisc.ac.in



S. Jadoun, N. Kumari Jangid (eds.), *Tuning of Conducting Polymers*, Springer Series on Polymer and Composite Materials,
https://doi.org/10.1007/978-3-032-00461-1_13

electropolymerization. Numerous strategies have been developed to modify and optimize CPs for seamless integration into biomedical applications, including biomaterials, biosensors, and regenerative medicine. These innovations hold significant promises for advancing biomedicine by fostering bioengineering and medical technologies breakthroughs.

Conducting polymers (CPs) have demonstrated remarkable potential in influencing various cellular mechanisms, significantly expanding their applications in the biomedical field. Their ability to respond to electrical stimuli from different tissues, such as muscle, connective tissue, epithelium, and nervous tissue, makes them ideal for improving biomedical devices' sensitivity, speed, stability, and interactions with biological systems. Additionally, CPs are highly regarded for their biocompatibility, ensuring safe interaction with biological samples and making them suitable for various medical applications.

CPs are materials in which electrons are delocalized along the polymer backbone [1], allowing these π-electrons to move freely within the unsaturated structure [2], creating an electrical pathway for charge carriers [3]. Some of the most widely used CPs include polyacetylene (PA), polythiophene (PT), poly[3,4-(ethylenedioxy)thiophene] (PEDOT), polypyrrole (PPy), polyphenylene, and polyaniline (PANI). The chemical structures of these materials are depicted in Fig. 13.1. This chapter focuses on the use of CPs in bioengineering applications [4] (Table 13.1).

Conducting polymers have been extensively studied for their potential in biomedical applications. ***Langer and colleagues*** were among the first to explore the interface between cells and conducting polymers by culturing mammalian cells on polypyrrole (PPy) thin films [5]. Building on this, ***Williams and Doherty***

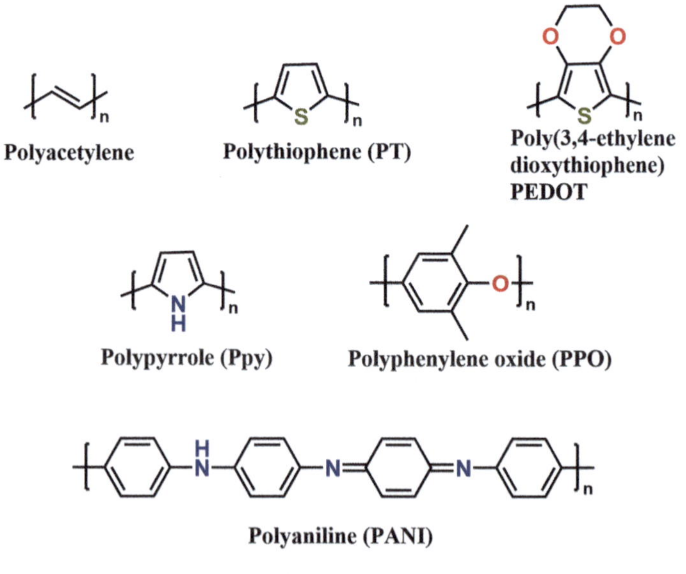


**Fig. 13.1** Chemical structures of π-conjugated polymers

**Table 13.1** Details of conductive polymers

| Conductive polymer | Formula | Electrical conductivity (S/cm) |
|---|---|---|
| Polyacetylene (PA) | $[C_2H_2]_n$ | $10^5$ |
| Polythiophene (PT) | $[C_4H_4S]_n$ | $10^0$–$10^3$ |
| Polypyrrole (PPy) | $[C_4H_2NH]_n$ | 2–100 |
| Poly(*p*-phenylene) (PPP) | $[C_6H_4]_n$ | $10^{-3}$–$10^2$ |
| Polyaniline (PANI) | $[C_6H_4NH]_n$ | $10^{-2}$–$10^0$ |

demonstrated the use of PPy in nerve tissue engineering, showing its cytocompatibility and dual functionality as a nerve guidance channel and a medium for electrical stimulation [6]. Polyaniline (PANi), another widely used CP, can be synthesized chemically or electrochemically from its monomer, aniline. ***Chen*** **et al.** developed aligned poly(ε-caprolactone)/PANi (PCL/PANi) nanofibers using electrospinning. They observed that the alignment significantly improved myoblast orientation and myotube formation, with an approximately 80% increase compared to random PCL scaffolds [7]. Similarly, ***Abdul Rahman*** **et al.** used electrospun poly(l-lactic acid)/poly(aniline-co-m-aminobenzoic acid) (PLLA/P(ANi-co-m-ABA)) nanofibers, culturing human adipose-derived stem cells (hASCs) on these scaffolds. The carboxylic acid functionalization of PANi copolymers enhanced solubility, acted as a self-dopant to eliminate the need for external doping, and improved biocompatibility. After 1 week, hASCs exhibited strong adhesion and proliferation on the nanofibers, with abundant focal adhesions observed [8].

## 13.2 Synthesis of Conducting Polymers

Conducting polymers (CPs) can be synthesized through chemical or electrochemical methods, each with distinct advantages and limitations. Chemical synthesis employs either condensation polymerization (step-growth polymerization) or addition polymerization (chain-growth polymerization). Condensation polymerization involves the elimination of small molecules, such as water or hydrochloric acid, during the reaction. Addition polymerization, on the other hand, proceeds through radical, cationic, or anionic intermediates, with the active end of the polymer chain determining the specific mechanism. This method allows various routes to produce CPs and enables scalability, which remains challenging for electrochemical synthesis.

Electrochemical synthesis offers a relatively straightforward approach, first demonstrated in 1968 when "pyrrole black" was deposited on a platinum electrode by applying an oxidative potential to an aqueous pyrrole solution with sulfuric acid. Modern electrochemical polymerization uses a three-electrode system (working, counter, and reference electrodes) immersed in a solution containing the monomer, a suitable solvent, and an electrolyte (dopant). When the current passes through the system, monomers at the positively charged working electrode undergo oxidation to form reactive radical cations. These cations react with monomers or other radicals to produce insoluble polymer chains deposited on the electrode surface. Several factors, including deposition time, temperature, solvent system, electrolyte composition, and electrode setup, influence the resulting CP films' morphology, mechanics, and conductivity, which are crucial for biomedical applications. For instance, non-protic, non-nucleophilic solvents yield more vigorous and conductive CPs, as protic and nucleophilic solvents can interfere with chain growth through side reactions [9, 10].

The primary distinction between chemical and electrochemical synthesis lies in the resulting CPs. While chemical methods typically produce powders or thick films, electrochemical techniques can generate ultra-thin films as fine as 20 nm. However, electrochemical synthesis is limited to monomers that can be oxidized to form radical ion intermediates. Standard CPs, such as polypyrrole (PPy), polythiophene (PT), polyaniline (PANI), and poly(3,4-ethylenedioxythiophene) (PEDOT), can be synthesized by both methods. In contrast, some novel CPs with modified monomers are restricted to chemical polymerization.

### *13.2.1 Poly(pyrrole)*

One of the notable conducting polymers is poly (pyrrole). This polymer has a heteroaromatic structure and π-conjugated chain backbone, contributing to its conductive nature. Despite the conjugation in the backbone, poly(pyrrole) always requires tuning or modification to exhibit sufficient conductivity [11]. This polymer is generally made using the chemical oxidative polymerization or Electrochemical polymerization methods [12]. The pyrrole monomer is oxidized to give radical cations in the chemical oxidative polymerization method. This radical cation couples with another radical cation, resulting in dihydromer dication formation. Then, the dimer is formed after deprotonation [13]. The exact mechanism (i.e., re-oxidation, coupling, and deprotonation) is followed to prepare the polymer poly(pyrrole). This reaction can be carried out in aqueous media or organic solvents. Ferric chloride and ammonium persulfate are oxidizing agents for the pyrrole polymerization reaction [14, 15]. If the $FeCl_3$ is added more than monomer molar concentration, excess counter chlorine anions will be left over in the solution. If the $FeCl_3$ said is lower than the monomer, the conversion of monomer to polymer will be less than 100% [16]. The mechanism of polymerization is shown in Fig. 13.2.

### *13.2.2 Poly(thiophene)*

Polythiophene is another conducting polymer with a heteroatomic structure, which is widely used in several applications when doped [17]. Polythiophenes were first prepared by metal-catalyzed polycondensation polymerization of dibromo

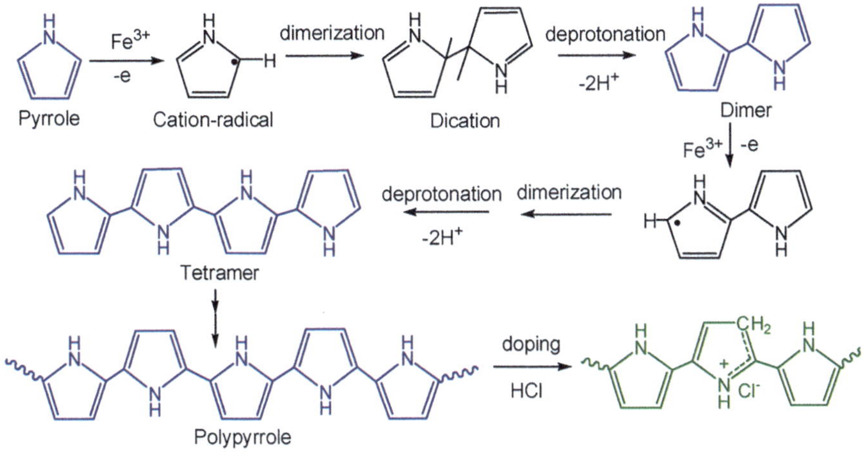


**Fig. 13.2** Chemical oxidative polymerization of poly(pyrrole) [14]

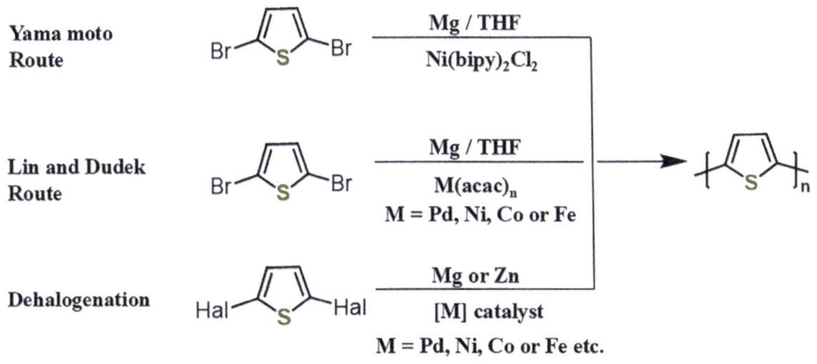


**Fig. 13.3** Synthesis of poly(thiophene) from various routes [21]

thiophene in the early 1980s. There are two types of poly(thiophene) synthesis routes: (i)Yamamoto's route and (ii) the Lin and Dudek route [18, 19] and another route based on dehalogenation [20]. The Yamamoto's route utilizes nickel(bipyridine) dichloride catalyst. In contrast, the Lin and Dudek route utilizes Palladium, nickel, cobalt, ferrous acetyl acetonate, and magnesium metal in THF solvent, which is evident from Fig. 13.3. In these routes, the magnesium/zinc reacts with one of the bromides and forms magnesium/zinc bromide, whereas another bromide reacts with a transition metal catalyst, forming a dimer. This complex undergoes condensation and propagates to create a low molecular weight poly(thiophenes) [17]. We also have other synthesis methods like Wurtz coupling reaction and electrochemical polymerization through radical coupling [11].

#### 13.2.2.1 Derivatives of PTs

Poly(3,4 ethylene dioxythiophene) is an alkoxyl-substituted polymer that falls under the class/derivatives of poly(thiophene) polymers [11]. PEDOT is highly crystalline, brittle, and insoluble in solvents. Nonetheless, it can form thin, transparent oxidized films with a conductivity of 300 S $cm^{-1}$ [22]. Because of its insolubility in water, it is often synthesized with poly(styrene sulfonate) PSS, resulting in PEDOT: PSS. Here, PSS acts as a dopant and a stabilizer for the PEDOT polymer [17]. PEDOT: PSS polymer is produced via (i) electrochemical, (ii) oxidative, and (iii) transition metal mediated coupling polymerizations in the presence of a counterion, which is shown in Fig. 13.4.

### *13.2.3 Poly(aniline)*

Polyaniline is one of the most famous and widely explored conducting polymers. It is also abbreviated as PANI. Conductivity depends upon the concentration of the dopant. When the pH is less than 3, it exhibits metal-like conductivity [24]. PANI exists in three different states depending on the mechanism (oxidation or reduction) used. During the polymerization of aniline monomer through oxidation, it can take: (i) Leucoemeraldine (clear/white), (ii) emeraldine (base-blue/salt-green), (iii) pernigraniline (violet/blue). The leucoemeraldine base is completely reduced PANI, and the Pernigraniline base is completely oxidized PANI, whereas the emerald base is half-oxidized and half-reduced [25]. PANI will be an insulator in its fully oxidized state, whereas it exhibits conductivity in the moderately oxidized state. In the emeraldine base (half-oxidized state), the polymer has an equal concentration of quinoid and benzoid rings. However, the fully oxidized state consists

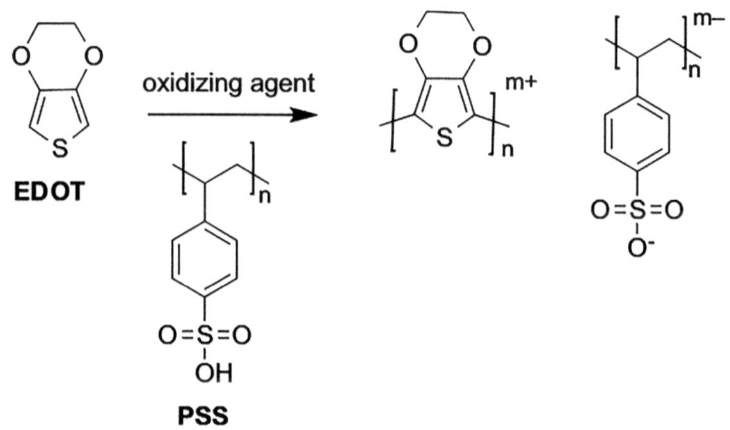


**Fig. 13.4** Synthesis of PEDOT:PSS [23]

of only benzoid rings, and the fully reduced state consists only of quinoid rings [11, 26]. For the synthesis of PANI, we have the chemical oxidation polymerization method and the electrochemical polymerization method. The chemical oxidation method is commonly used to prepare PANI, where the aniline monomer precursor reacts with an oxidizing agent and the dopant acid of their choice (doping is done during the polymerization reaction itself) to give polymer. Ammonium persulfate, ammonium peroxy disulfate, ceric nitrate, ceric sulfate, and potassium bichromate are some of the oxidizing agents used in the polymerization of PANI. If the color of the reaction medium changes to green, it confirms the formation of poly(aniline), Fig. 13.5 [27].

### *13.2.4 Poly(para-phenylene)*

Poly(para-phenylene) is a conducting polymer comprised of benzoid aromatic molecules, which are rigid, brittle, and strong due to the benzene rings' presence in the chain backbone. This polymer also needs tuning or modification to exhibit sufficient electrical conductivity. The solubility improves when flexible side chains are substituted with this polymer. These poly(para-phenylene) based polymers generally show high thermal and mechanical stability [29]. The synthesis of PPP is done by polymerizing benzene through chemical oxidation, a widely accepted and used method. Here, the polymerization reaction mechanism is based on the dehydrocoupling of benzenes and a catalyst-oxidant system. Here, the oxidation takes place using cupric chloride and auric chloride, a binary system where one acts as a lewis acid and another acts as an oxidizing system. We can also use a single reagent system like FeC13, which acts as a lewis acid and an oxidizing agent. This method was developed in the 1960s by the Kovacic group [30]. This reaction method comes under the school reaction [31], evident from Fig. 13.6. The Wurtz-Fittig reaction proposed by Riese in 1872 was the first proposed reaction for poly(para-phenylene)

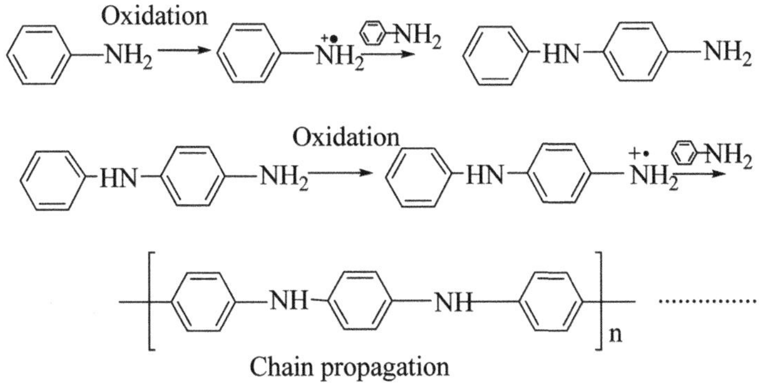


..........................5

**Fig. 13.5** Mechanism proposed for synthesis of PANI [28]

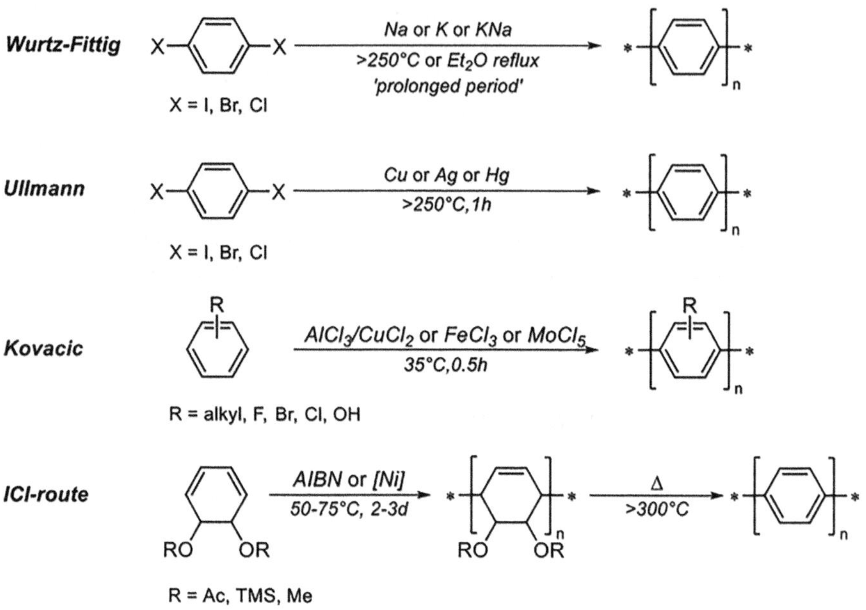


**Fig. 13.6** Synthesis routes for Poly(para-phenylene) conducting polymer [35]

preparation. Here, the reaction is forwarded by creating a carbanion using sodium metal, which reacts with the halide. Then this carbanion reacts with another monomer and forms a dimer and polymer [32]. These polymers can also be prepared using Ullmann coupling reaction [33] through which we can get substituted poly(paraphenylenes). We can also use the ICI route to prepare unsubstituted poly(paraphenylenes) [34].

### *13.2.5 Polyacetylene*

The discovery of poly(acetylene) opened the gates of conducting polymers until people thought that polymers were insulators. This is the first electroactive (conducting) polymer synthesized by Shirakawa in 1970 [36]. Poly(acetylene) can be synthesized through various methods like catalytic polymerization, non-catalytic polymerization, high-pressure polymerization, and so on. Ziegler Natta and Luttinger catalysts are commonly used in catalytic polymerization techniques [28]. We also have various other types of catalysts employed for poly (acetylene) polymerization, which can be referred to in Fig. 13.7. High-molecular weight poly(acetylene) has been synthesized by Natta et al. [37] after the advent of the Ziegler Natta catalyst in 1958. These polymers produced through the Ziegler Natta catalytic method are highly crystalline with high stereospecificity, and the catalysts are soluble in organic solvents. Luttinger catalysts are also used to synthesize

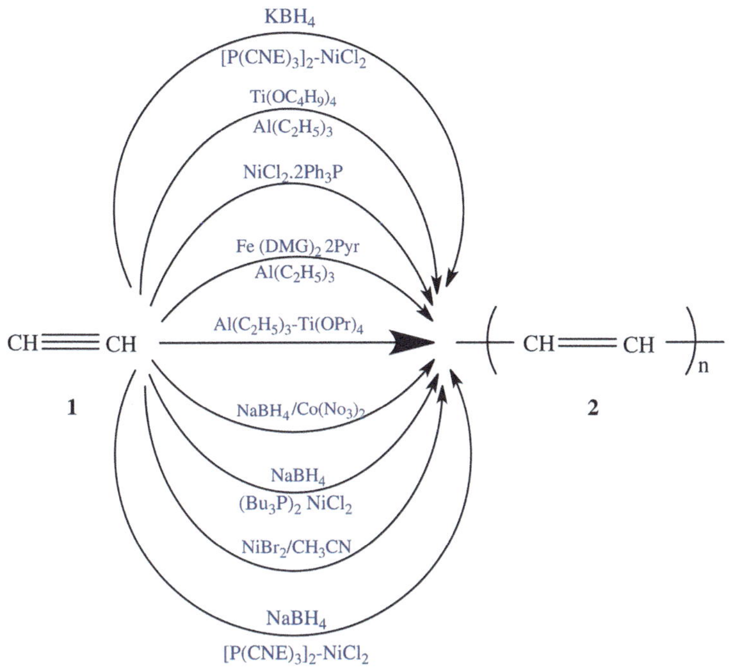


**Fig. 13.7** Different synthesis routes for catalytic polymerization of Poly(acetylene) [39]

Poly(acetylene), which comprises a hybrid reducing agent and halide complex of a group [VIII] metal. This Luttinger catalytic method utilizes hydrophilic solvents, and their activity is relatively lower than that of the Ziegler-Natta catalysts [38]. We have other means of synthesis, such as electrochemical polymerization, light-induced synthesis, and so on, which are not explored in detail in this book [28].

## 13.3 Tuning and Modification of Conducting Polymers

Conducting polymers, in their pristine form, act as insulators to semiconductors. The conjugation in these structures results in the delocalization of $\pi$ electrons, which helps develop charge carriers such as polarons, bipolarons, solitons, and so on [40]. These $\pi$ electrons delocalization results in decrease in energy gap between highest occupied molecular orbital (HOMO) and lowest unoccupied molecular orbital (LUMO). In conducting polymers, the orbital which is filled with electrons with high energy is considered to be HOMO, whereas the orbital which is has vacant molecular orbital with low energy is referred to as LUMO [41]. While tuning or modifying these polymeric structures, the energy levels are physically or chemically altered. These modifications altering the energy levels attributes to the change or enhancement in their electronic properties, which makes the polymers to be

**Table 13.2** Electrical conductivity values of various p and n doped conducting polymers

| Polymer | Dopant | Doping type | Temperature (°C) | Conductivity ($S\ cm^{-1}$) | Reference |
|---|---|---|---|---|---|
| Poly(pyrrole) | $FeCl_3$ | p-type | RT | $10^{-3}$ to $10^{-2}$ | [45] |
| Poly(thiophene) | $FeCl_3$ | p-type | RT | 10–25 | [46] |
| PEDOT:PSS | MAI/DMF | p-type | 140 | 1660 | [47] |
| PEDOT:PSS | Methyl red | n-type | RT | 51 | [48] |
| Poly(aniline) | $C_{10}H_{16}O_4S$ | p-type | RT | 300 | [49] |
| Poly(aniline) | NaH | n-type | RT | $10^{-1}$–$10^{-2}$ | [50] |
| Poly(para-phenylene vinylene) | I2 | p-type | RT | 200 | [51] |
| Poly(p-phenylene vinylene) | $C_{10}H_7Na$ | n-type | RT | $2 \times 10^{-4}$ | [52] |
| Polyacetylene | $I_2$ | p-type | RT | $10^5$ | [53] |
| Polyacetylene | $Na^+$ | n-type | RT | 80 | [54] |

capable of electron conduction, that is, the modified/doped polymer now becomes conducting in nature than the pristine polymer as such by decreasing the energy gap between HOMO and LUMO due to tuning and modification [42].

Chemical modification can be done during the synthesis process by co-polymerization, functionalization, and so on, whereas doping is a widely established and interesting process for conducting polymers. Dopants are generally used (either lewis acid or lewis base), which will either donate or withdraw electrons and alter the energy levels of the polymer and enhance its conductivity. Despite increasing the conductivity, doping improves stability, processability, and solubility [41, 42]. As in inorganics, we also have the same p-type and n-type doping in organic semiconductors. In p-type doping, the electrons from HOMO get transferred to the dopant creating a hole whereas in n-type doping, the electrons are transferred into LUMO from the dopant increasing electron density. This doping improves the delocalization of the charge carriers by doping enhancing the electron conduction [43, 44]. Some examples of the doped polymers and their conductivity are reported in Table 13.2.

## 13.4 Applications of Conducting Polymers

Conductive polymers have been employed in many biological applications due to their biocompatibility and tunability, especially with non-cytotoxic dopants [55–58]. Such flexibility allowed it to be used in human rated applications especially in in vivo materials such as tissue engineering like cardiac tissue architecture [59, 60], bio-actuators like artificial muscle [61], neural devices such as neural probes [62], neural imaging systems [63, 64], and so on, and mixed applications where both in vivo and in vitro such as bio-sensors [10, 65] and drug delivery [55, 57] which can be either supra dermal or implantable in nature.

As mentioned in the previous section, doping and tuning these polymers can be p or n. These dopants usually either increase or decrease the inherent Fermi levels of the material, thereby improving the workability and sensitivity of the material. This allows specific binding for cells of different types and simulates the actual bodily current flow, thus enabling more accurate process either growth modulation of particular tissue, growth modulation of neurons and nerves, muscle actuation and specific binging for drug delivery and sensing applications [1, 10, 55–59].

### *13.4.1 Drug Delivery*

Advancement in minimally invasive surgery techniques has forced the development of drug carriers that can be implanted, targeted toward the affected part, and released at will in a controlled fashion. Conducting polymers, due to their electromodulation from either electric current externally supplied or by using the body's natural electrical and pH changes, can be made into efficient drug delivery systems that can deliver drugs at will [66].

Ezazi et al. loaded a small molecule 3i-1000, a small molecule inhibitor of GATA4-NKX2-5 transcriptional synergy, which can inhibit cardiomyocytes hypertrophic response and promote myocardial repair. This was loaded onto a poly (glycerol sebacate) blended with collagen type 1 and polypyrolle. Polypyrolle (PPy) induces cell signaling and promotes cell adhesion while providing geometrical support to the patch. It was found that their system was not cytotoxic for cardiac cells and showed the importance of the conductive polymer for efficient protein binding, wettability, and cell proliferation [67].

Li et al. proposed an aldehyde-substituted polypyrrole made by graft copolymerizing polypyrrole with poly(ethylene glycol)monomethacrylate followed by oxidation in DMSO and Propionic acid for effective delivery of Heparin molecules loaded onto PVA (Fig. 13.8). They reported that electric modulation of 3.5 mA showed twice the drug release amount of non-modulated delivery systems. The aldehyde group effectively immobilized the Heparin patch onto the PPy substrate [68]. Wadhwa et al. synthesized a dexamethasone-loaded PPy-based coating containing a neural electrode array to reduce the swelling caused by implants. The coating was of 50 nm thickness, and the release studies showed 0.5 μg/cm2 of drug was eluded for each CV cycle of −0.8 V to +1.4 V at a scan rate of 100 mV/s, thus showcasing the on-demand drug elucidation ability of such a system [69]. George et al. doped PPy with biotin and decorated it with streptavidin for drug delivery application. They used the Nerve growth factor as the drug and tested their theory of controlled release in a PBS buffer solution. The release of the growth factor is controlled by applying 3 V (over which there was hydrolysis) for about 150 s to obtain a statistically significant value over other voltages. The release was measured using the PC 12 neurite outgrowth assay. They concluded that biotin doping provides better tunability to PPy in drug binding and release [70]. Pernaut et al. exploited the pH sensitivity of PPy and studied the drug release profiles of the Adenosine triphosphate

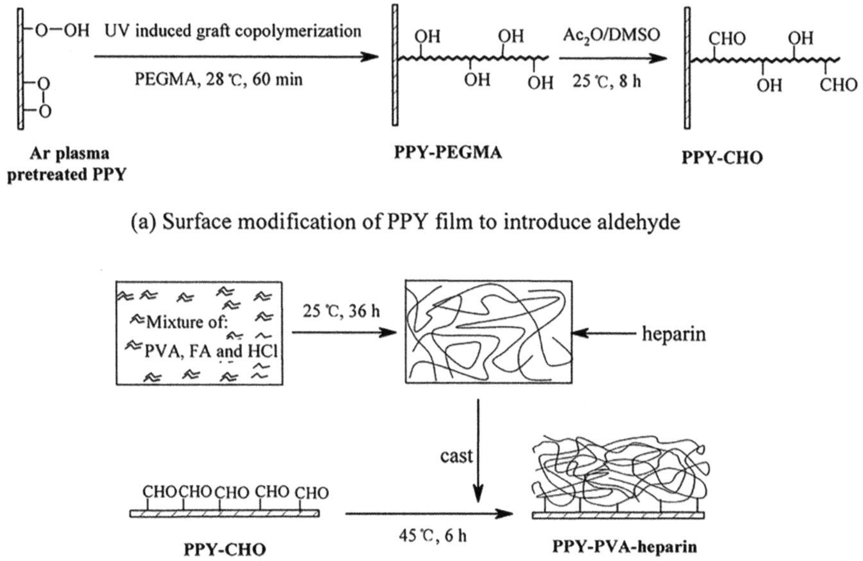


(a) Surface modification of PPY film to introduce aldehyde

(b) Immobilization of PVA or PVA-heparin on the surface modified PPY film

**Fig. 13.8** Surface modification of PVA, PPY, and PVA Heparin [68]

drug model over chemical and electrochemical triggers. They employed 0.1M Hydrazine solution and 0.1M NaOH solution as the Acid-Base chemical triggers and −0.7 V potential as the electrochemical trigger with elution rates up to 20 μg $cm^{-2}$ $min^{-1}$ [71]. Abidian et al. synthesized PEDOT nanotubes using PLGA nanofibers as a template. The templates were coated with EDOT monomer and electropolymerized around the fibers. The template was then removed by selective dissolution, and the tubes were loaded with Dexamethasone. The release of the drug was controlled at 1 V for 10 s to obtain statistically significant drug release [72]. A bioink made of poly(vinyl alcohol) (PVA) and κ-carrageenan was used to create a 3D conductive composite dressing by Chai et al., PEDOT: PSS was used as the conductive medium, and (+)-catechin-loaded mesoporous ZnO (CmZnO) was used as an antibacterial and anti-inflammatory agent. The composite, which was created by extrusion 3D printing, demonstrated good adhesiveness, strong blood clotting capacity, and appropriate conductivity, and the dressing showed potent antibacterial activity. Also, the electrical stimulation accelerated wound healing in a full-thickness skin defect model and encouraged in vivo blood clotting [73].

## *13.4.2 Tissue Engineering*

Tissue engineering requires precise growth of cells over scaffolds to generate a piece of tissue that can potentially change the quality of life of a human being. Conducting polymers, owing to their electrical modulation and conductivity, if used as scaffolds, will promote better cell selectivity and proliferation and can also mimic the natural electric signal of the body, thus making the tissue more natural-like and sensible [59].

Li et al. electrospunned poly aniline with gelatin and tested the cell cytotoxicity of the system. They found that the H9c2 rat cardiac myoblast cells proliferated seamlessly over the matrix. Depending upon the amount and distribution of PANI, the cell morphology differed initially, but after a weeks' time, all confluenced to a dame degree [74]. Gharbi et al. developed a series of methoxysilane end functionalized urethane prepolymers made of oil of castor and ricinoleic methyl ester with methoxysilane functionalised aniline tetramer to make polyurethane/siloxane cross-linked systems through sol gel process. The system showed optimum mechanical and electrical properties with excellent antibacterial properties and allowed cell proliferation of fibroblast cells. They studied the material with different ratios of starting materials, different solutions for preparation, and with camphor sulfonic acid doping and without doping. It was found that the electroactive dressing promoted the wound healing process, rate of collagen deposition, and vascularization [75]. Sanchez et al. reported an electrospun matrix made of Poly lactic acid and hydroxyapatite. The spun fabric was then coated with iodine-doped pyrrole, the plasma polymerised over the matrix. This doped PPy coating improved the osteogenic cell adhesion and proliferation. The in vitro studies proved that high cell proliferation rates were confirmed through the MTT assay. The porosity of the electrospun fiber matric allowed easy reach of nutrients and growth agents to the cell, thus increasing its survivability and ultimately leading to osteogenesis [76]. Song et al. designed a spring-like architecture with poly(glycolic acid)coated with DOPA followed by PPy to make the already strong and flexible spring conductive. When injected into the Myocardial infarction site, this material significantly improved left ventricular shortening and reduced infarct size. The tangled chains form a 3D network to modulate cardiac function, and the coating improved the growth rates of cardiomyocytes. They also concluded that the proposed material may be used as sensors to measure cardiomyocytes' contractile forces [77]. Feiner et al. proposed a similar system with PCL/Gelatin as their base matrix polymers for electrospinning, which was then coated with SU-8, Gold, and Titanium Nitride to make it conductive (Fig. 13.9). This material also showed properties required to reduce myocardial infarction [78].

Broda et al. employed Carbothane PC-3585A PU resin, mixed it with pyrrole, and polymerized in situ within the PU emulsion to form PPy nanoparticle dispersed PU and casted for TE applications. They tried to grow C2C12 cells over this substrate. They found good cell adhesion toward the substrate due to inherent conductivity that backed the cell signaling and allowed proliferation to form myotubes [79]. To address the reactive oxygen species that cause bone inflammation, Huang et al. designed polyorganophosphazene-based scaffolds to scavenge the reactive oxygen species. They employed PATGP type of polyorganophospazenes with aniline tetramer side group with doping of glycine ethyl ester and compared it against the conventional system made of PLGA. The made blends and core shell structures to analyze the effect of the materials and conductivity and found that the core shell architecture of PLGA/PATGP effectively scavenges the oxygen species and promotes proliferation of osteogenic differentiation and improve bone forming

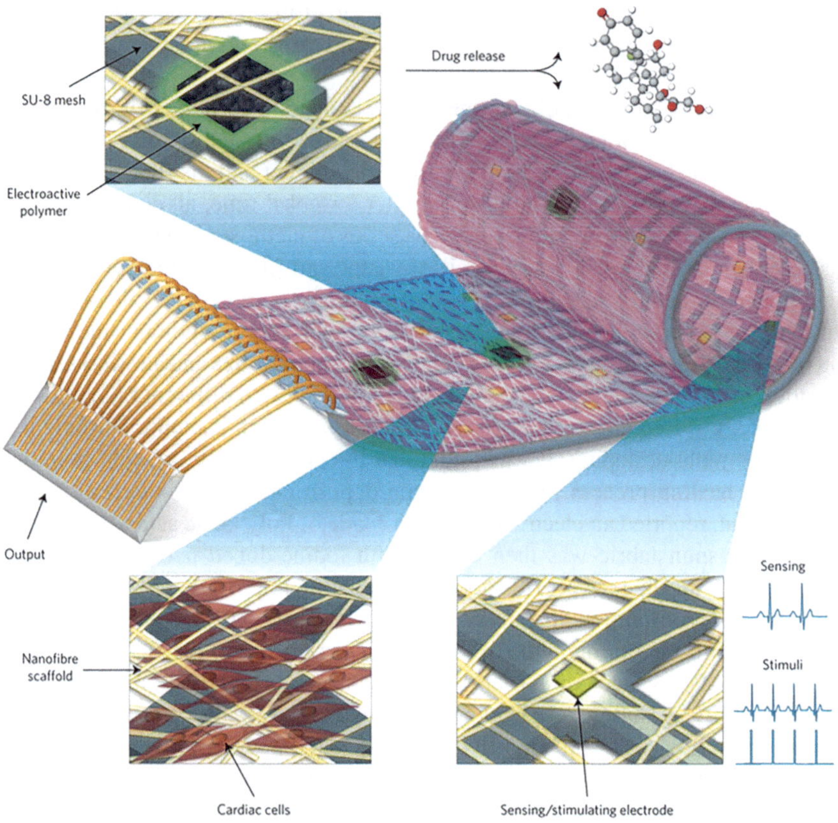


**Fig. 13.9** Schematics of novel cardiac patch [78]

capability. The distribution of PATGP decided the proliferation rates and uniformity of growths [80].

## *13.4.3 Nerve Tissue Engineering (NTE)*

NTE is a subfield of tissue engineering that deals with synthesizing or growing functional neural tissue that can be implanted to treat damaged neural tissues. This subfield takes a special place in this chapter due to the complexity of choosing the materials and its subsequent fabrication process onto a scaffold. Usually, NTE requires 3D constructs that mimic the geometry and function of nerve tissue. These artificial nerves are generally employed in treating Alzheimer's, Parkinson's, dementia, and chrons, monitoring sports performances, mental health, extracellular neural recording, and understanding the brain and brain–computer interfaces [63, 64, 81].

Schmidt et al. were the first to explore using conductive polymers for nerve tissue regeneration. They used oxidized polypyrolle as the substrate and tested it with PC 12 and chicken Sciatic nerve explants. The cell proliferation was excellent in this material even without electrical stimulation, and with electrical stimulation, neurite length increased [82]. Oribe et al. developed a soft subdural electrode with an array of PEDOT modified carbon fabric over a PVA hydrogel substrate. The electrode performance was analyzed using electrocorticography in various media, including ex vivo and vivo brain models. The sizeable double-layer capacitance of the material ring low impedance at frequencies (0.1 Hz–1000 Hz) of brain wave including epileptic seizure. The material also had better adhesion to the brain, even in structurally complex areas. A promising ECoG electrode for intraoperative monitoring during epilepsy surgery and long-term brain stimulation or monitoring is the hydrogel-based PEDOT-CF electrode [83]. Borah et al. developed poly [2 methoxy 5 (2 ethyl hexyloxy) 1,4 phenylene vinylene] based nanofibers for electrically simulated neuronal growth. They studied blends of MEH PPV with PCL and core sheath architecture by coaxial electrospinning and found that the core sheath has better electrical properties. The materials showed good cell adhesion and proliferation with 3T3 fibroblast and a neuronal rat pheochromocytoma PC12 cell line when coated with collagen 1. A Beta (III) tubulin immunochemistry assay proved the proliferation and differentiation into the sympathetic nerve system of PC12. The neurite growth was observed at 50 mV $cm^{-1}$ for 2 h $day^{-1}$ even without NGF [84]. Lin et al. used a spinal cord injury model in rats to check the viability of CP in nerve regeneration in traumatic conditions and observed the rehabilitation of the model. They used poly (ε-caprolactone-co-ethyl ethylene phosphate) as the base material for electrospinning, and the spunned matrix was coated with rat tail collagen 1 and neurotrophins 3 NGF. They observed that the rat model developed more axons and anti-inflammatory cells during rehabilitation, followed by recovery of gait movement and significant ankle flexion force, thus showing better recovery characteristics [85].

Das et al. used electrospun silk fibroin-polyaniline conduits as a nerve regeneration scaffold and modeled it on a rat sciatic nerve injury model. The SF was electrospunned, coated with Polyaniline, and then rolled into a conduit. The cell proliferation and adhesion studies with the Schwann cells of rats showed positive observations (Fig. 13.10). In rat models, this conduit showed excellent nerve regeneration support, and the model showed complete recovery of nerve conduction velocity, compound muscle action potential, and motor unit potential, regaining complete gait movements [86]. Abidian et al. used PEDOT-deposited PLLA nanofiber coating over neural microelectrodes and tested their cell adhesion and biosignaling. The modified electrodes showed improved signal quality, lower impedance, and reduced response time when tested as an implant in the barrel cortex of a rat for 7 days. Not to mention, the ability of this material to act as a drug delivery medium enables it to be a smart system that can record and treat diseases [87]. Chen et al. added a PEDOT layer between the microneedle array dry electrode to lower the contact impedance and allow high-quality data acquisition [88].

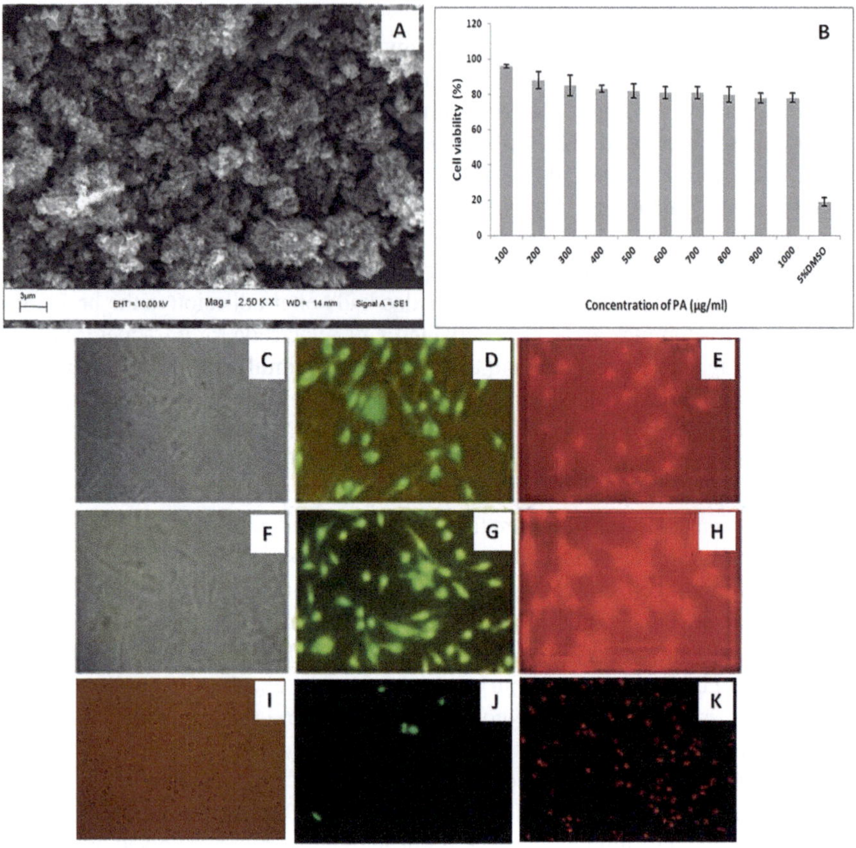


**Fig. 13.10** Scanning electron microscope (SEM) image of polyaniline (PA) powder and viability of rat Schwann cells (SCTM41) treated with PA measured by MTT assay compared to the 5% DMSO treated group [86]

## *13.4.4 Sensors*

Conducting polymers are employed in biosensors that are used to detect a specific biological effluent to quantify or identify certain diseases or to draw conclusions regarding some tests. These materials are designed to bind to specific functional groups in the target molecule, changing the polymer's electrical conductivity or the photochrome effect. The sensors can also be designed to be implanted inside the body for precise targeting of diseases or act as a drug delivery system for certain stimulus [89, 90].

Yoon et al. designed a polyaniline-based flexible pH sensor architected as a nanopillar array. The device was fabricated with two electrodes, a PAN array working electrode and an Ag/AgCl reference electrode. The electrodes were fabricated using soft lithography and electrodeposition techniques for working and reference.

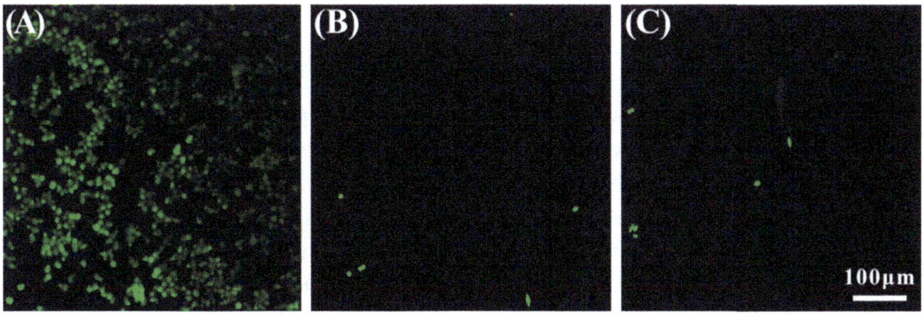


Fig. 13.11 Fluorescence images for anti-cell attachment performance of (a) PANI-NWs, (b) BSA/PANI-NW-modified surfaces, and (c) biosensors. All the modified surfaces were co-incubated with HEK 293 cells for 24 h at 37 °C [92]

The sensor exhibited a near Nernstian response of almost 60.3 mV/pH and covers a detection range of 2.38–11.61 even at its bent state. The fabricated device had a sensing performance on par with a commercial pH meter [91]. Li et al. proposed crosslinked BSA protein with polyaniline nanowires as an effective biosensor with excellent antifouling properties and electrochemical quantification toward IgG. With the immobilization of peptide aptamer for IgG recognition, the sensor showed excellent selectivity (Fig. 13.11). The sensors' detection range was 1 ng/mL to 10 μg/mL with a lower detection limit of 0.27 ng/mL, thus making its working range and selectivity comparable to ELISA [92].

Song et al. designed a unique immobilized peptide with an amine-rich end that can effectively anchor, antifouling, and recognize moiety. The designer peptide has a head rich in amine that can bind to the electroactive polymer PEDOT-citrate film. At the same time, the tail is designed to bind to aminopeptidase N or human hepatocellular carcinoma cells. The middle part was intended to have antifouling properties. The lower detection limit of this sensor was found to be 0.4 ng/mL or 20 cells/mL [93]. Taylor et al. designed a dopamine detection sensor of carbon fiber microelectrode electrodeposited with PEDOT and graphene oxide mixture. The coated electrode showed very high sensitivity to dopamine, and the sensitivity increases with an increase in layer thickness, but the total uptake and diffusion process will lag. The 25s deposited system showed optimal thickness and an 880% increase in sensitivity with a 50% decrease in the detection limit. This device was tested on both solutions and rat model dorsal striatum and performed similarly in both cases [94]. A similar work was presented by Serrano et al. They fabricated PEDOT: PSS CP eutectogel for body surface potential mapping and electrocardiogram-assisted imaging. Measurements from manufactured devices using simultaneous electrode arrays show that different ECG signals may be acquired from each electrode, including temporal and spatial information such as speed and direction of propagation, spectrum information, and correlations. Results confirming the correctness of the optimal array design parameter selection have been given after the recorded BSPM was interpolated without aliasing aberrations [95]. Santos et al. fabricated poly(9,9-dioctylfluorenyl-2,7-diyl-co-(1,4-benzo-thiadiazole)) (PFBT) nano particles with stabilizing agent as polystyrene co N acrloxysuccininmide. Due to its structural similarity, this stabilizing agent has an affinity toward collagen type molecules. Thus, formed nanoparticles were prevented from self-trimerization, maintaining collagen-binding activity without pre-activation. The particles specifically bound denatured collagen in histological mouse cornea tissue, revealing collagen fibril organization through fluorescence imaging [96]. Wu et al. molecularly imprinted o-phenylenediamine and polymerized it onto the surface of a gold electrode. The MIP electrode was doped with biocompatible sulfonated graphene to improve current response. This modified MIP electrode has a detection limit of 0.11 mg/L with a working range from 0.5 to 7.0 mg/L for Dopamine. The sensor has very high selectivity and is never bound to ascorbic acid model molecule against Dopamine [97].

Garma et al. inkjet-printed PEDOT: PSS over polyimide film with glycerol as a viscosity modifier. Air plasma treatment was done to improve the wettability of the surface of polyimide sheets. PEDOT:PSS tracks were then connected to copper tracks with silver, thus completing the fabrication process. The electrodes showed a low impedance of 19.5 kΩ at 1 kHz, with the cell viability of cardiomyocyte cells being 94.9% [98]. Mugo et al. fabricated a composite system of nanoparticles and conductive polymers for sweet sensing. They deposited CNT/CNC mixed in the pNIPAM layer over the PVA substrate. This layer is coated with lactate imprinted poly(aniline/phenylboronic acid). The final layer substantially reduces the ion selectivity of pNIPAM, thus regulating the taste sensors to sense sweetness and not salt. With a limit of detection (LOD) of 0.23 mM NaCl, the pNIPAM @CNC/CNT sensor's linear concentration range in response to NaCl was 3–80 mM. An order of magnitude higher than the non-imprinted variant, the lactate imprinted pANI/PBA-pNIPAM @CNC/CNT sensor produced a linear range of 1 mM–25 mM and a LOD of 0.10 mM. The sensor precision for both analytes varied between 0.5% and 5.0%, indicating general validity and dependability. It was shown that the manufactured flexible sensors successfully measured lactate and electrolytes (mainly NaCl) in an actual sweat sample [99].

### *13.4.5 Actuators*

Soft tissues like muscle tissues are vital for day-to-day bodily function. They are force generation, conversion and responsible for movements. Regenerating muscle tissue requires substrates that allow the muscle cells to grow and orient themselves in a specific direction, as most muscles are fibrillar. The substrate must also assist the cell in achieving optimum strength. Conductive polymers can do the job as they can modulate and orient cells and allow them to gain the strength required for its viability [61].

Otero et al. electrogenerated PPy over SS electrodes with controlled morphology during polymerization. The polymer is coated only on one side of the ss plate, thus making it a bilayer system. Therefore, when the conducting polymer is immersed in $LiClO_4$ electrolyte, the polymer expands due to absorption of counterions; hence, oxidation and the vice versa causes contraction or reduction, mimicking the muscle movement. 3 mg PPy layer demonstrated about 180° under 30-mA current, and the device can lift to 1000 times its weight [60]. Chen et al. electrospunned PCL and PANI with an external magnetic field at the collector region. This blended mat showed about 63.6 mS/cm electrical conductivity even at 3 wt% of PANI (Fig. 13.12). The mat was then subjected to random culturing of C2C12 myoblast for differentiation, and it showed good cell adhesion and proliferation. The aligned nanofibers showed better proliferation and promoted myotube formation. These findings indicated a possible application of A-PCL/PANI nanofibers for skeletal muscle regeneration. They showed that the combined effect of both guidance cues was more successful than each alone [7]. A similar work was reported by Ahadian et al. where they assessed C2C12 myoblast proliferation in PdMGSMWs, or palladium-based metallic glass sub-micron wires, which were used to improve the mechanical strength and conductivity of gels made of gelatin methacryloyl (GelMA). They observed myotube formation in the material and concluded that this material can be used for skeletal muscle repair [100]. Azcon et al. used GelMA as the base material with CNT dispersed dielectrophoretically. C2C12 myoblast was used as the model cell line for cell differentiation and proliferation studies. It was found that at an electrical stimulation of 8 V, 1 Hz, there was promotion of muscle myotube alignment and contractility [101].

**Fig. 13.12** SEM micrographs of the electrospun nanofibers: (**a**) pure PCL, (**b**) PCL/PANi-1, (**c**) PCL/PANi-2 and (**d**) PCL/PANi-3 nanofibers [7]

## 13.5 Conclusion and Future Scope

Conducting polymers (CPs) are advanced materials in biomedical engineering, recognized for their biocompatibility, tunable conductivity, and versatility. Their conjugated backbones allow efficient electron conduction, comparable to metals, while their organic nature ensures excellent compatibility with biological systems. CPs are highly adaptable, with straightforward synthesis and modification processes, making them well-suited for various applications, including biosensors, drug delivery, tissue engineering, and neural interfaces. Beyond biomedicine, their flexibility and conductivity enable their use in photovoltaic devices and nerve regeneration.

In the biomedical field, CPs have shown significant promise in biosensing, controlled drug release, and guiding cellular behavior. Although the field has seen notable progress, opportunities remain to develop advanced, responsive systems and explore hybrid materials for improved functionality. Continued innovations in biofabrication techniques, biodegradable CPs, and wearable technologies are anticipated to expand their applications in regenerative medicine and bioelectronics. CPs can potentially drive substantial healthcare and biomedical technology advancements through collaborative research across disciplines.

## References

1. Gerard M, Chaubey A, Malhotra BD (2002) Application of conducting polymers to biosensors. Biosens Bioelectron 17(5):345–359. https://doi.org/10.1016/s0956-5663(01)00312-8
2. Green RA, Baek S, Poole-Warren LA, Martens PJ (2010) Conducting polymer-hydrogels for medical electrode applications. Sci Technol Adv Mater 11(1):014107. https://doi.org/10.1088/1468-6996/11/1/014107
3. Park AR, Kim JS, Kim KS, Zhang K, Park J, Park JH, Lee JK, Yoo PJ (2014) Si–Mn/reduced graphene oxide nanocomposite anodes with enhanced capacity and stability for lithium-ion batteries. ACS Appl Mater Interfaces 6(3):1702–1708. https://doi.org/10.1021/am404608d
4. Jin B, Song H, Jiang R, Song J, Zhao Q, Xie T (2018) Programming a crystalline shape memory polymer network with thermo- and photo-reversible bonds toward a single-component soft robot. Sci Adv 4(1):eaao3865. https://doi.org/10.1126/sciadv.aao3865
5. Jin G, Li K (2014) The electrically conductive scaffold as the skeleton of stem cell niche in regenerative medicine. Mater Sci Eng C 45:671–681. https://doi.org/10.1016/j.msec.2014.06.004
6. Williams RL, Doherty PJ (1994) A preliminary assessment of poly(pyrrole) in nerve guide studies. J Mater Sci Mater Med 5(6):429–433. https://doi.org/10.1007/BF00058978
7. Chen M-C, Sun Y-C, Chen Y-H (2013) Electrically conductive nanofibers with highly oriented structures and their potential application in skeletal muscle tissue engineering. Acta Biomater 9(3):5562–5572. https://doi.org/10.1016/j.actbio.2012.10.024
8. Abdul Rahman N, Feisst V, Dickinson ME, Malmström J, Dunbar PR, Travas-Sejdic J (2013) Functional polyaniline nanofibre mats for human adipose-derived stem cell proliferation and adhesion. Mater Chem Phys 138(1):333–341. https://doi.org/10.1016/j.matchemphys.2012.11.065
9. Diaz AF, Kanazawa KK, Gardini GP (1979) Electrochemical polymerization of pyrrole. J Chem Soc Chem Commun No. 14:635. https://doi.org/10.1039/c39790000635

10. Guimard NK, Gomez N, Schmidt CE (2007) Conducting polymers in biomedical engineering. Prog Polym Sci 32(8):876–921. https://doi.org/10.1016/j.progpolymsci.2007.05.012
11. Skotheim, T.A., & Reynolds, J. (Eds.). (2007). Handbook of Conducting Polymers, 2 Volume Set (3rd ed.). CRC Press. https://doi.org/10.1201/b12346
12. Wallace, G.G., Teasdale, P.R., Spinks, G.M., & Kane-Maguire, L.A.P. (2002). Conductive Electroactive Polymers: Intelligent Materials Systems, Second Edition (2nd ed.). CRC Press. https://doi.org/10.1201/9781420031898
13. Bjorklund RB (1987) Kinetics of pyrrole polymerisation in aqueous iron chloride solution. J Chem Soc Faraday Trans 1 Phys Chem Condens Phases 83(5):1507. https://doi.org/10.1039/f19878301507
14. Chen Y, Wang F, Dong L, Li Z, Chen L, He X, Gong J, Zhang J, Li Q (2019) Design and optimization of flexible polypyrrole/bacterial cellulose conductive nanocomposites using response surface methodology. Polymers 11(6):960. https://doi.org/10.3390/polym11060960
15. Yussuf A, Al-Saleh M, Al-Enezi S, Abraham G (2018) Synthesis and characterization of conductive polypyrrole: the influence of the oxidants and monomer on the electrical, thermal, and morphological properties. Int J Polym Sci 2018:1–8. https://doi.org/10.1155/2018/4191747
16. Patil BH, Bulakhe RN, Lokhande CD (2014) Supercapacitive performance of chemically synthesized polypyrrole thin films: effect of monomer to oxidant ratio. J Mater Sci Mater Electron 25(5):2188–2198. https://doi.org/10.1007/s10854-014-1858-6
17. McCullough RD (1998) The chemistry of conducting polythiophenes. Adv Mater 10(2):93–116. https://doi.org/10.1002/(SICI)1521-4095(199801)10:2<93::AID-ADMA93>3.0.CO;2-F
18. Yamamoto T, Sanechika K, Yamamoto A (1980) Preparation of thermostable and electric-conducting poly(2,5-thienylene). J Polym Sci Polym Lett Ed 18(1):9–12. https://doi.org/10.1002/pol.1980.130180103
19. Lin JW, Dudek LP (1980) Synthesis and properties of poly(2,5-thienylene). J Polym Sci Polym Chem Ed 18(9):2869–2873. https://doi.org/10.1002/pol.1980.170180910
20. Kobayashi M, Chen J, Chung T-C, Moraes F, Heeger AJ, Wudl F (1984) Synthesis and properties of chemically coupled poly(thiophene). Synth Met 9(1):77–86. https://doi.org/10.1016/0379-6779(84)90044-4
21. Livi F, Carlé JE, Bundgaard E (2014) Thiophene in conducting polymers: synthesis of poly(thiophene)s and other conjugated polymers containing thiophenes, for application in polymer solar cells. In: Joule JA (ed) Thiophenes. Topics in heterocyclic chemistry, vol 39. Springer International Publishing, Cham, pp 203–226. https://doi.org/10.1007/7081_2014_128
22. Groenendaal L, Jonas F, Freitag D, Pielartzik H, Reynolds JR (2000) Poly(3,4-ethylenedioxythiophene) and its derivatives: past, present, and future. Adv Mater 12(7):481–494. https://doi.org/10.1002/(SICI)1521-4095(200004)12:7<481::AID-ADMA481>3.0.CO;2-C
23. Elschner, A.; Kirchmeyer, S. PEDOT-type materials in organic solar cells. In Organic photovoltaics; Brabec, C., Dyakonov, V., Scherf, U., Eds.; Wiley, 2008; pp. 211–242. doi:https://doi.org/10.1002/9783527623198.ch7
24. Wang Y, Levon K (2012) Influence of dopant on electroactivity of polyaniline. Macromol Symp 317–318(1):240–247. https://doi.org/10.1002/masy.201200008
25. Krishna JBM, Saha A, Okram GS, Soni A, Purakayastha S, Ghosh B (2009) Electrical properties of polyaniline doped with metal ions. J Phys Appl Phys 42(9):095404. https://doi.org/10.1088/0022-3727/42/9/095404
26. Gospodinova N, Terlemezyan L (1998) Conducting polymers prepared by oxidative polymerization: polyaniline. Prog Polym Sci 23(8):1443–1484. https://doi.org/10.1016/S0079-6700(98)00008-2
27. Yang L, Yang L, Wu S, Wei F, Hu Y, Xu X, Zhang L, Sun D (2020) Three-dimensional conductive organic sulfonic acid co-doped bacterial cellulose/polyaniline nanocomposite films for detection of ammonia at room temperature. Sens Actuators B Chem 323:128689. https://doi.org/10.1016/j.snb.2020.128689

28. Namsheer K, Rout CS (2021) Conducting polymers: a comprehensive review on recent advances in synthesis, properties and applications. RSC Adv 11(10):5659–5697. https://doi.org/10.1039/D0RA07800J
29. Shacklette LW, Chance RR, Ivory DM, Miller GG, Baughman RH (1980) Electrical and optical properties of highly conducting charge-transfer complexes of poly(p-phenylene). Synth Met 1(3):307–320. https://doi.org/10.1016/0379-6779(80)90020-X
30. Kovacic P, Oziomek J (1964) P-polyphenyl from benzene—Lewis acid catalyst—oxidant. Reaction scope and investigation of the benzene—aluminum chloride—cupric chloride system[1]. J Org Chem 29(1):100–104. https://doi.org/10.1021/jo01024a023
31. Kovacic P, Jones MB (1987) Dehydro coupling of aromatic nuclei by catalyst-oxidant systems: poly(p-phenylene). Chem Rev 87(2):357–379. https://doi.org/10.1021/cr00078a005
32. Riese F (1872) Ueber Die Einwirkung von Natrium Auf Krystallinisches Dibrombenzol. Justus Liebigs Ann Chem 164(2):161–176. https://doi.org/10.1002/jlac.18721640202
33. Schlüter A-D, Wegner G (1993) Palladium and nickel catalyzed polycondensation—the key to structurally defined polyarylenes and other aromatic polymers. Acta Polym 44(2):59–69. https://doi.org/10.1002/actp.1993.010440201
34. Ballard DGH, Courtis A, Shirley IM, Taylor SC (1983) A biotech route to polyphenylene. J Chem Soc Chem Commun No. 17:954. https://doi.org/10.1039/c39830000954
35. Kleybolte ME, Vagin SI, Rieger B (2023) A polymer lost in the shuffle: the perspective of poly(para)phenylenes. Macromol Chem Phys 224(3):2200441. https://doi.org/10.1002/macp.202200441
36. Shirakawa H (1995) Synthesis and characterization of highly conducting polyacetylene. Synth Met 69(1–3):3–8. https://doi.org/10.1016/0379-6779(94)02340-5
37. Natta G, Mazzanti G, Corradini P (1967) Stereospecific polymerization of acetylene. In: Stereoregular polymers and stereospecific polymerizations. Elsevier, pp 463–465. https://doi.org/10.1016/B978-1-4831-9883-5.50084-2
38. Luttinger LB, Colthup EC (1962) Hydridic reducing agent—group VIII metal compound. A new catalyst system for the polymerization of acetylenes and related compounds. II. J Org Chem 27(11):3752–3756. https://doi.org/10.1021/jo01058a002
39. Poddar AK, Patel SS, Patel HD (2021) Synthesis, characterization and applications of conductive polymers: a brief review. Polym Adv Technol 32(12):4616–4641. https://doi.org/10.1002/pat.5483
40. Le T-H, Kim Y, Yoon H (2017) Electrical and electrochemical properties of conducting polymers. Polymers 9(4):150. https://doi.org/10.3390/polym9040150
41. Kar P (2013) Doping in conjugated polymers. Polymer science and plastics engineering. Wiley, Hoboken. https://doi.org/10.1002/9781118816639
42. Graham MDA, Mammone RJ, Kaner RB, Lord P (1985) The concept of 'doping' of conducting polymers: the role of reduction potentials. Philos Trans R Soc Lond Ser Math Phys Sci 314(1528):3–15. https://doi.org/10.1098/rsta.1985.0004
43. Nollau A, Pfeiffer M, Fritz T, Leo K (2000) Controlled *n* -type doping of a molecular organic semiconductor: naphthalenetetracarboxylic dianhydride (NTCDA) doped with bis(ethylenedithio)-tetrathiafulvalene (BEDT-TTF). J Appl Phys 87(9):4340–4343. https://doi.org/10.1063/1.373413
44. Bajpai M, Srivastava R, Dhar R, Tiwari RS (2016) Review on optical and electrical properties of conducting polymers. Indian J Mater Sci 2016:1–8. https://doi.org/10.1155/2016/5842763
45. Chitte HK, Shinde GN, Bhat NV, Walunj VE (2011) Synthesis of polypyrrole using ferric chloride ($FeCl_3$) as oxidant together with some dopants for use in gas sensors. J Sens Technol 01(02):47–56. https://doi.org/10.4236/jst.2011.12007
46. Dillingham TR, Cornelison DM, Townsend SW (1996) Structural and chemical characterization of vapor-deposited polythiophene films. J Vac Sci Technol Vac Surf Films 14(3):1494–1498. https://doi.org/10.1116/1.579975
47. Yu Z, Xia Y, Du D, Ouyang J (2016) PEDOT:PSS films with metallic conductivity through a treatment with common organic solutions of organic salts and their application as a transparent electrode of polymer solar cells. ACS Appl Mater Interfaces 8(18):11629–11638. https://doi.org/10.1021/acsami.6b00317

48. Huseynova G, Boampong AA, Yu KM, Lee Y-S, Lee J, Kim M-H, Lee J-H (2023) Highly conductive and low-work-function polymer electrodes for solution-processed *n* -type oxide thin-film transistors. J Inf Disp 24(1):47–56. https://doi.org/10.1080/15980316.2022.2126017
49. Holland ER, Pomfret SJ, Adams PN, Monkman AP (1996) Conductivity studies of polyaniline doped with CSA. J Phys Condens Matter 8(17):2991–3002. https://doi.org/10.1088/0953-8984/8/17/011
50. Hua M-Y, Hwang G-W, Chuang Y-H, Chen S-A, Tsai R-Y (2000) Soluble N-doped polyaniline: synthesis and characterization. Macromolecules 33(17):6235–6238. https://doi.org/10.1021/ma000376x
51. Murase I, Ohnishi T, Noguchi T, Hirooka M (1987) Highly conducting poly(phenylene vinylene) derivatives via soluble precursor process. Synth Met 17(1–3):639–644. https://doi.org/10.1016/0379-6779(87)90812-5
52. Gagnon DR, Capistran JD, Karasz FE, Lenz RW, Antoun S (1987) Synthesis, doping, and electrical conductivity of high molecular weight poly(p-phenylene vinylene). Polymer 28(4):567–573. https://doi.org/10.1016/0032-3861(87)90471-X
53. Tsukamoto J, Takahashi A, Kawasaki K (1990) Structure and electrical properties of polyacetylene yielding a conductivity of $10^5$ S/Cm. Jpn J Appl Phys 29(1R):125. https://doi.org/10.1143/JJAP.29.125
54. Chiang CK, Druy MA, Gau SC, Heeger AJ, Louis EJ, MacDiarmid AG, Park YW, Shirakawa H (1978) Synthesis of highly conducting films of derivatives of polyacetylene, (CH)x. J Am Chem Soc 100(3):1013–1015. https://doi.org/10.1021/ja00471a081
55. Ravichandran R, Sundarrajan S, Venugopal JR, Mukherjee S, Ramakrishna S (2010) Applications of conducting polymers and their issues in biomedical engineering. J R Soc Interface 7(suppl_5):S559–S579. https://doi.org/10.1098/rsif.2010.0120.focus
56. Bettucci O, Matrone GM, Santoro F (2022) Conductive polymer-based bioelectronic platforms toward sustainable and biointegrated devices: a journey from skin to brain across human body interfaces. Adv Mater Technol 7(2):2100293. https://doi.org/10.1002/admt.202100293
57. Idumah CI (2022) Recent advancements in conducting polymer bionanocomposites and hydrogels for biomedical applications. Int J Polym Mater Polym Biomater 71(7):513–530. https://doi.org/10.1080/00914037.2020.1857384
58. Maziz A, Özgür E, Bergaud C, Uzun L (2021) Progress in conducting polymers for biointerfacing and biorecognition applications. Sens Actuators Rep 3:100035. https://doi.org/10.1016/j.snr.2021.100035
59. Yu R, Zhang H, Guo B (2021) Conductive biomaterials as bioactive wound dressing for wound healing and skin tissue engineering. Nano-Micro Lett 14(1):1. https://doi.org/10.1007/s40820-021-00751-y
60. Otero TF, Sansieña JM (1998) Soft and wet conducting polymers for artificial muscles. Adv Mater 10(6):491–494. https://doi.org/10.1002/(SICI)1521-4095(199804)10:6<491::AID-ADMA491>3.0.CO;2-Q
61. Dong R, Ma PX, Guo B (2020) Conductive biomaterials for muscle tissue engineering. Biomaterials 229:119584. https://doi.org/10.1016/j.biomaterials.2019.119584
62. Kiyotake EA, Martin MD, Detamore MS (2022) Regenerative rehabilitation with conductive biomaterials for spinal cord injury. Acta Biomater 139:43–64. https://doi.org/10.1016/j.actbio.2020.12.021
63. Dimov IB, Moser M, Malliaras GG, McCulloch I (2022) Semiconducting polymers for neural applications. Chem Rev 122(4):4356–4396. https://doi.org/10.1021/acs.chemrev.1c00685
64. Berggren M, Głowacki ED, Simon DT, Stavrinidou E, Tybrandt K (2022) In vivo organic bioelectronics for neuromodulation. Chem Rev 122(4):4826–4846. https://doi.org/10.1021/acs.chemrev.1c00390
65. Borges MHR, Nagay BE, Costa RC, Souza JGS, Mathew MT, Barão VAR (2023) Recent advances of polypyrrole conducting polymer film for biomedical application: toward a viable platform for cell-microbial interactions. Adv Colloid Interf Sci 314:102860. https://doi.org/10.1016/j.cis.2023.102860

66. Svirskis D, Travas-Sejdic J, Rodgers A, Garg S (2010) Electrochemically controlled drug delivery based on intrinsically conducting polymers. J Control Release 146(1):6–15. https://doi.org/10.1016/j.jconrel.2010.03.023
67. Zanjanizadeh Ezazi N, Ajdary R, Correia A, Mäkilä E, Salonen J, Kemell M, Hirvonen J, Rojas OJ, Ruskoaho HJ, Santos HA (2020) Fabrication and characterization of drug-loaded conductive poly(glycerol sebacate)/nanoparticle-based composite patch for myocardial infarction applications. ACS Appl Mater Interfaces 12(6):6899–6909. https://doi.org/10.1021/acsami.9b21066
68. Li Y, Neoh KG, Kang ET (2005) Controlled release of heparin from polypyrrole-poly(vinyl alcohol) assembly by electrical stimulation. J Biomed Mater Res A 73A(2):171–181. https://doi.org/10.1002/jbm.a.30286
69. Wadhwa R, Lagenaur CF, Cui XT (2006) Electrochemically controlled release of dexamethasone from conducting polymer polypyrrole coated electrode. J Control Release 110(3):531–541. https://doi.org/10.1016/j.jconrel.2005.10.027
70. George PM, LaVan DA, Burdick JA, Chen C-Y, Liang E, Langer R (2006) Electrically controlled drug delivery from biotin-doped conductive polypyrrole. Adv Mater 18(5):577–581. https://doi.org/10.1002/adma.200501242
71. Pernaut J-M, Reynolds JR (2000) Use of conducting electroactive polymers for drug delivery and sensing of bioactive molecules. A redox chemistry approach. J Phys Chem B 104(17):4080–4090. https://doi.org/10.1021/jp994274o
72. Abidian MR, Kim D-H, Martin DC (2006) Conducting-polymer nanotubes for controlled drug release. Adv Mater 18(4):405–409. https://doi.org/10.1002/adma.200501726
73. Chai X, Lou Y, Nie L, Shavandi A, Yunusov KE, Sun Y, Jiang G (2025) A three-dimensional printable conductive composite dressing for accelerating wound healing under electrical stimulation. Colloids Surf B Biointerfaces 245:114264. https://doi.org/10.1016/j.colsurfb.2024.114264
74. Li M, Guo Y, Wei Y, Macdiarmid A, Lelkes P (2006) Electrospinning polyaniline-contained gelatin nanofibers for tissue engineering applications. Biomaterials 27(13):2705–2715. https://doi.org/10.1016/j.biomaterials.2005.11.037
75. Gharibi R, Yeganeh H, Rezapour-Lactoee A, Hassan ZM (2015) Stimulation of wound healing by electroactive, antibacterial, and antioxidant polyurethane/siloxane dressing membranes: in vitro and in vivo evaluations. ACS Appl Mater Interfaces 7(43):24296–24311. https://doi.org/10.1021/acsami.5b08376
76. Flores-Sánchez MG, Islas-Arteaga NC, Raya-Rivera AM, Esquiliano-Rendon DR, Morales-Corona J, Uribe-Juarez OE, Vivar-Velázquez FI, Ortiz-Vázquez GP, Olayo R (2021) Effect of a plasma synthesized polypyrrole coverage on polylactic acid/hydroxyapatite scaffolds for bone tissue engineering. J Biomed Mater Res A 109(11):2199–2211. https://doi.org/10.1002/jbm.a.37205
77. Song C, Zhang X, Wang L, Wen F, Xu K, Xiong W, Li C, Li B, Wang Q, Xing MMQ, Qiu X (2019) An injectable conductive three-dimensional elastic network by tangled surgical-suture spring for heart repair. ACS Nano 13(12):14122–14137. https://doi.org/10.1021/acsnano.9b06761
78. Feiner R, Engel L, Fleischer S, Malki M, Gal I, Shapira A, Shacham-Diamand Y, Dvir T (2016) Engineered hybrid cardiac patches with multifunctional electronics for online monitoring and regulation of tissue function. Nat Mater 15(6):679–685. https://doi.org/10.1038/nmat4590
79. Broda CR, Lee JY, Sirivisoot S, Schmidt CE, Harrison BS (2011) A chemically polymerized electrically conducting composite of polypyrrole nanoparticles and polyurethane for tissue engineering. J Biomed Mater Res A 98A(4):509–516. https://doi.org/10.1002/jbm.a.33128
80. Huang Y, Du Z, Li K, Jing W, Wei P, Zhao B, Yu Y, Cai Q, Yang X (2022) ROS-scavenging electroactive polyphosphazene-based core–shell nanofibers for bone regeneration. Adv Fiber Mater 4(4):894–907. https://doi.org/10.1007/s42765-022-00153-8

81. Bianchi M, De Salvo A, Asplund M, Carli S, Di Lauro M, Schulze-Bonhage A, Stieglitz T, Fadiga L, Biscarini F (2022) Poly(3,4-ethylenedioxythiophene)-based neural interfaces for recording and stimulation: fundamental aspects and in vivo applications. Adv Sci 9(12):2104701. https://doi.org/10.1002/advs.202104701
82. Schmidt CE, Shastri VR, Vacanti JP, Langer R (1997) Stimulation of neurite outgrowth using an electrically conducting polymer. Proc Natl Acad Sci 94(17):8948–8953. https://doi.org/10.1073/pnas.94.17.8948
83. Oribe S, Yoshida S, Kusama S, Osawa S, Nakagawa A, Iwasaki M, Tominaga T, Nishizawa M (2019) Hydrogel-based organic subdural electrode with high conformability to brain surface. Sci Rep 9(1):13379. https://doi.org/10.1038/s41598-019-49772-z
84. Borah R, Ingavle GC, Sandeman SR, Kumar A, Mikhalovsky S (2018) Electrically conductive MEH-PPV:PCL electrospun nanofibres for electrical stimulation of rat PC12 pheochromocytoma cells. Biomater Sci 6(9):2342–2359. https://doi.org/10.1039/C8BM00559A
85. Lin J, Anopas D, Milbreta U, Lin PH, Chin JS, Zhang N, Wee SK, Tow A, Ang WT, Chew SY (2019) Regenerative rehabilitation: exploring the synergistic effects of rehabilitation and implantation of a bio-functional scaffold in enhancing nerve regeneration. Biomater Sci 7(12):5150–5160. https://doi.org/10.1039/C9BM01095E
86. Das S, Sharma M, Saharia D, Sarma KK, Muir EM, Bora U (2017) Electrospun silk-polyaniline conduits for functional nerve regeneration in rat sciatic nerve injury model. Biomed Mater 12(4):045025. https://doi.org/10.1088/1748-605X/aa7802
87. Abidian MR, Ludwig KA, Marzullo TC, Martin DC, Kipke DR (2009) Interfacing conducting polymer nanotubes with the central nervous system: chronic neural recording using poly(3,4-ethylenedioxythiophene) nanotubes. Adv Mater 21(37):3764–3770. https://doi.org/10.1002/adma.200900887
88. Chen Y, Pei W, Chen S, Wu X, Zhao S, Wang H, Chen H (2013) Poly(3,4-Ethylenedioxythiophene) (PEDOT) as interface material for improving electrochemical performance of microneedlesaArray-based dry electrode. Sens Actuators B Chem 188:747–756. https://doi.org/10.1016/j.snb.2013.07.075
89. Mei X, Ye D, Zhang F, Di C (2022) Implantable application of polymer-based biosensors. J Polym Sci 60(3):328–347. https://doi.org/10.1002/pol.20210543
90. Wang C-C, Wei S-C, Luo S-C (2022) Recent advances and biomedical applications of peptide-integrated conducting polymers. ACS Appl Bio Mater 5(5):1916–1933. https://doi.org/10.1021/acsabm.1c01194
91. Yoon JH, Hong SB, Yun S-O, Lee SJ, Lee TJ, Lee KG, Choi BG (2017) High performance flexible pH sensor based on polyaniline nanopillar array electrode. J Colloid Interface Sci 490:53–58. https://doi.org/10.1016/j.jcis.2016.11.033
92. Li Y, Han R, Chen M, Zhang L, Wang G, Luo X (2021) Bovine serum albumin-cross-linked polyaniline nanowires for ultralow fouling and highly sensitive electrochemical protein quantification in human serum samples. Anal Chem 93(9):4326–4333. https://doi.org/10.1021/acs.analchem.1c00089
93. Song Z, Chen M, Ding C, Luo X (2020) Designed three-in-one peptides with anchoring, antifouling, and recognizing capabilities for highly sensitive and low-fouling electrochemical sensing in complex biological media. Anal Chem 92(8):5795–5802. https://doi.org/10.1021/acs.analchem.9b05299
94. Taylor IM, Robbins EM, Catt KA, Cody PA, Happe CL, Cui XT (2017) Enhanced dopamine detection sensitivity by PEDOT/graphene oxide coating on in vivo carbon fiber electrodes. Biosens Bioelectron 89:400–410. https://doi.org/10.1016/j.bios.2016.05.084
95. Serrano RR, Velasco-Bosom S, Dominguez-Alfaro A, Picchio ML, Mantione D, Mecerreyes D, Malliaras GG (2024) High density body surface potential mapping with conducting polymer-eutectogel electrode arrays for ECG imaging. Adv Sci 11(27):2301176. https://doi.org/10.1002/advs.202301176

96. Santos JL, Li Y, Culver HR, Yu MS, Herrera-Alonso M (2014) Conducting polymer nanoparticles decorated with collagen mimetic peptides for collagen targeting. Chem Commun 50(95):15045–15048. https://doi.org/10.1039/C4CC06056C
97. Wu D, Li H, Xue X, Fan H, Xin Q, Wei Q (2013) Sensitive and selective determination of dopamine by electrochemical sensor based on molecularly imprinted electropolymerization of O-phenylenediamine. Anal Methods 5(6):1469. https://doi.org/10.1039/c3ay26200f
98. Garma LD, Ferrari LM, Scognamiglio P, Greco F, Santoro F (2019) Inkjet-printed PEDOT:PSS multi-electrode arrays for low-cost *in vitro* electrophysiology. Lab Chip 19(22):3776–3786. https://doi.org/10.1039/C9LC00636B
99. Mugo SM, Dhanjai, Alberkant J (2020) A biomimetric lactate imprinted smart polymers as capacitive sweat sensors. IEEE Sensors J 20(11):5741–5749. https://doi.org/10.1109/JSEN.2020.2974851
100. Ahadian S, Banan Sadeghian R, Yaginuma S, Ramón-Azcón J, Nashimoto Y, Liang X, Bae H, Nakajima K, Shiku H, Matsue T, Nakayama KS, Khademhosseini A (2015) Hydrogels containing metallic glass sub-micron wires for regulating skeletal muscle cell behaviour. Biomater Sci 3(11):1449–1458. https://doi.org/10.1039/C5BM00215J
101. Ramón-Azcón J, Ahadian S, Estili M, Liang X, Ostrovidov S, Kaji H, Shiku H, Ramalingam M, Nakajima K, Sakka Y, Khademhosseini A, Matsue T (2013) Dielectrophoretically aligned carbon nanotubes to control electrical and mechanical properties of hydrogels to fabricate contractile muscle myofibers. Adv Mater 25(29):4028–4034. https://doi.org/10.1002/adma.201301300

# Chapter 14
# Modification of Conducting Polymers for Smart Textile and Wearable

**Astha Kamboj, Anshul Yadav, Chetna Kumari, Nirmala Kumari Jangid, and Kavita Poonia**

**Abstract** Conductive polymers are organic polymers that possess the ability to conduct electricity, and they have alternate single and double bonds between the carbon atoms. To improve the conductivity, stability, and flexibility of conductive polymers, a number of modification techniques have been used, including chemical vapor deposition (CVD), in situ polymerization, grafting, doping, and many more, making them more crucial for the creation of E-textiles. E-textiles are fabrics with special properties, such as the ability to sense and respond to environmental conditions and stimuli. These fabrics are important in various fields, that is, aerospace, automotive, sportswear, healthcare, and many more. Conductive polymers are a great option for textile electrodes because they have the ability to combine the electrical properties of semiconductors and metals with the benefits associated with conventional polymers, such as affordability, structural variety, adaptability, and longevity. This chapter explains the modification methods of conducting polymers in order to improve its functions in different types of smart textile applications.

**Keywords** Conductive polymers · Smart textiles · In situ polymerization · Grafting · Doping · Health care · Sports wear

## 14.1 Introduction

E-textiles, or smart textiles, are designed to detect and react to environmental stimuli in an intelligent manner. Smart textiles deal with sensors that are physically and mechanically incorporated into textiles [1, 2], and by giving people real-time information about their surroundings and the ability to sense various stimuli such as strain, pressure, temperature, and humidity, they assist them managing their everyday activities in better way [1]. The goal of the Smart textiles is to address

A. Kamboj · A. Yadav · C. Kumari · N. K. Jangid · K. Poonia (✉)
Department of Chemistry, Banasthali Vidyapith, Banasthali, Rajasthan, India



S. Jadoun, N. Kumari Jangid (eds.), *Tuning of Conducting Polymers*, Springer Series on Polymer and Composite Materials,
https://doi.org/10.1007/978-3-032-00461-1_14

cutting-edge uses in the field of sports, fitness, safety, healthcare, and the military or defense. They can also serve as carriers for autonomously developed sensors or support the sensor that is an incorporated component of a material [2].

Additionally, certain textile qualities should be maintained when chemical sensors are non-invasively incorporated into clothing. It should be a lightweight device and able to absorb low power [3]. In the recent years, a number of wearable sensor-based devices have been created and modified to track the health of people. These comprise the heart rate [4], breath, metabolism [5], vocalization, facial expression, wrist pulse [6], and many more. In addition to creating opportunities for advancements in biomonitoring, telemedicine, and rehabilitation, these wearable textile-based sensors also lead to advancements in a number of new technological domains, including wireless communication systems, ergonomics, virtual augmented reality, tele-assistance, and people telecontrol [7].

Additionally, there are three types of E-textiles: very smart, active, and inactive or passive. The first category of smart textiles that can provide additional functionality without responding to environmental changes is referred to as passive smart textiles. Bulletproof, anti-odor, antistatic, and anti-microbial textiles are the examples of passive smart textiles or inactive smart textiles. Active smart textiles, which fall under the second group, can sense and respond to external cues. And these materials can be used as actuators and sensors. The third group of textile that has a recognition, reasoning, and actuation unit is very smart textiles. These textiles, including space suits and health monitoring devices, sense, respond, and adjust to their surroundings or stimuli [8]. The development of smart textiles heavily relies on conductive polymers, which are organic polymers that have the ability to conduct electricity and display either conductive or semiconductive behavior. Because of their affordability, ease of preparation, biocompatibility, relatively high electrical conductivity flexibility, and ability to be customized for sensing and actuation functions, these polymers have gained popularity as conductive materials. Extrinsically conductive polymers and electroactive conjugated polymers, sometimes referred to as intrinsically conductive polymers (ICPs), are the essential components of actuators and sensor systems. To improve smart monitoring, these materials have desirable optical and electrical qualities that were previously exclusive to inorganic materials [9]. They can be used in fuel cells, super capacitors, energy storage, conversion devices, conducting inks, adsorbents, heterogeneous catalysts, antistatic packaging, and many more. The creation of light-smart materials without the use of metals is possible with conductive polymers. Conductive polymers are essential to the creation of commercially viable smart textile products, including protective clothing, medical textiles, touch screen displays, flexible fabric keyboards, and sensors for a range of applications [10].

## 14.2 Conducting Polymers

A vast array of complex structures can be created by molding polymers. Conducting polymers, like metals, exhibit a broad spectrum of electrical conductivity while retaining their mechanical properties [11]. Due to their electrical conductivity and conductive or semiconductive qualities, conducting polymers also referred to as organic polymers are essential to the development of smart textiles [12]. Polyacetylene, polyurethane, polyaniline, and more than 25 additional conducting polymers have been discovered [13]. In polymers, these materials show both mechanical and electrical properties [14]. Due to their relatively high regulating electrical conductivity, adaptability, biocompatibility, ease of manufacture, flexibility, light weight, and affordability, these polymers have become widely used conductive materials. The key characteristics of metals and conventional polymers were merged into conductive polymers. Two subgroups of conducting polymers are extrinsically conductive polymers and intrinsically conductive polymers (ICPs) [15].

### 14.2.1 Extrinsically Conductive Polymers (ECPs)

ECPs are lightweight, flexible, and can be easily integrated into the fabrication of textile; they have increasingly being recognized as the main materials for smart textiles and wearable electronics [16]. Conductive polymers have low conductivity due to low charge mobility, but by adding external conductive fillers like metal (such as stainless steel and silver), carbon (such as carbon nanotube and carbon black), and oxide/non-organic (such as CuS and indium tin oxide) or conductive polymer polypyrrole [17], the conductivity is greatly increased compared to conventional conductive materials, such as intrinsically conductive polymers [18]. When conductive fillers are dispersed uniformly in a polymer matrix, a conductive network is created that facilitates the flow of electrical charge, which is crucial for applications in electronic textiles, such as flexible electronic devices, sensors, and energy storage devices integrated into fabrics and wearable technologies. The use of electrically conductive polymer enables fibers to maintain the unique mechanical behaviors (flexibility, stretchability, and lightweight) of the polymer matrix with tunable electrical conductivity. Conductive polymers can make flexible and stretchable sensors that monitor a variety of physiological signals, such as heart rate, muscle activity, body temperature, and environmental factors like humidity and pressure. And for health monitoring through wearable technology, these sensors can easily be integrated into clothing or accessories such as shirts, socks, or wristbands, for medical evaluations or fitness tracking. Electrically conductive polymer can be applied in energy harvesting devices to convert mechanical energy associated with human body movements into electricity that is used for the power supply of small wearable electronics [19].

### 14.2.2 Intrinsically Conductive Polymers (ICPs)

Since their backbone chains contain lengthy conjugated double bonds that combine to generate intrinsically conductive polymers, and drawn interest as promising polymer materials. These double bonds are an indication of increased polymer conductivity. There are two straightforward ways to manufacture them: chemical oxidation and electrochemistry. The most encouraging compounds within these polymer groups include polyaniline, polyacetylene, polypyrrole, and poly (3, 4-ethylene dioxythiophene), a derivative of polythiophene. Their electrical properties are weak, but they are easily manufactured and exhibit excellent electrical conductivity and good environmental stability [17].

## 14.3 Most Popular ICPs in Smart Sensors

Smart sensors are made using a variety of ICPs, but the following polymers are frequently used as shown in Fig. 14.1.

### *14.3.1 Polyacetylene*

The simplest structure of all the polymers makes polyacetylene a conducting polymer. Doping the synthetic polymer polyacetylene with electron donors or acceptors results in high electrical conductivity. Trans-Cisoid polyacetylene and Cis-Transoid polyacetylene are two examples of the various configurations that can be created by the arrangement of its long chain of alternating double bonds. Polyacetylene is an insulator in normal circumstances, but it becomes conductive when doped with the right substances (such iodine or an oxidizing agent). According to the research, $AsF_5$ can be used to chemically oxidize methyl acetylene and acetylene copolymer films in order to increase their conductivities within the range of 50 S/cm [20]. Additionally, polyacetylene is relatively easy to process and can be synthesized in a variety of forms, including films, fibers, and composites, increasing its versatility

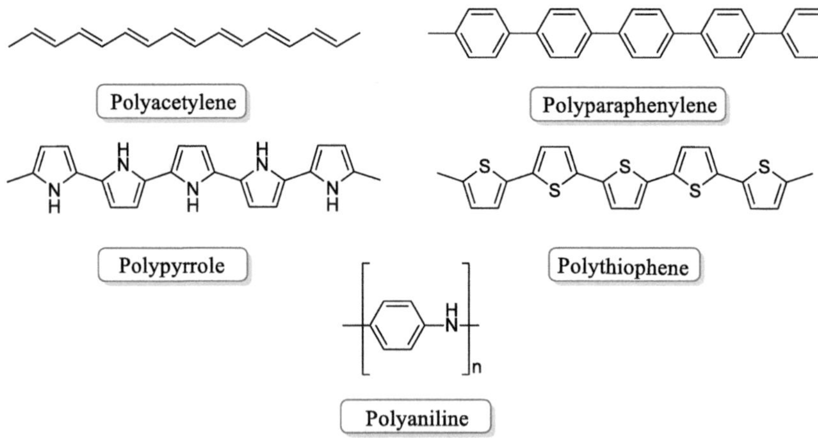


**Fig. 14.1** Chemical structure of different types of intrinsically conductive polymers

for various textile applications [21]. Polyacetylene maintains its elasticity, which is crucial when incorporating it into textiles, in contrast to metals or ceramics. Because of this property, materials based on polyacetylene can stretch and flex without losing their ability to conduct electricity. Additionally, it is significantly lighter than metals or carbon-based materials that are commonly used in electronics, which is essential for textile applications that are wearable and portable. By altering the synthesis conditions, doping agents, and processing methods, polyacetylene electrical conductivity, mechanical qualities, and optical traits can be adjusted [22].

### 14.3.2 Polyparaphenylene

Another member of the conducting polymer family is poly (p-phenylene); Polyparaphenylene was synthesized for the first time in 1979. To put it simply, poly (p-phenylene) is created when benzene rings connected with a single bond to form lengthy chains [23]. N-type poly (p-phenylene) can be produced by using alkali metals as an agent for reduction of Poly (p-phenylene) oligomers, biphenyl, terphenyl, and tetraphenyl react with hexafluoroarsenate in an immediate oxidative polymerization to form conducting poly (p-phenylene). When $AsF_5$ interacts with a single terphenyl crystal plate, the polymer PPP exhibits significant optical and electrical anisotropy [24]. Furthermore, Polyparaphenylene-containing textiles can be used as sensors that users wear to track their body temperature and heart rate as well as their muscular activity. Polyparaphenylene made fibers, for instance, can be woven into textiles to make soft, flexible wearable health monitoring device [22].

### 14.3.3 Polypyrrole

The importance of polypyrrole (PPy) has increased due to its excellent conductivity and stability in oxidized states. Its low cost and straightforward oxidation reaction are two important features that set it apart from other conducting polymers. But pyrrole monomers are more expensive than aniline monomers when compared to PANI, which makes them less appealing for some possible uses, Oxidized PPy is more chemically stable at room temperature than polyacetylene [25]. Only at temperatures higher than 150–300 °C PPy start to degrade [26]. Depending on whether counter ions are present, Polypyrrole electrical conductance can vary from 10 to 1000 S/cm. While acetonitrile, propylene carbonate, and dichloromethane [27] are examples of aqueous and nonaqueous media that can be utilized for the electrochemical polymerization of polypyrrole [28]. Furthermore, numerous wearable electronics applications exist for conductive polypropylene. These textiles have the required conductivity to enable sensor integration for health monitoring systems. For example, conducting polymer-based textiles can be used as electrodes for electrocardiogram (ECG) monitoring, without the need for large, cumbersome hardware, enabling real-time heart rate tracking and conductive fabric based on propylene alter the color or texture of textile when they come in contact with external stimuli like pressure, temperature, humidity, and other related factors [22].

### 14.3.4 Polythiophene

Although pyrrole and polythiophene have structural similarities, polythiophene contains sulfur instead of pyrrole's amine group. However, Polythiophene particles produced through chemical polymerization have been reported to have low electrical conductivity [29]. Nonetheless, electrochemically polymerized polythiophene (PT) has electrical conductivity that is similar to that of PPy. Because of their similar structures, thiophene and pyrrole can be prepared similarly to PPy. Polythiophene is electrochemically polymerized using a nonaqueous medium. Benzonitrile [30], propylene carbonate [31], and acetonitrile [27] are the electrolytes most frequently employed for polymerization. Because of its electrical conductivity and stability, Poly (3, 4-ethylene dioxythiophene) is most researched polymers of our time. It exhibits more conductivity than PPy in the doped condition and post-treatment or blending techniques can enhance this conductivity even more.

It is also a great material for an actuator's electrode because of its extremely stable oxidized state, which allows it to maintain conductivity for a long period even at high temperatures. In addition to this, PEDOT is more beneficial when it is synthesized using straightforward techniques such chemical reactions or electrochemical processes [32]. And they are used as a sensor for muscle activity and heart rate and also integrated with energy harvesting system that generates electrical energy from movement [22].

### 14.3.5 Polyaniline (PANI)

The most popular conductive polymer (PANI) is polyaniline. Excellent chemical stability, low cost monomers, a high capacitance value, customizable characteristics, and simplicity of synthesis are just a few of polyaniline noteworthy attributes. These characteristics provide this polymer a significant advantage over other conducting polymers. At temperatures above 200 °C, PANI doped with counter ions always undergoes thermal degradation, and PANI has good electrical conductivity in the 10–100 S/cm range electrical conductivity is believed to be tuned by the fraction of Crystallinity, molecular weight, oxidation level, and doping degree. Both reduction and oxidation can transform PANI into a hydrogenated or dehydrogenated state [33]. If the pH level of the solution is extremely high, PANI will create side effects during the electrochemical polymerization process. As a result, in an acidic environment, PANI is always polymerized. The final product's conductivity decreases. Additionally, PANI is being studied for tissue engineering, controlled medication administration, and brain probes [34].

## 14.4 Technique for Preparing Intrinsically Conducting Polymers

Polymers can generally be prepared and modified by a number of ways, including curing, grafting, and blending and many more.

### *14.4.1 Blending*

In blending to achieve the desired properties, two or more polymers are physically mixed together as shown in Fig. 14.2 [21].

### *14.4.2 Grafting*

It involves the covalent bonding of monomers onto the polymer chain (Fig. 14.3a). Grafting makes conductive polymers easier to produce and offers the chance to get the specific use characteristics needed for a certain application. Water-soluble PPy is produced via grafting procedures on self-doped conductive polymers. Additionally, aniline was polymerized in a poly (p-amino styrene) aqueous solution to produce a graft copolymer that is soluble in specific solvents, making PANI soluble. Grafting can also make the polymer chain less stiff, which allows the solvents to solvate it and give it solubility [21].

### *14.4.3 Curing*

An oligomer mixture is polymerized during curing to create a covering that clings to the substrate through physical forces (Fig. 14.3b) [21].

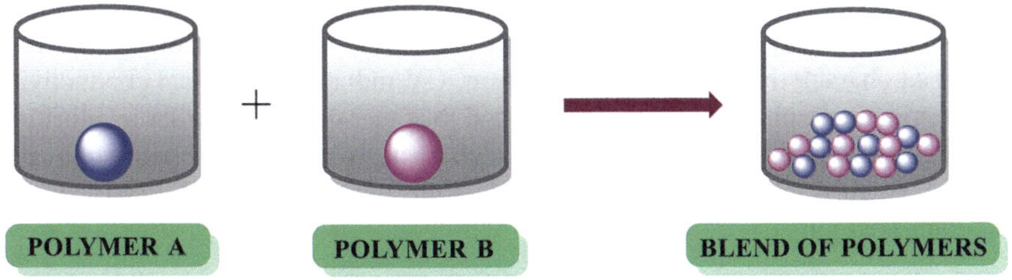


**Fig. 14.2** Diagrammatic representation of blending

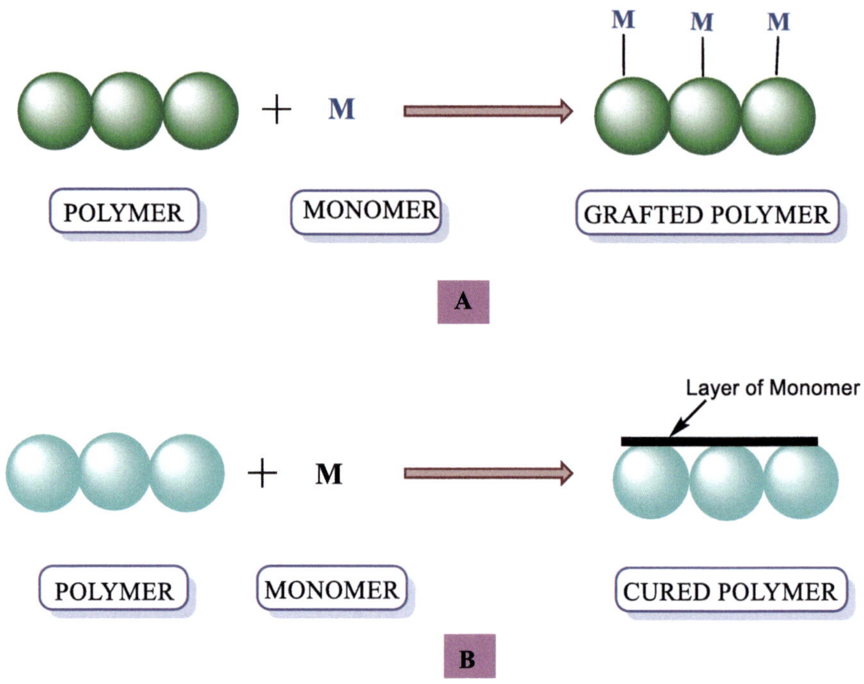


**Fig. 14.3** Diagrammatic representation of (**a**) grafting (**b**) curing

### *14.4.4 Doping*

A suitable dopant can be used to achieve high electrical conductivity in conductive polymers, such as polyvinyl chloride and polystyrene. Conducting polymer-coated textiles are prepared with the aid of these polymers molecular differences. These physical and mechanical characteristics of these polymer are influenced by dopant, monomers, anions, oxidants, and manufacturing parameters. The production of side products is decreased by low-temperature polymerization. Dopant counter ions size and mobility can be controlled to quantify the benefits of conductive polymer-based sensors, which include elasticity and flexibility [35]. Conductive polymer-based sensors designed to integrate seamlessly with textile structures have two main advantages: elasticity and potential flexibility. Furthermore, all of these material characteristics, including moisture, UV radiation, electrical conductivity, and temperature change, are observable and thus measurable. In order to adjust the mechanical, optical, and electrochemical characteristics of the doped conjugated polymers, this approach is useful for adjusting the size and mobility of dopant counter ions [36].

### 14.4.5 Electrochemical Polymerization

Conducting polymers can be prepared in a variety of methods, but the most used method is electrochemical synthesis since it is simple and reproducible [37]. This technique creates a thin polymer coating and enables accurate control of the film viscosity by the manipulation of electrochemical polymerization parameters, including the electro-polymerization duration and current density [38]. Typically, the electrochemical coating is applied using an electrochemical cell. The three electrodes that make up this cell are the working electrode, which is used to coat the substrate, the reference electrode, which is made of inert materials such as Saturated Calomel Electrode (SCE), and the counter electrode, which is made of platinum. Polymerization solution is made consisting of monomer, solvent, and supporting electrolyte [39]. According to further research simplicity, suitability for practical uses and ability to vary the thickness of the polymer film through modification of the polymerization time make this technique beneficial [40].

## 14.5 Textile and Clothing

Nowadays, textiles are used in every aspect of human activity. They also have a number of advantages over other materials, such as ductility, strength, and flexibility. And these factors have made the recent incorporation of multifunctional qualities into such a common material an inventive step toward the next generation of textiles. Innovations in materials science and nanotechnology have enabled new methods of system design for applications in industries like work wear, sports training, healthcare, and the military [41].Additionally, conducting polymer containing textiles can store energy in supercapacitors or use piezoelectric mechanism to transform motion into energy. This feature makes it possible to create clothes that can generate electricity, which might be used to charge wearable [42]. Furthermore, textiles are pliable and permeable fabrics created by braiding, knitting, nonweaving, or synthetic weaving, natural, or synthetic fibers such as polyester or cotton. Fabrics are composed of hierarchically structured fibrous materials. The initial levels of integration known as fibers, that are entangled to form a thread have a high thickness to length ratio [43]. The process of twisting thread to form yarn signifies the second level of integration and in the third level of integration, the yarn is subsequently transformed into fabric using various methods, such as knitting and weaving [44], the introduction of portable electronics and soft computing has sparked a lot of interest in fabrics, which are novel silicon wafers. In terms of structure, electronic apparel that can be worn for observation was made using knitted and woven materials. The primary benefit of textiles in relation to a few different levels of wearing comfort is their flexibility. Knitted materials are fairly elastic and malleable, which makes them useful in situations where the cloth must fit closely to the body, like in sportswear [45]. Because of their nature, fabrics are perfect for creating sensors that come into direct contact with persons. Natural, regenerated, or synthetic fiber unit has a big impact on the characteristics of the fabric. Because synthetic fibers absorb less water, they are more hydrophobic than natural fibers, and they have excellent mechanical agreement and are hydrophilic. Numerous fiber types, such as natural fibers, polyester, nylon, and acrylic, have been used in textile production through a variety of methods [46].

## 14.6 Methods of Coating Textiles

Applying a prepared substance or viscous liquid (liquid) to a textile substrate is known as coating. To create textiles that are electrically conductive, it can be applied to fibers surface, yarns or the fabrics. This technique can be used on a variety of fiber types that can provide high conductivity levels without changing the substrates inherent handling, flexibility, or density characteristics. Typically, a conductive coating paste is used to cover polymers [47].

### *14.6.1 In Situ Polymerization*

Chemical polymerization inside the bulk solution is achieved through the sequential addition of reagents and fiber components in in situ polymerization (Fig. 14.4). This process can result in spontaneous deposition on the surface of fiber or precipitation as a substance that are insoluble [48]. The ideal reaction conditions, temperature, and material treatment are chosen to accomplish bulk polymerization on the fabric surface for coating [49]. Other benefits of in situ polymerization include ease of mechanization, the use of inexpensive materials, and the ability to integrate with a number of other curing and heating techniques. Furthermore, this technique offers excellent miscibility with almost all polymer types and can be utilized to create polymer composites that are insoluble and thermally unstable enough to be treated by melt and solution processing and for certain applications, the chemical structure of the polymer can be precisely controlled using in situ polymerization to attain the required degree of conductivity, mechanical strength, and stability [50].

### *14.6.2 Two-Step Polymerization*

Certain compounds are first absorbed by fiber materials and then the remaining reagents are added to start the polymerization reaction. To promote the reagent's penetration into the textiles structure, the first step usually involves heating, agitating, padding, or ultrasonic vibration treatment. Oxidizing agents or monomers

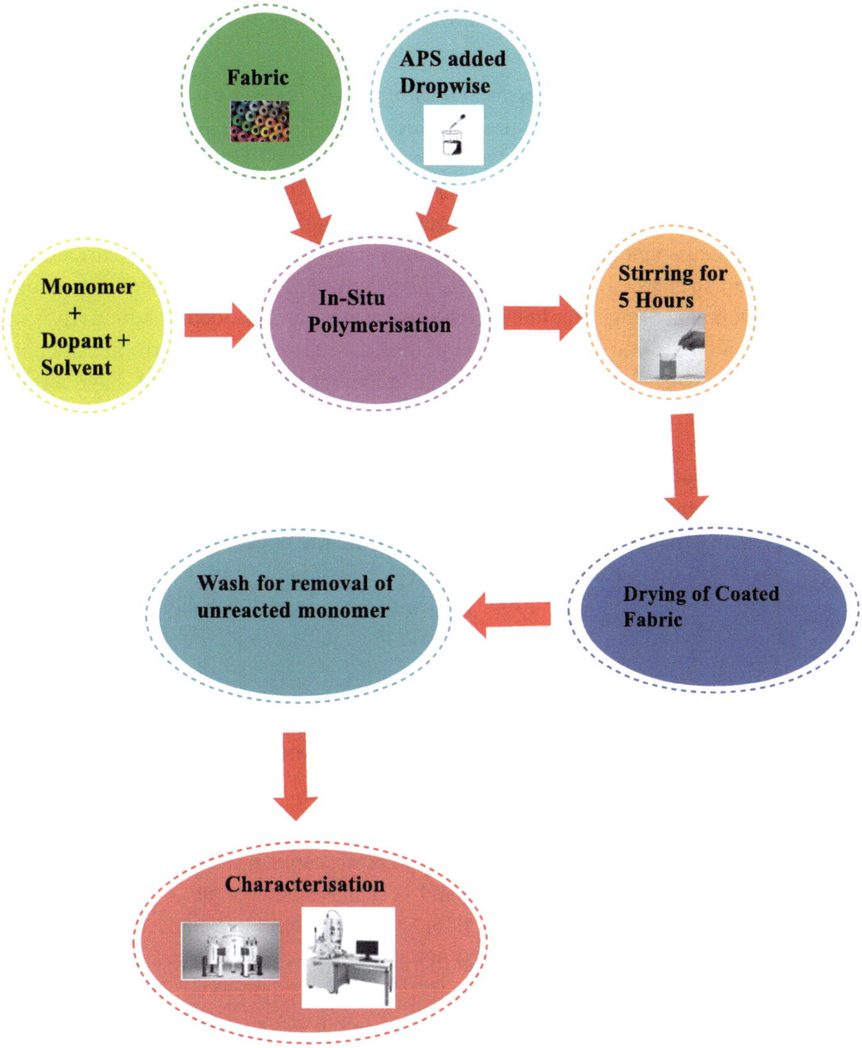


**Fig. 14.4** Flow diagram showing the preparation of conducting textile

should be added to the surface undercoating procedure. The main feature of these techniques is that the polymerization only occur at the surface of fiber, potentially even inside its structure. However, nevertheless, this procedure is not appropriate for certain textile materials that have not been adequately enhanced with monomers or oxidants in a separate stage before the surface polymerization. Because there are wide variety of conducting polymers and their molecular variations, certain polymerization conditions of basic monomers can be modified to create textiles coated with conductive polymers that can be used for a variety of purposes [51].

### *14.6.3 Chemical Vapor Deposition (CVD)*

Conducting polymers can also be coated using CVD. There are two steps involved in coating polymers to create electro conductive materials. The economical CVD method makes it simple to recover unreacted monomers, oxidants, and dopant solutions. This technique ensures uniform application of the oxidant and dopant on the fiber surface during polymerization, producing a uniform layer on the fabric. This method produces nanoparticles with minimal aggregation, enhancing surface area coverage and strengthening the bond between polymeric chains. Therefore, in contrast to aqueous coating, conducting polymers exhibit enhanced conductivity [52]. The interaction between the polymers and the conducting fibers determines the polymer's capacity to coat the substrate. Because animal fibers, like wool, have a lot of functional groups which make it easier for dopant anions and polymer molecules to connect through hydrogen or ionic bonds, they can also be an excellent choice for substrate material for the surface attachment of conductive polymers. Furthermore, conductive polymers have the ability to adhere to the surface shape of wool fibers [9]. This approach offers several advantages, including the deposition of a wide range of chemicals with high purity, which also offers a fast deposition rate and also allows for uniform coating. Since multiple pieces can be coated simultaneously, CVD is cost-effective in production because it does not require a high vacuum process [53].

## 14.7 Application of Smart Sensors in Smart Textiles

The essential parts of wearable electronic systems in monitoring applications are sensors, which gather data from the wearer and transmit it to a processor. Through advancements in monitoring, clothing and vests have been created with a variety of sensors that send location information and physiological data including body temperature, heart rate and skin conductivity via satellite-based services [54]. Many conventional sensors devices have been produced based on their specifications and dimensions. Sensors for smart textiles can use conducting elements, such as metals or carbons. However, because textile materials properties change depending on the environment, such as changes in temperature, pressure, and moisture, modern research interests are using the novel array of polymers to create flexible sensors. As a result, new flexible sensors have been specially designed for integration into textiles and smart clothing [46]. As a sensory nervous system, sensors in smart clothes pick up impulses and send information to the processor. When exposed to volatile compounds, temperature fluctuations, or ambient radiation, conductive polymers are adept at electrically reacting to changes in their molecular structure [55]. The Life Shirt system (which monitors physiological parameters), an electronic inner wear for breast cancer detection, and baby pajamas helps to prevent cot death are contemporary examples of textile incorporated sensors. Such items or clothing have implications for a variety of industries, including the military, rescue, sports, clinical, and healthcare, where it is essential to identify critical indicators. Smart monitoring clothes are used in the medical field for ongoing, continuously moniter patient condition. They also offer dormant support in preventing infant cot death [34].

These wearable sensory devices allow for the real-time recording and monitoring of a person's physiological state and activities. They can include various flexible sensors that can be attached directly to the body or integrated into clothing, textile fibers, and elastic bands. These sensors are capable of accurately measuring a range of physiological signals like electrocardiogram (ECG), respiration rate (RR), electromyogram (EMG), heart rate (HR), arterial oxygen saturation (SpO2), blood pressure (BP), electro-dermal activity (EDA), and body temperature [56]. Continuous monitoring of physiological signals aids in the identification and diagnosis a number of diseases in their early stages, including neurological, cardiovascular, and pulmonary problems. Additionally, it is quite possible to monitor person actions in real time using these sensory devices, which are very useful for fall detection, posture analysis, gait monitoring, and sleep assessment [29].

## 14.8 Types of Pressure Sensors in Smart Textiles and Their Mechanisms

### *14.8.1 Monitoring Postures and Movements*

Numerous templates have been established for the purpose of monitoring posture and movement, which aids in rehabilitation, improves body posture, and lowers the risk of sports-related and other injuries. These devices were not well-known in the past due to a number of drawbacks, including their weight, mass, and difficulty in removing, as they needed to be affixed to the body directly in order to collect data. The majority of flexible sensors work on the premise that stretching alters the electrical resistance of the sensors. With the advancement of wearable sensing instruments, flexible wearable sensors are becoming increasingly popular. Another textile-based polymer-coated sensor was made for this search. First, aniline was polymerized in situ to produce polyaniline (PANI). Then, using the deposition process, PANI was applied to cotton knit fabric to create conductive fabric. With a strong linearity and maximum gauge factor, this conductive knit fabric provides satisfactory performance. Thus, by examining the variation in resistance of the fabric sensor when various human body parts are moving, including the neck, elbows, and knees, this device can be utilized to monitor human body signals and movement [40].

### *14.8.2 Stretch and Strain Detection*

Wearable strain gauges are an alternative to traditional strain gauges for measuring movement across a range of strains. These strain gauges are composed of textile fabrics covered with conducting polymers. In 2009, De Rossi and colleagues conducted the first investigation of flexible strain gauges utilizing conducting polymer-coated textiles. Another study is concentrating on creating wearable strain gauges made from nylon Lycra fabric covered with conductive polypyrrole so that biomechanical monitoring can be done, an example of this would be a smart knee sleeve. Made from conducting polypyrrole-coated Lycra strip, this knee sleeve responds to the player by emitting an auditory tone. Consequently, various audio tones are released in accordance with the strain generated by the coated fabric, giving the user real-time feedback. Moreover the primary focus in a number of application domains is on posture and gesture analysis as well as body kinematics monitoring. A leotard made of electrical polymer-coated elastic fabric (Lycra) was created in 2000. This sensor has been used to track the posture and movement of the trunk as well as several body sections, including the scapular segment, glen humeral, vertebral column, and elbow joints of the spine [57].

And a flexible strain sensor was created by using polypyrrole-coated textiles in 2005 by combining in situ polymerization and chemical vapor deposition. Flexible strain sensors with superior sensitivity and stability are made from these materials. The sensitivity and stability of the sensors have been enhanced by a number of techniques, including (a) low-temperature pyrrole polymerization, (b) reinforcing fabrics with conductive polypyrrole coatings, (c) applying thin coats of polypyrrole to the fabric's surface using the chemical vapor deposition techniques, (d) the polypyrrole film is enriched with a large anion, such as dodecyl benzene sulfonate.

Additionally, polypyrrole-coated fabric is used in wearable hardware, rehabilitation, sensing clothing, and other fields. A conducting polypyrrole-coated fabric developed was developed on the basis of nylon Lycra. In situ chemical polymerization was employed to create this sensor fabric. The sensor was created using $FeCl_3$ as an oxidizing agent, white nylon Lycra fabric, and Tetra-hydrate of 1, 5-naphthalene sulfonic acid as a dopant. After immersing a piece of cloth in an aqueous mixture of a dopant and pyrrole monomer, and then the flask was filled with $FeCl_3$ solution to initiate the polymerization process over the fabric surface. The resulting fabric may take on the shape of different body parts to examine the movement of various human body parts, making it ideal for use as a biomechanical wearable sensor. And this innovative approach has broad applications in medical care, rehabilitation, injury prevention, and sports skill enhancement. In situ chemical polymerization was also used to create polyaniline-coated conductive cloth, according to PSG College of Technology in India. GSM 86, a polyester fabric with a plain weave, was used for the coating substrate. These coated fabrics function as intelligent sensors when the body is stretched [58].

### *14.8.3 Electrocardiogram Monitoring*

Physiological factors such as, cardiac rate variability, electrocardiograph (ECG), and arterial pressure must be monitored during all medical procedures. Additionally, new advancements in smart sensors for ECGs also improve comfort and ease of living. According to the research, these sensors were made to be worn close to the body in a variety of shapes and sizes, including jackets, vests, sleeves, wrist bands, and more. A fabric-based smart sensor was created by Cetiner and his colleagues to track an ECG or watch the electrical activity of a human heart. 3, 4-ethylene dioxythiophene was polymerized in situ to fabricate the sensor. Acetonitrile, p-toluene sulfonic acid monohydrate, and iron (III) chloride ($FeCl_3$) were also employed as polymerizing agents. However, the substrate was made of woven polyethylene terephthalate (PET) materials. The development of these textile-based sensors could potentially replace commercial ECG monitoring devices and offer a solution for extended physical signal monitoring without the use of hydrogel. These sensors offer adaptability in terms of material integration, sensor size and sensor design they are also lightweight, washable, and appropriate for clothing incorporation.

A smart sensor that detects ECG signals by employing polyamide as a conducting material. On ordinary cotton cloth, bands of polyamide threads were stitched between two electrodes. These bands were covered with thermoplastic polyurethane films for protection. During prolonged ECG monitoring, the embroidered bands have the ability to record electrical signals. Bio-signal detection is greatly aided by biomedical conducting devices. These high-performance sensing devices are essential for acquiring biosignals for medical diagnostics. In this case, these devices were made using polypyrrole (PPy) and polyvinyl chloride (PVC). These polymer-based devices are highly effective for gathering various types of biosignals, and the PPy/PVC sensors performed better than commercially available sensors [22].

## 14.9 Future Perspective

The potential of conducting polymers (CPs) in wearables and smart textiles is enormous, due to developments in fabrication methods, nanotechnology, and material science. The development of next-generation smart fabrics is expected to be heavily reliant on conductive polymer as customer demand for wearable electronics becomes more functional, flexible, and sustainable.

Investigation and advancement of producing single components, such as microprocessors, actuators, sensors, and many more, is becoming more and more popular in industries. The industries keen interest in this domain drives innovation. Moreover, research on interactive clothing design can be utilized in various fields such as clinical health monitoring, sports activities, defense (including injury and vital signs tracking), location, communication systems and position tracking. And a small number of systems have been heavily commercialized, and much more work needs to be done to address all of the technological issues, particularly with regard to: (i) The system's robustness, which includes water resistance and long-term performance accuracy and the effectiveness of power management, ease of handling, and the capacity to put on and utilize clothing without help; (ii) Movement, including weight reduction; (iii) Comfort and attractiveness, including handling, breathability, draping, absorption, and other factor; (iv) cost in comparison to the product's longevity or durability. Given the numerous benefits of smart textiles, it is clear that medical and healthcare applications are expanding. To meet customer expectations, more end-user apps must be introduced to the market with a strong emphasis on these areas. The objective of smart textiles should focus on creating innovative applications, particularly for smart home textiles, given the quick pace of life in many regions.

## 14.10 Conclusion

Conducting polymer (CP) integration has become a game-changing strategy for creating functional fabrics for wearable's and smart textiles. This chapter has examined the several approaches to altering conducting polymers to improve their characteristics and make them appropriate for usage in demanding and dynamic applications. Approaches include chemical vapor deposition, doping, in situ polymerization, grafting, blending, and many more combining with other materials, and the use of sophisticated production methods like 3D printing and electrospinning. With these changes, the wearer's comfort, mechanical flexibility, electrical conductivity, and environmental stability are all designed to be perfectly balanced. For monitoring applications, the technological and wearable sensor systems functional elements have also been extensively researched and are still developing. From energy-harvesting textiles and responsive clothing to electronic skin these modified conducting polymers have a wide range of potential uses in wearables and smart textiles.

These applications mainly depend on conducting polymers capacity to adapt to the intricate and dynamic interactions between humans and their surroundings without sacrificing functionality. Wearable health monitoring devices, like strain sensors and biosensors, require conducting polymers to be flexible, washable, and highly sensitive to physiological changes in addition to conducting electricity. The fields of textile and clothing research were the sole focus of invention at first, but as time went on, advancements were made in response to industry and consumer demands. Wearable monitoring clothing is expected to find more uses for individual consumers as consumers become more conscious for lifestyle products. These clothing systems could be crucial in facilitating the freedom of the elderly and disabled by allowing them to continue their regular routine, which can be observed by internet monitoring systems, given the significant elderly population.

## References

1. Gonçalves C, Da Silva AF, Gomes J, Simoes R (2018) Wearable e-textile technologies: a review on sensors, actuators and control elements. Inventions 3(1):14. https://doi.org/10.3390/inventions3010014
2. Luo N, Zhang J, Ding X, Zhou Z, Zhang Q, Zhang Y, Chen S, Hu J, Zhao N (2017) Textile-enabled highly reproducible flexible pressure sensors for cardiovascular monitoring. Adv Mater Technol 3(1). https://doi.org/10.1002/admt.201700222

3. Park S, Jayaraman S (2003) Smart textiles: wearable electronic systems. MRS Bull 28(8):585–591. https://doi.org/10.1557/mrs2003.170
4. Güder F, Ainla A, Redston J, Mosadegh B, Glavan A, Martin TJ, Whitesides GM (2016) Paper-based electrical respiration sensor. Angew Chem Int Ed 55(19):5727–5732. https://doi.org/10.1002/anie.201511805
5. Imani S, Bandodkar AJ, Mohan AMV, Kumar R, Yu S, Wang J, Mercier PP (2016) A wearable chemical–electrophysiological hybrid biosensing system for real-time health and fitness monitoring. Nat Commun 7(1). https://doi.org/10.1038/ncomms11650
6. Nassar JM, Mishra K, Lau K, Aguirre-Pablo AA, Hussain MM (2017) Recyclable nonfunctionalized paper-based ultralow-cost wearable health monitoring system. Adv Mater Technol 2(4). https://doi.org/10.1002/admt.201600228
7. Carpi F, DeRossi D (2005) Electroactive polymer-based devices for e-textiles in biomedicine. IEEE Trans Inf Technol Biomed 9(3):295–318. https://doi.org/10.1109/titb.2005.854514
8. Patwary MSSU (2015) Smart textiles and nano-technology: a general overview. J Textile Sci Eng 05(01). https://doi.org/10.4172/2165-8064.1000181
9. Huynh T, Haick H (2018) Autonomous flexible sensors for health monitoring. Adv Mater 30(50). https://doi.org/10.1002/adma.201802337
10. Grancarić AM, Jerković I, Koncar V, Cochrane C, Kelly FM, Soulat D, Legrand X (2017) Conductive polymers for smart textile applications. J Ind Text 48(3):612–642. https://doi.org/10.1177/1528083717699368
11. Burr GW, Kurdi BN, Scott JC, Lam CH, Gopalakrishnan K, Shenoy RS (2008) Overview of candidate device technologies for storage-class memory. IBM J Res Dev 52(4.5):449–464. https://doi.org/10.1147/rd.524.0449
12. Majumder S, Mondal T, Deen M (2017) Wearable sensors for remote health monitoring. Sensors 17(1):130. https://doi.org/10.3390/s17010130
13. Benhamou SM, Hamouni M (2014) Determination of reflection loss, absorption loss, internal reflection and shielding effectiveness of a double electromagnetic shield of conductive polymer. J Mater Environ Sci 5(6):1982–1987
14. Lekpittaya P, Yanumet N, Grady BP, O'Rear EA (2004) Resistivity of conductive polymer–coated fabric. J Appl Polym Sci 92(4):2629–2636. https://doi.org/10.1002/app.20270
15. Cristian I, Nauman S, Cochrane C, Koncar V (2011) Electro-conductive sensors and heating elements based on conductive polymer composites in woven structures. In: InTech eBooks. https://doi.org/10.5772/18274
16. Ates M, Sarac AS (2009) Conducting polymer coated carbon surfaces and biosensor applications. Prog Org Coat 66(4):337–358. https://doi.org/10.1016/j.porgcoat.2009.08.014
17. Langereis G, Bouwstra S, Chen W (2012) Sensors, actuators and computing systems for smart textiles for protection. In: Elsevier eBooks, pp 190–213. https://doi.org/10.1533/9780857097620.1.190
18. Heinze J, Frontana-Uribe BA, Ludwigs S (2010) Electrochemistry of conducting polymers—persistent models and new concepts. Chem Rev 110(8):4724–4771. https://doi.org/10.1021/cr900226k
19. Yan J, Chen A, Liu S (2023) Flexible sensing platform based on polymer materials for health and exercise monitoring. Alex Eng J 86:405–414. https://doi.org/10.1016/j.aej.2023.11.085
20. Ishii K, Sato K, Oaki Y, Imai H (2018) Highly porous polymer dendrites of pyrrole derivatives synthesized through rapid oxidative polymerization. Polym J 51(1):11–18. https://doi.org/10.1038/s41428-018-0115-x
21. Ankhili A, Zaman SU, Tao X, Cochrane C, Končar V, Coulon D (2019) Washable embroidered textile electrodes for long-term electrocardiography monitoring. Textile Leather Rev 2(3):126–135. https://doi.org/10.31881/tlr.2019.27
22. Zahid M, Rathore HA, Tayyab H, Rehan ZA, Rashid IA, Lodhi M, Zubair U, Shahid I (2021) Recent developments in textile based polymeric smart sensor for human health monitoring: a review. Arab J Chem 15(1):103480. https://doi.org/10.1016/j.arabjc.2021.103480
23. Chow PCY, Someya T (2019) Organic photodetectors for next-generation wearable electronics. Adv Mater 32(15). https://doi.org/10.1002/adma.201902045

24. Dias O a T, Konar S, Leão AL, Sain M (2019) Flexible electrically conductive films based on nanofibrillated cellulose and polythiophene prepared via oxidative polymerization. Carbohydr Polym 220:79–85. https://doi.org/10.1016/j.carbpol.2019.05.057
25. Shi X, Chen W, Zhang T, Zou J, Chen Z (2020) Fiber-based thermoelectrics for solid, portable, and wearable electronics. Energy Environ Sci 14(2):729–764. https://doi.org/10.1039/d0ee03520c
26. Fernandez FDM, Khadka R, Yim J (2021) A comparative study between vapor phase polymerized PPy and PEDOT—thermoplastic polyurethane composites for ammonia sensing. Polymer 217:123463. https://doi.org/10.1016/j.polymer.2021.123463
27. Tamburri E, Orlanducci S, Toschi F, Terranova ML, Passeri D (2008) Growth mechanisms, morphology, and electroactivity of PEDOT layers produced by electrochemical routes in aqueous medium. Synth Met 159(5–6):406–414. https://doi.org/10.1016/j.synthmet.2008.10.014
28. She C, Li G, Zhang W, Xie G, Zhang Y, Li L, Yue F, Liu S, Jing C, Cheng Y, Chu J (2020) A flexible polypyrrole/silk-fiber ammonia sensor assisted by silica nanosphere template. Sensor Actuat A Phys 317:112436. https://doi.org/10.1016/j.sna.2020.112436
29. Ameen MA, Liu J, Kwak K (2010) Security and privacy issues in wireless sensor networks for healthcare applications. J Med Syst 36(1):93–101. https://doi.org/10.1007/s10916-010-9449-4
30. Sapurina IY, Shishov MA (2012) Oxidative polymerization of aniline: molecular synthesis of polyaniline and the formation of supramolecular structures. In: New polymers for special applications, vol 740(7), p 272
31. Balint R, Cassidy NJ, Cartmell SH (2014) Conductive polymers: towards a smart biomaterial for tissue engineering. Acta Biomater 10(6):2341–2353. https://doi.org/10.1016/j.actbio.2014.02.015
32. Xu Z, Song J, Liu B, Lv S, Gao F, Luo X, Wang P (2021) A conducting polymer PEDOT: PSS hydrogel based wearable sensor for accurate uric acid detection in human sweat. Sensor Actuat B Chem 348:130674. https://doi.org/10.1016/j.snb.2021.130674
33. Liu A, Lv S, Jiang L, Liu F, Zhao L, Wang J, Hu X, Yang Z, He J, Wang C, Yan X, Sun P, Shimanoe K, Lu G (2021) The gas sensor utilizing polyaniline/MoS2 nanosheets/SnO2 nanotubes for the room temperature detection of ammonia. Sensor Actuat B Chem 332:129444. https://doi.org/10.1016/j.snb.2021.129444
34. Shakir HMF, Tariq A, Afzal A, Rashid IA (2019) Mechanical, thermal and EMI shielding study of electrically conductive polymeric hybrid nano-composites. J Mater Sci Mater Electron 30(18):17382–17392. https://doi.org/10.1007/s10854-019-02088-0
35. Hebeish A, Farag S, Sharaf S, Shaheen TI (2016) Advancement in conductive cotton fabrics through in situ polymerization of polypyrrole-nanocellulose composites. Carbohydr Polym 151:96–102. https://doi.org/10.1016/j.carbpol.2016.05.054
36. Scholes DT, Yee PY, Lindemuth JR, Kang H, Onorato J, Ghosh R, Luscombe CK, Spano FC, Tolbert SH, Schwartz BJ (2017) The effects of crystallinity on charge transport and the structure of sequentially processed F4TCNQ-doped conjugated polymer films. Adv Funct Mater 27(44). https://doi.org/10.1002/adfm.201702654
37. Hsiao S, Liao W, Liou G (2017) A comparative study of redox-active, ambipolar electrochromic triphenylamine-based polyimides prepared by electrochemical polymerization and conventional polycondensation methods. Polym Chem 9(2):236–248. https://doi.org/10.1039/c7py01897e
38. Wustoni S, Hidalgo TC, Hama A, Ohayon D, Savva A, Wei N, Wehbe N, Inal S (2019) In situ electrochemical synthesis of a conducting polymer composite for multi metabolite sensing. Adv Mater Technol 5(3). https://doi.org/10.1002/admt.201900943
39. De Leon A, Advincula RC (2014) Conducting polymers with superhydrophobic effects as anticorrosion coating. In: Elsevier eBooks, pp 409–430. https://doi.org/10.1016/b978-0-12-411467-8.00011-8
40. Zhou X, Hu C, Lin X, Han X, Zhao X, Hong J (2021) Polyaniline-coated cotton knitted fabric for body motion monitoring. Sensor Actuat A Phys 321:112591. https://doi.org/10.1016/j.sna.2021.112591

41. Gorgutsa S, Bélanger-Garnier V, Ung B, Viens J, Gosselin B, LaRochelle S, Messaddeq Y (2014) Novel wireless-communicating textiles made from multi-material and minimally-invasive fibers. Sensors 14(10):19260–19274. https://doi.org/10.3390/s141019260
42. Libertino S, Plutino MR, Rosace G (2018) Design and development of wearable sensing nanomaterials for smart textiles. AIP Conf Proc. https://doi.org/10.1063/1.5047770
43. Castano LM, Flatau AB (2014) Smart fabric sensors and e-textile technologies: a review. Smart Mater Struct 23(5):053001. https://doi.org/10.1088/0964-1726/23/5/053001
44. Shim BS, Chen W, Doty C, Xu C, Kotov NA (2008) Smart electronic yarns and wearable fabrics for human biomonitoring made by carbon nanotube coating with polyelectrolytes. Nano Lett 8(12):4151–4157. https://doi.org/10.1021/nl801495p
45. Dhawan A, Seyam AM, Ghosh TK, Muth JF (2004) Woven fabric-based electrical circuits: part I: evaluating interconnect methods. Text Res J 74(10):913–919. https://doi.org/10.1177/004051750407401011
46. Tang SLP (2007) Recent developments in flexible wearable electronics for monitoring applications. Trans Inst Meas Control 29(3–4):283–300. https://doi.org/10.1177/0142331207070389
47. Munro BJ, Campbell TE, Wallace GG, Steele JR (2007) The intelligent knee sleeve: a wearable biofeedback device. Sensor Actuat B Chem 131(2):541–547. https://doi.org/10.1016/j.snb.2007.12.041
48. Sasso C, Beneventi D, Zeno E, Chaussy D, Petit-Conil M, Belgacem N (2011) Polypyrrole and polypyrrole/wood-derived materials conducting composites: a review. Bioresources 6(3)
49. Qu H, Skorobogatiy M (2015) Conductive polymer yarns for electronic textiles. In: Elsevier eBooks, pp 21–53. https://doi.org/10.1016/b978-0-08-100201-8.00003-5
50. Rashid IA, Tariq A, Shakir HF, Afzal A, Ali F, Abuzar M, Haider T (2021) Electrically conductive epoxy/polyaniline composite fabrication and characterization for electronic applications. J Reinf Plast Compos 41(1–2):34–45. https://doi.org/10.1177/07316844211023991
51. Malinauskas A (2001) Chemical deposition of conducting polymers. Polymer 42(9):3957–3972. https://doi.org/10.1016/s0032-3861(00)00800-4
52. Cucchi I, Boschi A, Arosio C, Bertini F, Freddi G, Catellani M (2008) Bio-based conductive composites: preparation and properties of polypyrrole (PPy)-coated silk fabrics. Synth Met 159(3–4):246–253. https://doi.org/10.1016/j.synthmet.2008.09.012
53. Achard J, Jacques V, Tallaire A (2020) Chemical vapour deposition diamond single crystals with nitrogen-vacancy centres: a review of material synthesis and technology for quantum sensing applications. J Phys D Appl Phys 53(31):313001. https://doi.org/10.1088/1361-6463/ab81d1
54. Goulev P, Stead L, Mamdani E, Evans C (2004) Computer aided emotional fashion. Comput Graph 28(5):657–666. https://doi.org/10.1016/j.cag.2004.06.005
55. Zubair K, Ashraf A, Gulzar H, Shakir MF, Nawab Y, Rehan ZA, Rashid IA (2021) Study of mechanical, electrical and EMI shielding properties of polymer-based nanocomposites incorporating polyaniline coated graphene nanoparticles. Nano Express 2(1):010038. https://doi.org/10.1088/2632-959x/abe843
56. Nemati E, Deen M, Mondal T (2012) A wireless wearable ECG sensor for long-term applications. IEEE Commun Mag 50(1):36–43. https://doi.org/10.1109/mcom.2012.6122530
57. Das TK, Prusty S (2012) Review on conducting polymers and their applications. Polym-Plast Technol Eng 51(14):1487–1500. https://doi.org/10.1080/03602559.2012.710697
58. Li Y, Cheng X, Leung M, Tsang J, Tao X, Yuen M (2005) A flexible strain sensor from polypyrrole-coated fabrics. Synth Met 155(1):89–94. https://doi.org/10.1016/j.synthmet.2005.06.008

# Chapter 15
# Emerging Applications of Conducting Polymer in Modern Architecture

**Hiba Gul**

**Abstract** The integration of advanced materials into architectural design has gained significant momentum in recent years, with conducting polymers emerging as a promising frontier. This chapter explores the evolving applications of conducting polymers in modern architecture, focusing on their unique properties, including electrical conductivity, flexibility, and environmental responsiveness. Through a review of the current research and case studies, this chapter examines how these materials are transforming building technologies, enabling innovations in energy efficiency, smart surfaces, and dynamic façades. Conducting polymers present sustainable solutions for responsive environments, positioning them at the intersection of architectural aesthetics and functionality. This chapter highlights their potential to revolutionize urban design, offering architects new avenues for creating intelligent and eco-friendly structures in the context of a rapidly urbanizing world. Introducing Polymers in architecture.

**Keywords** Conducting polymers · Modern architecture · Advanced materials · Energy efficiency · Smart surfaces · Dynamic façades · Sustainable solutions

## 15.1 Introducing Conducting Polymers in Architecture

The progress in materials science has greatly impacted architectural design, leading to structures that excel in creativity, sustainability, and practicality [1]. The most widely used polymers in architecture include paints, waterproofing membranes, sealants, roofing and floor coatings, and all kinds of materials we can use in day-to-day architecture life. With the development of thousands of polymers on the market in the laboratory environment, products used for new applications are always emerging.

H. Gul (✉)
Department of Architecture and Planning, National Institute of Technology Patna, Patna, India



S. Jadoun, N. Kumari Jangid (eds.), *Tuning of Conducting Polymers*, Springer Series on Polymer and Composite Materials,
https://doi.org/10.1007/978-3-032-00461-1_15

Among these advancements, conducting polymers stand out as game-changing materials [2]. They are valued for their ability to conduct electricity, adapt to different environments, and be easily shaped or processed [3]. These unique qualities make them ideal for transforming modern architecture. This chapter highlights the various new ways conducting polymers are being used in architecture, focusing on their ability to create smart, energy-efficient, and eco-friendly buildings [3]. By combining materials science with architecture, we can design structures that are practical, environmentally friendly, and suited to the changing demands of urban life [4].

The integration of advanced materials into architectural design has gained more significance in recent years, with the emerging role of conducting polymers in the field of contemporary architecture as well as retaining sustainability [5]. This chapter explores the evolving applications of conducting polymers in the field of sustainable architecture, focusing mainly on their unique properties like electrical conductivity, flexibility, and environmental responsiveness [6]. Reviewing current research and case studies, the study examines how these materials are transforming building technologies, enabling innovations in energy efficiency, smart surfaces, and dynamic façades [7]. Conducting polymers also provides sustainable solutions for responsive environments, positioning them at the intersection of architectural aesthetics and functionality [8–10]. This chapter highlights their potential to revolutionize urban design, offering architects new avenues for designing intelligent buildings and eco-friendly cities in the rapidly urbanizing world.

## 15.2 Significance of Conducting Polymers in Modern Architecture

Conducting polymers, such as polyaniline (PANI), polypyrrole (PPy), and poly(3,4-ethylene dioxythiophene) (PEDOT), exhibit a unique combination of properties that make them suitable for architectural applications [11–18].

### *15.2.1 Sustainability*

Many conducting polymers can be synthesized using environmentally friendly methods; aligning with the principles of green architecture [19, 20]. Advances in green chemistry have facilitated the production of polymers with reduced ecological footprints [21, 22] (Fig. 15.1).

Also, the focus is on advanced building controls where polymers play a big role, which can be applied to new buildings or retrofitted in old buildings to improve their overall energy efficiency, increase integration of clean energy sources, and coordinate electricity consumption within buildings and with the power grid [23, 24]. This involves integrating technology that automates operational functions, such as ventilation, heating, cooling, and lighting systems, according to schedules and other energy-saving adjustment parameters [25–27].

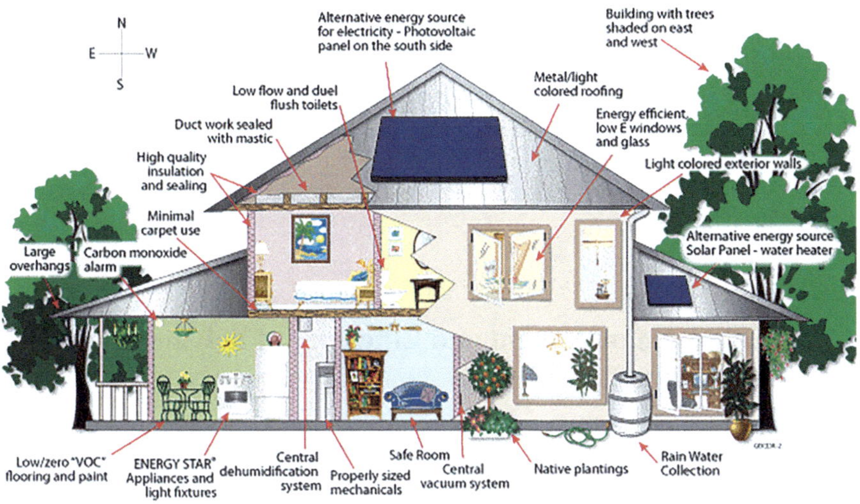


**Fig. 15.1** Integration of polymers in a house

### 15.2.2 *Electrical Conductivity*

Unlike traditional polymers, conducting polymers can conduct electricity, enabling their use in energy management systems and smart building technologies [28, 29]. This property is particularly relevant in the context of developing integrated electronic systems within building structures [30, 31].

### 15.2.3 *Flexibility and Process Ability*

Polymers can be used after processed into different usable films, coatings, and composites, allowing their integration into various architectural design components [31, 32]. Conducting polymers are much compatible with a variety of fabrication techniques, including printing and extrusion, which support scalable production [33–36].

### 15.2.4 *Environmental Responsiveness*

Conducting polymers respond to environmental stimuli such as light, heat, and humidity, making them ideal for adaptive architecture [37, 38]. Their responsiveness can be fine-tuned by modifying their chemical structures, enabling highly specialized applications [39, 40].

### 15.2.5 Aesthetically Pleasing

As polymers are available and can be easily moulded in different forms and colours, they could be used to make each and every aesthetic design on the building facade [41, 42]. They give architects design flexibility. The customization options extend to optical transparency and texture, allowing seamless incorporation into diverse architectural styles [43].

## 15.3 Energy Management and Efficiency of Polymers in Buildings

One of the most significant applications of conducting polymers in architecture is in enhancing energy efficiency:

### 15.3.1 Smart Windows

Smart windows incorporating conducting polymers can modulate light and heat transmission in response to environmental conditions [44, 45]. For example, electro chromic windows using PEDOT can switch between transparent and opaque states, reducing reliance on artificial lighting and heating/cooling systems [46, 47]. Advanced designs may include multi-layered structures where conducting polymers work synergistically with nanomaterials to improve energy control [48]. These windows also enhance occupant comfort by reducing glare and optimizing natural light [49] (Fig. 15.2).

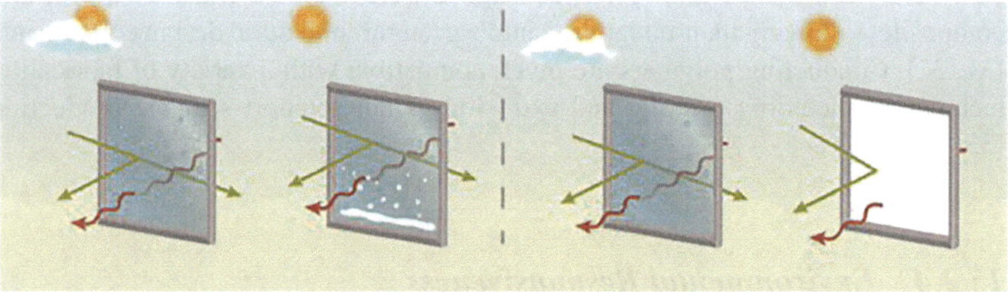

**Fig. 15.2** Smart window using photonic crystal for energy saving

### 15.3.2 Solar Energy Harvesting

Conducting polymers is integral to the development of lightweight and flexible organic photovoltaic (OPV) cells [50]. These cells can be integrated into building surfaces such as roofs, façades, and even windows, enabling structures to generate renewable energy [51, 52]. New developments in tandem solar cells, combining conducting polymers with perovskites, offer enhanced efficiency and broader application potential in urban environments [53, 54]. Building-integrated photovoltaic (BIPVs) represent a promising area for widespread adoption, where entire facades double as energy generators [55].

### 15.3.3 Energy Storage

Architectural components such as walls and façades can incorporate energy storage systems based on conducting polymers [56]. Super capacitors and batteries utilizing polyaniline or polypyrrole offer high energy density and durability, enabling efficient building energy management [57, 58]. Hybrid systems combining energy storage and solar energy generation within the same polymer-based structure are also being explored [59]. These systems contribute to the development of self-sustaining buildings capable of operating off-grid [60, 61].

## 15.4 Adaptive Building Systems

The adaptability of conducting polymers enables their use in dynamic architectural systems (Fig. 15.3):

Adaptive facades are of different types which are used in various designs facades of buildings [62].

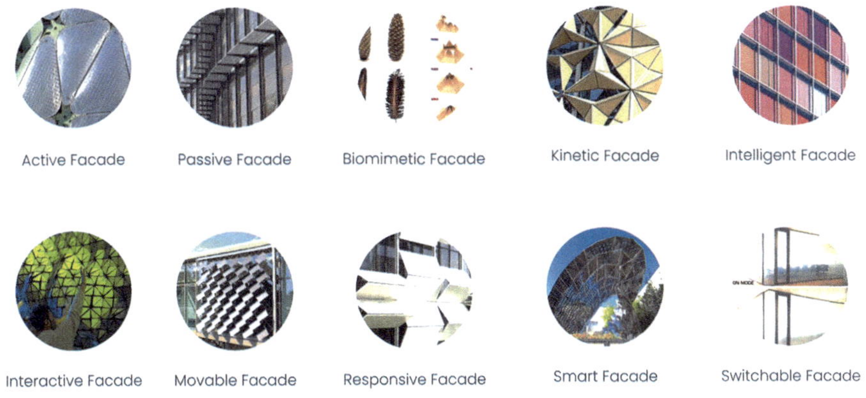


**Fig. 15.3** Various kinds of adaptive facades

### 15.4.1 Responsive Facades

Polymers used in building facades can adjust their properties in real time with respect to changes in orientation, sunlight, temperature, or air quality [63]. Such systems contribute to passive climate control, enhancing occupant comfort while reducing energy consumption. Research in integrating these facades with artificial intelligence (AI) systems promises further optimization of building performance [64]. The multivariate properties extend to aesthetic aspects, where facades can change colour and texture to suit environmental needs. Sometimes they are used in such a way that it goes with the design and sustainability of the built environment (Fig. 15.4).

### 15.4.2 Self-Healing Materials

Conducting polymers with self-healing capabilities can repair microcracks or damage in building components, prolonging their lifespan and reducing maintenance costs. This is particularly relevant for infrastructure exposed to harsh environmental conditions. Innovations in multi-functional self-healing polymers combine conductivity with structural integrity, broadening their applications. For instance, polymer blends incorporating graphene or carbon nanotubes enhance these materials' robustness and healing capabilities [65, 66].

**Fig. 15.4** Al Bahar Twin Towers with kinetic façade at Abu Dhabi

### 15.4.3 Smart Sensors and Actuators

Embedded sensors and actuators based on conducting polymers enable buildings to monitor and respond to environmental changes [67]. Applications include air quality monitoring, structural health assessments, and automated climate control systems. For instance, polypyrrole-based sensors are being used to detect toxic gases, ensuring safe indoor environments. Such systems are integral to predictive maintenance strategies, reducing downtime and operational costs [68].

## 15.5 Smart Technologies and Their Integration with the Internet of Things

Conducting polymers plays a pivotal role in the integration of Internet of Things (IoT) technologies into architecture. They are used in communities as well as in individual buildings (Fig. 15.5).

### 15.5.1 Smart Surfaces

Interactive surfaces made from conducting polymers can detect touch or proximity, enabling applications such as interactive walls and control panels in smart homes and offices [70]. Advanced versions include augmented surfaces that display real-time data or respond to voice commands. These surfaces can also incorporate energy harvesting capabilities, enabling dual functionality [71]. Conducting polymers are used in smart surfaces widely in mosques also, in mosques of Dubai, they used interactive prayer mats where one can easily make their prayers more focused [72].

**Fig. 15.5** Application of IOT in making smart cities [69]

### 15.5.2 Wearable Architecture

Flexible, conductive materials enable the concept of wearable architecture, where structural components can adapt their shape and function based on user input or environmental conditions [73]. This futuristic application aligns with the increasing demand for modular and customizable living spaces. Wearable components also integrate seamlessly with portable power generation and storage systems, creating self-sustaining structures [74].

### 15.5.3 Data Transmission and Connectivity

Conducting polymers are being explored for use in flexible antennas and communication devices, facilitating seamless connectivity within smart buildings [75]. Such integration ensures reliable data flow across IoT devices, enhancing the overall functionality of intelligent architectural systems. These polymers also support the miniaturization of electronic components, making them ideal for compact designs [69].

## 15.6 Sustainability and Environmental Impact

The sustainable characteristics of conducting polymers make them ideal for green architecture:

### 15.6.1 Eco-Friendly Synthesis

Recent advancements in the synthesis of conducting polymers emphasize the use of renewable feedstocks and low-energy processes, reducing their environmental footprint [22]. For instance, biosynthesized conducting polymers from natural precursors like lignin have gained attention [76]. These processes also produce fewer hazardous by-products, making them suitable for large-scale applications.

### 15.6.2 Reduction of Construction Waste

The adaptability and longevity of conducting polymers reduce material wastage, aligning with circular economy principles in construction [77]. Modular components using these materials allow for easy disassembly and reuse. Incorporating recyclable composites ensures minimal waste generation during renovation or demolition activities [78].

### 15.6.3 Recyclability

Many conducting polymers can be recycled or reprocessed, minimizing their impact on the environment [79]. Emerging technologies are exploring solvent-based recycling methods that recover high-purity polymers for reuse. Efforts are also being directed towards closed-loop systems, ensuring that materials are reused without degradation in quality [80].

## 15.7 Case Studies

### 15.7.1 The Adaptive Building Envelope

A prominent example of conducting polymers in architecture is the development of adaptive building envelopes. These systems use conducting polymers to regulate thermal and light transmission, demonstrating significant energy savings in various climates [81]. Plastic is also a very good example of application of polymer. According to one study plastic is composed of organic compounds with a high molecular weight that can be liquefied and poured into certain moulds [82] (Fig. 15.6).

Projects like the "Chameleon Façade" prototype showcase how these materials can integrate seamlessly into high-rise buildings. This innovation has inspired new designs for skyscrapers in metropolitan areas aiming for net-zero energy certifications [83].

**Polymers Are Widely Used in Polymer Concrete**
According to study of conducting polymers the following could be derived

- Which Repairs to corrosion-damaged concrete,
- All Marine works,
- Prestressed concrete,
- Electrical works and industrial construction,
- Used in the construction of roads,
- Nuclear power plant construction,
- Irrigation works,
- Floor Waterproofing.

Seco et al. prepared PPC building products and characterized their durability based on the damage and mechanical strength losses after freezing and thawing. Results showed that no surface damage existed in PPC building products after

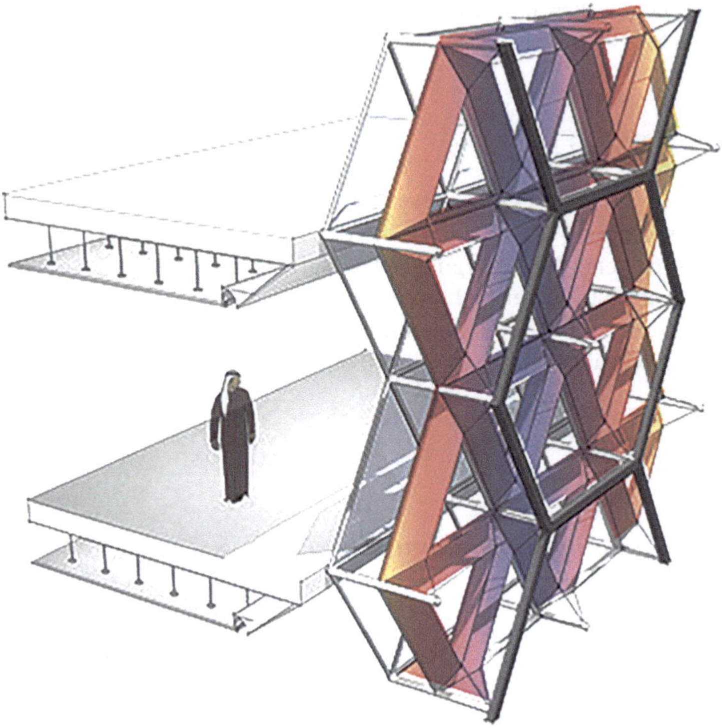

**Fig. 15.6** Chameleon façade concept in a proposed project in Dubai

25 cycles of freezing followed by thawing in water according to European Standard EN 14617-5 [84].

Polyester polymer concrete (PPC) has been widely used in constructions due to its advantages of fast setting and hardening time, high mechanical strength, low permeability, and good chemical resistance [85].

## *15.7.2 Smart Flooring Systems*

Flooring materials incorporating conducting polymers can generate electricity through piezoelectric effects, providing a sustainable energy source for smart buildings [86]. For example, polyaniline-based composites have been tested in high-traffic areas to harness kinetic energy from footfalls. Such systems are particularly effective in commercial spaces like airports and train stations, where foot traffic is consistent and high.

### 15.7.3 Organic Photovoltaic Integration

Several modern buildings have integrated OPV panels based on conducting polymers, showcasing their potential in urban renewable energy generation. Examples include urban skyscrapers with fully photovoltaic glass façades, offering both aesthetic appeal and energy functionality [87]. These applications also highlight the role of conducting polymers in retrofitting existing structures with renewable energy solutions [88].

In addition to the waterproofing of concrete bridge decks or roofs, many other building elements, including walls, facades, and cultural heritage sites, also require waterproofing. Water penetration in these building elements seriously impacts their durability. A common approach to protect these building elements is to use waterproof coatings to prevent the transport of water into the interior [89].

## 15.8 Challenges in Architecture Techniques

While conducting polymers offers immense potential, several challenges must be addressed to realize their full integration into architecture:

### 15.8.1 Durability

Ensuring the long-term stability of conducting polymers in outdoor environments remains a critical challenge [19]. Advances in protective coatings and hybrid composites are being explored. Future efforts may involve embedding self-repairing functionalities to enhance resilience.

### 15.8.2 Scalability

Developing cost-effective manufacturing processes for large-scale applications is essential [90]. Innovations in 3D printing with conducting polymers hold promise for mass production. Automated manufacturing lines integrated with quality control systems can further streamline production [91].

### 15.8.3 *Standardization*

Establishing industry standards for the use of conducting polymers in architecture is necessary to facilitate widespread adoption [92]. Collaborative efforts between regulatory bodies and researchers are crucial. International cooperation on certification processes will ensure uniformity in application [93].

Over the past few years, polymers have had a growing impact on the development of new structures and the upkeep of deteriorating infrastructure [94–96]. Polymers are used in a variety of structural and non-structural applications in the building industry; around 18% of the United States' plastic use occurs in the building and construction industry [97, 98]. Plastic smart windows are quickly becoming one of the primary components of the next generation of highly energy-efficient buildings [99–101].

## 15.9 Concluding the Case in Terms of Solution

This research concludes about the use of conducting polymers as a paradigm shift in contemporary architecture, offering solutions that amalgamate aesthetics, functionality, form and sustainability. Polymers unique properties enable a wide range of applications, from energy management systems to adaptive building technologies.

Conducting polymers are also used as the binder, substrate, and active materials in construction process. Conducting polymer-based electrodes shows better electrochemical performance than other. Non-conducting polymers increase the "dead volume" of electrodes [102]. Most of the electrodes which are customized design are wavy, spring, and kirigami-like structures [103].

Engineering fields such as communication, acoustics, architecture, and structure are now included in the broad and multidisciplinary study of sustainable construction. Owners, contractors, suppliers, and building users all participate in it [104]. The past several years have seen a significant amount of research on sustainable construction [105]. In this context, polymer-based CBMs have seen increased application in recent years in the field of construction engineering [106].

As research and development continue, conducting polymers is set to play a pivotal role in shaping the future of architectural design, addressing the challenges of urbanization, climate change, and resource efficiency [107–109]. By including these materials in architecture and built structures, architects use to create intelligent, creative, eco-friendly, and sustainable structures that redefine the built environment [110].

## 15.10 Recommending for Future Research in Polymers

Future research should focus on enhancing the performance of conducting polymers through nanocomposites, hybrid materials, and advanced fabrication techniques. Collaborative efforts between material scientists, architects, designers, and engineers will drive the development of innovative applications. Exploring synergistic integrations with other advanced materials, such as shape-memory alloys and graphene, will expand their utility.

## References

1. Bechthold M, Weaver JC (2017) Materials science and architecture. Nat Rev Mater:2. https://doi.org/10.1038/natrevmats.2017.82
2. Kenry LB (2018) Recent advances in biodegradable conducting polymers and their biomedical applications. Biomacromolecules 19:1783–1803
3. Gish L (1990) Criteria for design and testing of sealants used in single-ply roofing and waterproofing systems. In: Buildings sealants: materials, properties, and performance. ASTM International, West Conshohocken, pp 237–260
4. Groenendaal L, Zotti G, Aubert PH, Waybright SM, Reynolds JR (2003) Electrochromic polymers: recent developments in viewing systems, sensors and wearables. Adv Mater 15(11):855–879. https://doi.org/10.1002/adma.200300373
5. Bubnova O, Crispin X (2012) Towards polymer-based organic thermoelectric generators. Energy Environ Sci 5(11):9345–9362. https://doi.org/10.1039/C2EE22470D
6. Irimia-Vladu M (2014) "Green" electronics: biodegradable and biocompatible materials and devices for sustainable future. Chem Soc Rev 43(2):588–610. https://doi.org/10.1039/c3cs60235d
7. Ates M (2013) A review study of (bio)sensor systems based on conducting polymers. Mater Sci Eng C 33(4):1853–1863. https://doi.org/10.1016/j.msec.2013.01.035
8. Pacheco-Torgal F, Jalali S (2011) Nanotechnology: advantages and drawbacks in the field of construction and building materials. Constr Build Mater 25(2):582–590. https://doi.org/10.1016/j.conbuildmat.2010.07.009
9. Ramakrishna S, Mayer J, Wintermantel E, Leong KW (2001) Biomedical applications of polymer-composite materials: a review. Compos Sci Technol 61(9):1189–1224. https://doi.org/10.1016/S0266-3538(00)00247-2
10. Sánchez C, Julián B, Belleville P, Popall M (2005) Applications of hybrid organic–inorganic nanocomposites. J Mater Chem 15(35–36):3559–3592. https://doi.org/10.1039/B509097K
11. Chiang CK, Fincher CR, Park YW, Heeger AJ, Shirakawa H, Louis EJ et al (1977) Electrical conductivity in doped polyacetylene. Phys Rev Lett 39(17):1098–1101. https://doi.org/10.1103/PhysRevLett.39.1098
12. Skotheim TA, Reynolds JR (2007) Conjugated polymers: theory, synthesis, properties, and characterization.3rd edn. CRC Press, Boca Raton
13. Bredas JL, Street GB (1985) Polarons, bipolarons, and solitons in conducting polymers. Acc Chem Res 18(10):309–315. https://doi.org/10.1021/ar00118a004
14. Yuen MC, Hwang JY, Kan CW (2016) A review of PEDOT: PSS as a conducting polymer in electronic textiles. Fibers Polym 17(5):763–773. https://doi.org/10.1007/s12221-016-5764-1
15. Stejskal J, Gilbert RG (2002) Polyaniline. Preparation of a conducting polymer. Pure Appl Chem 74(5):857–867. https://doi.org/10.1351/pac200274050857
16. Wallace GG, Spinks GM, Kane-Maguire LA, Teasdale PR (2008) Conductive electroactive polymers: intelligent materials systems.3rd edn. CRC Press

17. Martins P, Lopes AC, Lanceros-Mendez S (2014) Electroactive phases of poly(vinylidene fluoride): determination, processing and applications. Prog Polym Sci 39(4):683–706. https://doi.org/10.1016/j.progpolymsci.2013.07.006
18. Ghosh S, Inganäs O (2000) Electrochemical characterisation of poly(3,4-ethylene dioxythiophene) based conducting hydrogel networks. Synth Met 110(1):47–56
19. Skotheim TA, Reynolds JR (2007) Handbook of conducting polymers.3rd edn. CRC Press, Boca Raton
20. Le T, Dittmer A, Dittmer W (2019) Green synthesis of conducting polymers. Polymers (Basel) 11(2):362
21. Güner A, Erkan S, Toppare L (2012) Conducting polymers as environmentally benign materials: recent developments. Prog Polym Sci 37(10):1264–1290
22. Groenendaal L, Jonas F, Freitag D, Pielartzik H, Reynolds JR (2000) Poly(3,4-ethylenedioxythiophene) and its derivatives: past, present, and future. Adv Mater 12(7):481–494
23. Zhang J, Huang Y, Hu J, Xie J (2020) Application of functional polymer materials in smart building energy systems. Renew Sust Energ Rev 117:109493
24. Li H, Wang X, Yu Y, Liu Y (2021) Smart polymers and their applications in smart building technologies. Adv Mater Sci Eng 2021:8893185
25. Baetens R, Jelle BP, Gustavsen A (2010) Phase change materials for building applications: a state-of-the-art review. Energ Buildings 42(9):1361–1368
26. Lu N, Du P (2012) Demand response in smart grids: a review. Proc IEEE 100(1):198–207
27. Kilkis B (2011) Advanced HVAC control strategies and their impact on building performance. Energ Buildings 43(2–3):418–424
28. Bhadra S, Khastgir D, Singha NK, Lee JH (2009) Progress in preparation, processing and applications of polyaniline. Prog Polym Sci 34(8):783–810
29. Kumar R, Singh RK, Sharma S, Dutta V (2021) Conducting polymers for energy applications: recent advances and challenges. J Energy Chem 59:315–337
30. Yoon H (2013) Current trends in sensors based on conducting polymer nanomaterials. Nano 3(3):524–549
31. Guo Y, Zhang X, Zheng Q, Li H (2016) Smart materials in architecture: applications of conductive polymers. Smart Mater Struct 25(9):093001
32. Tiwari A, Advincula RC (2018) Handbook of polymer nanocomposites for industrial applications. Elsevier, Amsterdam
33. Inzelt G (2012) Conducting polymers: a new era in electrochemistry.2nd edn. Springer, Berlin
34. Sekine C, Tsubata Y, Yamada T, Kitano M, Doi S (2014) Recent progress of high performance polymer-based flexible electronics for organic transistors, organic photovoltaics, and organic light emitting diodes. Sci Technol Adv Mater 15(3):034203
35. Molina-Lopez F, Briand D, de Rooij NF (2012) All additive inkjet printed humidity sensors on plastic substrate. Sens Actuators B Chem 166–167:212–222
36. Torrisi F, Carey T, Sun Z (2014) Inkjet printing of graphene and 2D materials: technology, applications and prospects. Nanoscale 6(15):8284–8295
37. Kim DH, Ghaffari R, Lu N, Rogers JA (2012) Flexible and stretchable electronics for biointegrated devices. Annu Rev Biomed Eng 14:113–128
38. Wang J, Wang Y, Yang M, Tian H, Zhao Y (2020) Adaptive materials and structures for building applications. Smart Mater Struct 29(6):063002
39. Groenendaal L, Zotti G, Aubert PH, Waybright SM, Reynolds JR (2003) Electrochromic conducting polymers for displays and microelectronic devices. Adv Mater 15(11):855–879
40. Malti A, Edberg J, Granberg H, Khan ZU, Jonsson MP, Berggren M et al (2016) An organic mixed ion-electron conductor for power electronics. Adv Sci (Weinh) 3(2):1500305
41. Mijenko A, Hargreaves M, Ganesh T (2018) Polymers in architecture: integration of functionality and form. Arch Mater Struct 3(1):25–34
42. Tzatzalos E, Xenidou E, Symeou E, Karapanagiotis I (2020) The aesthetic and functional contribution of smart materials to architecture. J Cult Herit 45:206–212

43. Chuangchote S, Sagawa T, Yoshikawa S (2009) Design of polymer-based transparent films for building integration. Sol Energy Mater Sol Cells 93(3):476–480
44. Granqvist CG (2014) Electrochromics for smart windows: oxide-based thin films and devices. Thin Solid Films 564:1–38
45. Llordés A, Garcia G, Gazquez J, Milliron DJ (2013) Tunable near-infrared and visible light transmittance in nanocrystal-in-glass composites. Nature 500(7462):323–326
46. Beaujuge PM, Reynolds JR (2010) Color control in π-conjugated organic polymers for use in electrochromic devices. Chem Rev 110(1):268–320
47. Wang Y, Runnerstrom EL, Milliron DJ (2016) Switchable materials for smart windows. Annu Rev Chem Biomol Eng 7:283–304
48. Sannicolo T, Lagrange M, Cabos A, Celle C, Simonato JP, Bellet D (2016) Metallic nanowire-based transparent electrodes for next generation flexible devices: a review. Small 12(44):6052–6075
49. Baetens R, Jelle BP, Gustavsen A (2010) Properties, requirements and possibilities of smart windows for dynamic daylight and solar energy control in buildings: a state-of-the-art review. Sol Energy Mater Sol Cells 94(2):87–105
50. Krebs FC (2009) Fabrication and processing of polymer solar cells: a review of printing and coating techniques. Sol Energy Mater Sol Cells 93(4):394–412
51. Li G, Zhu R, Yang Y (2012) Polymer solar cells. Nat Photonics 6(3):153–161
52. Kumar A, Pathak A, Tiwari GN (2020) Advancement in building integrated photovoltaic system: a review. Renew Sust Energ Rev 126:109841
53. Hou Y, Du X, Scheiner S, McMeekin DP, Wang Z, Khaja Nazeeruddin M et al (2018) Organic–inorganic hybrid tandem solar cells with advanced functional layers. Nat Mater 17(5):401–409
54. Saliba M, Correa-Baena JP, Grätzel M, Hagfeldt A, Abate A (2018) Perovskite solar cells: from the atomic level to film quality and device performance. Angew Chem Int Ed Engl 57(10):2554–2569
55. Jelle BP, Breivik C, Drolsum R (2012) Building integrated photovoltaic products: a state-of-the-art review and future research opportunities. Sol Energy Mater Sol Cells 100:69–96
56. Yang Z, Zhang J, Kintner-Meyer MCW, Lu X, Choi D, Lemmon JP et al (2011) Electrochemical energy storage for green grid. Chem Rev 111(5):3577–3613
57. Zhou G, Paek E, Hwang GS, Manthiram A (2015) Long-life Li/Polysulphide batteries with high sulfur loading enabled by lightweight three-dimensional nitrogen/sulfur-codoped graphene sponge. Nat Commun 6:7760
58. Snook GA, Kao P, Best AS (2011) Conducting-polymer-based supercapacitor devices and electrodes. J Power Sources 196(1):1–12
59. Pushparaj VL, Shaijumon MM, Kumar A, Murugesan S, Ci L, Vajtai R et al (2007) Flexible energy storage devices based on nanocomposite paper. Proc Natl Acad Sci USA 104(34):13574–13577
60. Wang Y, Song Y, Xia Y (2016) Electrochemical capacitors: mechanism, materials, systems, characterization and applications. Chem Soc Rev 45(21):5925–5950
61. Liu Z, Neale ZG, Cao G (2016) Understanding electrochemical potentials of cathode materials in rechargeable batteries. Mater Today 19(2):109–123
62. Kaner RB, Ouyang JY (2021) Design of multifunctional conducting polymer nanocomposites for energy and environment. Acc Chem Res 54(5):1097–1109
63. Addington DM, Schodek DL (2005) Smart materials and technologies in architecture. Architectural Press, Oxford
64. Müller C, Voss K (2020) Smart façades and AI integration in sustainable buildings: a review. Energ Buildings 224:110229
65. Cordier P, Tournilhac F, Soulié-Ziakovic C, Leibler L (2008) Self-healing and thermoreversible rubber from supramolecular materials. Nature 451(7181):977–980
66. Ma S, Ren L, Liu T, Yang W, Xu L, Liu Y (2020) Self-healing conductive polymer composites for smart electronic applications. Adv Mater Technol 5(2):1900665

67. Someya T, Bao Z, Malliaras GG (2016) The rise of plastic bioelectronics. Nature 540(7633):379–385
68. Wang X, Li X, Zhang L, Yao J, Zhang J, Guo Z (2021) Polypyrrole-based gas sensors for ammonia detection: a review. Synth Met 275:116750
69. Kumar S, Tiwari P, Zymbler M (2019) Internet of Things is a revolutionary approach for future technology enhancement: a review. J Big Data 6:111
70. Someya T, Amagai M (2019) Toward a new generation of smart skins. Nat Biotechnol 37(4):382–388
71. Wang C, Xia K, Wang H, Liang X, Yin Z, Zhang Y (2019) Advanced carbon for flexible and wearable electronics. Adv Mater 31(9):1801072
72. Al-Qaysi N, Al-Emran M, Al-Sharafi MA, Shaalan K (2020) Smart HCI-based systems in Islamic prayer practices: the case of interactive prayer mats. Appl Sci 10(21):7673
73. Menges A, Sheil B, Glynn R, Skavara M (2017) Fabricate: rethinking design and construction. UCL Press
74. Stoppa M, Chiolerio A (2014) Wearable electronics and smart textiles: a critical review. Sensors 14(7):11957–11992
75. Sekitani T, Someya T (2010) Stretchable, large-area organic electronics. Adv Mater 22(20):2228–2246
76. Zhan Y, Mei L, Wang W, Zhou J, Ma F, Wang X (2021) Lignin-based conducting polymers: from synthesis to application. Chem Eng J 406:126896
77. Radwan AB, Hassan M, Abd El-Aziz ME (2022) Conducting polymers as eco-friendly materials for sustainable construction. J Build Eng 52:104423
78. Sharma A, Bansal P (2020) Modular construction and the circular economy: integrating design for disassembly. J Clean Prod 276:123324
79. Song J, Wang L, Xu Z (2021) Recycling technologies for thermoset composite materials—a review. Waste Manag 120:570–586
80. Sayyed MI, Yasmeen T, Al-Hadeethi Y (2022) Closed-loop recycling systems for high-performance polymers: environmental and economic perspectives. Polymers (Basel) 14(2):379
81. Debele TA, Mekonnen YS, Tsai HC (2020) Polymers for daylighting applications: development of electrochromic smart windows. Polymers (Basel) 12(2):443
82. Andrady AL, Neal MA (2009) Applications and societal benefits of plastics. Philos Trans R Soc Lond Ser B Biol Sci 364(1526):1977–1984
83. Seco A, Ramírez F, Miqueleiz L, García B (2010) Durability of polyester polymer concrete made with recycled aggregates from construction and demolition waste. Constr Build Mater 24(10):1781–1789
84. Fowler DW (1999) Polymers in concrete: a vision for the 21st century. Cem Concr Compos 21(5–6):449–452
85. Almusallam AA, Al-Gahtani AS, Dakhil FH, Saadi MM (1997) Effect of temperature and humidity on the properties of polymer concrete. Constr Build Mater 11(6):353–356
86. Wang Z, Wang W, Zang X, Wang C, Li X, Pan L et al (2015) Piezoelectric polyaniline nanofiber matrix for energy harvesting and self-powered sensing. Nano Energy 14:364–371
87. Krebs FC (2009) Polymer solar cell modules prepared using roll-to-roll methods: knife-over-edge coating, slot-die coating and screen printing. Sol Energy Mater Sol Cells 93(4):465–475
88. Kaltenbrunner M, White MS, Głowacki ED, Sekitani T, Someya T, Bauer S et al (2012) Ultrathin and lightweight organic solar cells with high flexibility. Nat Commun 3:770
89. Van den Bossche N, Wouters P, Roels S (2013) The durability of building elements: waterproofing performance of façade joints. Build Environ 59:701–710
90. Tee BC-K, Wang C, Allen R, Bao Z (2012) An electrically and mechanically self-healing composite with pressure- and flexion-sensitive properties for electronic skin applications. Nat Nanotechnol 7(12):825–832
91. Ligon SC, Liska R, Stampfl J, Gurr M, Mülhaupt R (2017) Polymers for 3D printing and customized additive manufacturing. Chem Rev 117(15):10212–10290

92. Kousiatza C, Karalekas D (2016) In-situ monitoring of thermal strain fields in fused deposition modeling process. Mater Des 97:400–406
93. Malucelli G, Bongiovanni R, Priola A (2020) UV-curable polymeric materials for coating applications: a review. Polymers (Basel) 12(8):1790
94. Bank LC (2006) Composites for construction: structural design with FRP materials. Wiley
95. Hollaway LC (2010) A review of the present and future utilisation of FRP composites in the civil infrastructure with reference to their important in-service properties. Constr Build Mater 24(12):2419–2445
96. Benmokrane B, El-Salakawy E, El-Ragaby A, Lackey T (2006) Fiber-reinforced polymer bars for the concrete construction of bridges in Canada. Adv Struct Eng 9(6):721–733
97. De la Fuente A, Pujadas P, Blanco A, Aguado A (2012) Experiences in the use of synthetic fibres in concrete tunnel linings. Tunn Undergr Space Technol 27(1):60–71
98. Pilakoutas K, Guadagnini M, Dritsos S (2004) Rehabilitation of deteriorated concrete structures using fiber reinforced polymers. Mater Struct 37(5):288–296
99. American Chemistry Council. Plastics and polymer composites in construction. 2021
100. PlasticsEurope Plastics—the facts 2020: an analysis of European plastics production, demand and waste data. PlasticsEurope, Brussels, p 2020
101. Granqvist CG (2006) Electrochromic materials: out of a niche. Nat Mater 5(2):89–90
102. Wang X, Zhi L, Müllen K (2008) Transparent, conductive graphene electrodes for dye-sensitized solar cells. Nano Lett 8(1):323–327
103. Kim D, Lee H, Kwon J, Song YM (2019) Kirigami-inspired highly stretchable and conformable triboelectric energy harvester for self-powered wearable electronics. Nano Energy 60:398–406
104. Kibert CJ (2016) Sustainable construction: green building design and delivery.4th edn. Wiley
105. Ding G, Akintoye A (2000) Sustainable construction: the role of environmental assessment tools. Build Environ 35(3):339–346
106. Lu Y, Yuan H (2011) Exploring the eco-efficiency of construction and demolition waste recycling. J Clean Prod 19(10):1044–1052
107. Tam VWY, Tam CM, Zeng SX, Chan KKW (2006) Environmental performance measurement indicators in construction. Build Environ 41(2):164–173
108. Ahn YH, Kwon YJ, Jeon S (2019) Sustainability in construction engineering: a review of recent research trends. Sustainability 11(3):781
109. Zhang Y, Xie Y, Liu W, Xie J, Wang H, Liu Y (2022) Review of polymer-based building materials for sustainable construction. Constr Build Mater 321:126318
110. Finkenstadt VL, Liu LS (2010) Poly(lactic acid) and starch composites reinforced with natural fibers. J Polym Environ 18(2):316–324

# Chapter 16
# Conclusion and Future Aspects

**Anshul Yadav, Chetna Kumari, Astha Kamboj, Nirmala Kumari Jangid, and Kavita Poonia**

**Abstract** Conducting polymers (CPs) represent a novel class of materials that combine the advantages of polymer production with the electrical characteristics of metals. They are very adaptable for a wide range of applications due to their tunable electrical, optical, and mechanical characteristics. The potential applications of conducting polymers (CPs) in a variety of domains, including biological and environmental, have led to a great deal of interest in their tuning. Researchers have significantly improved conductivity, stability, and biocompatibility by varying the polymer structure, doping levels, and incorporation of composite materials. CPs are used in anti-corrosion coatings, energy storage devices such as supercapacitors and batteries, and sensors that detect pollutants in the environment. They are perfect for these uses because of their capacity to interact with different chemical species and conduct electricity. CPs are utilized in drug delivery systems, biosensors, tissue engineering, in the biomedical industry. Their adjustable characteristics and biocompatibility provide precise control over drug release mechanisms and cell interactions. This chapter concludes with a discussion of CP tuning and its applications in biomedicine and the environment and their future aspects. Overall, the tuning of conducting polymers opens up new possibilities for creative solutions in both environmental and biomedical applications, leading the way for advanced technologies that tackle pressing global issues.

**Keywords** Conducting polymers (CPs) · Electrical · Optical · Mechanical · Biocompatibility · Supercapacitors · Biosensors

---

A. Yadav · C. Kumari · A. Kamboj · N. K. Jangid · K. Poonia (✉)
Department of Chemistry, Banasthali Vidyapith, Banasthali, Rajasthan, India



S. Jadoun, N. Kumari Jangid (eds.), *Tuning of Conducting Polymers*, Springer Series on Polymer and Composite Materials,
https://doi.org/10.1007/978-3-032-00461-1_16

## 16.1 Introduction

Conducting polymers, once considered an obscure class of materials, have emerged at the forefront of material science due to their unique combination of electrical conductivity, flexibility, and process ability. These properties, which were traditionally associated with metals and semiconductors, are now found in organic materials, opening up new avenues for innovative applications across a broad range of fields [1]. As the demand for multifunctional materials grows, conducting polymers have gained significant attention for their versatility and adaptability in addressing contemporary challenges in both the environmental and biomedical sectors [2].

The concept of "tuning" conducting polymers refers to the ability to modify and enhance their properties to suit specific applications. By altering factors such as doping levels, molecular structure, and processing conditions, researchers can tailor the electrical, mechanical, optical, and chemical characteristics of these polymers to meet the demands of various industries [3].

In the environmental arena, conducting polymers have shown promise in applications such as sensors for pollutant detection, water treatment systems, and energy storage devices. Their ability to efficiently conduct electricity while remaining lightweight and flexible makes them ideal candidates for these sustainable technologies. Similarly, in the biomedical field, the biocompatibility, electroactivity, and mechanical properties of conducting polymers enable groundbreaking advancements in areas like tissue engineering, neural interfaces, and biosensors. These polymers are even being explored for the development of smart prosthetics and controlled drug release systems, demonstrating their potential to transform healthcare and improve quality of life [4].

By bringing together the latest research and development in both environmental and biomedical applications, this book highlights the interdisciplinary nature of the field and provides a comprehensive understanding of how conducting polymers are being utilized to solve some of the most pressing challenges of our time.

## 16.2 Tuning of Conducting Polymers According to Applications

The ability to tailor the properties of conducting polymers has made them one of the most versatile classes of materials in modern science and technology [5].

This book explored various methods and strategies for tuning conducting polymers, with a particular focus on how these modifications enable the polymers to excel in distinct applications. The tuning process involves altering a variety of material properties, including conductivity, mechanical flexibility, chemical stability, and biocompatibility, through methods such as doping, functionalization, molecular engineering, and hybridization with other materials [6]. By carefully adjusting these parameters, conducting polymers can be optimized for specific requirements in different sectors.

In environmental applications, conducting polymers can be engineered for the detection of pollutants, water purification, and energy harvesting, among other uses. For instance, by modifying their surface properties or incorporating nanoparticles, these materials can be tailored to selectively interact with specific environmental pollutants or to improve energy storage efficiency [7]. Similarly, in biomedical applications, the electrical properties of conducting polymers can be fine-tuned for use in bioelectronic devices, drug delivery systems, and tissue engineering scaffolds. The biocompatibility and electroactivity of these materials make them ideal candidates for interfacing with living tissues, facilitating advancements in areas such as neural stimulation, biosensing, and regenerative medicine [8, 9].

Ultimately, this chapter aims to provide a comprehensive understanding of how the strategic tuning of conducting polymers opens up innovative possibilities for their use across a variety of industries, pushing the boundaries of what is achievable with these remarkable materials.

## 16.3 Characterization of Tuned Conducting Polymers

This chapter will provide an overview of the key characterization methods used for tuned conducting polymers, focusing on how each technique contributes to a comprehensive understanding of the material's properties. By exploring these characterization tools, this chapter aim to provide readers with the necessary knowledge to assess and refine the performance of conducting polymers, ultimately enabling the development of more efficient and reliable applications across fields such as environmental monitoring, energy storage, and biomedicine [10].

The development and application of conducting polymers depend heavily on their ability to be finely tuned for specific uses, ranging from environmental sensing to biomedical devices [11].

This chapter delves into the various techniques used to characterize the properties of tuned conducting polymers. This process is essential for both optimizing performance and ensuring consistency in the polymer's functionality over time [12].

The characterization of conducting polymers is multi-faceted and typically involves a combination of analytical methods. Electrical characterization techniques, such as conductivity measurements, electrochemical analysis, and impedance spectroscopy, provide crucial information on the material's charge transport properties. Structural analysis, including X-ray diffraction (XRD), scanning electron microscopy (SEM), and atomic force microscopy (AFM), is used to assess the polymer's morphology and molecular arrangement, which are key factors influencing its electrical and mechanical performance [13]. Additionally, spectroscopic techniques such as UV-Vis, Fourier-transform infrared (FTIR), and Raman spectroscopy allow researchers to examine the chemical composition and interactions within the polymer, providing a deeper understanding of the doping mechanisms and functional groups that play a role in tuning the material's properties [14].

For polymers intended for biomedical applications, biocompatibility and biodegradability are also crucial factors. Thus, techniques like cytotoxicity assays, in vitro testing, and even in vivo assessments are often employed to evaluate the interaction between the conducting polymer and biological systems [15].

Through a combination of fundamental principles and practical insights, this chapter highlights the importance of accurate and thorough characterization in the ongoing development of conducting polymers.

## 16.4 Effect of Tuning and Post-treatment on Conducting Polymers

The performance of conducting polymers is highly sensitive to the methods used to tune and modify their properties, making the processes of tuning and post-treatment critical to achieving desired outcomes for specific applications [9]. Conducting polymers, by their nature, are flexible, lightweight, and easily processed, but their inherent conductivity and mechanical properties can be further enhanced—or, in some cases, tailored down—through a combination of chemical, physical, and environmental treatments [10, 16].

Tuning typically refers to the deliberate modification of a polymer's structure, chemical composition, or doping levels during the synthesis phase to enhance or optimize specific characteristics [17]. This is often followed by post-treatment procedures—such as thermal annealing, chemical crosslinking, solvent treatments, or exposure to various atmospheric conditions—that can further modify the polymer's final properties, including its morphology, adhesion to substrates, and overall performance in a given application [18].

Tuning methods such as doping with electron-rich or electron-deficient materials can significantly alter the electrical conductivity of a polymer. However, the effects of such modifications can be influenced by post-treatment steps, such as annealing or exposure to reactive gases, which may either stabilize or degrade the doped structure [19]. Similarly, post-treatment processes like surface functionalization or chemical modification can enhance the polymer's compatibility with other materials or improve its environmental stability, particularly for applications in harsh or fluctuating conditions [20].

In biomedical applications, post-treatment processes are crucial for ensuring that conducting polymers maintain their biocompatibility and functionality when interfacing with living tissues [21]. For instance, post-treatment strategies such as hydrophilic surface modification or the incorporation of biologically active molecules can improve the adhesion of conducting polymers to cells, enhance their resistance to biological degradation, and promote desirable interactions in tissue engineering or drug delivery systems [22].

The ability to precisely control and modify these materials' properties opens up vast opportunities for innovation, enabling the creation of smarter, more efficient materials for a wide array of technological and industrial uses.

## 16.5 Tailored Conducting Polymers for Photocatalytic Treatment of Water

Water pollution, driven by industrial effluents, agricultural runoff, and urban waste, continues to be one of the most pressing environmental challenges worldwide. Traditional water treatment methods often fall short in addressing complex pollutants. As the demand for cleaner water grows, the development of advanced materials for efficient, sustainable water treatment is of paramount importance. Among these, conducting polymers have emerged as promising candidates, making them well-suited for photocatalytic water treatment applications [23].

Conducting polymers, with their ability to absorb light, facilitate electron transfer, and interact with a wide range of chemical species, are ideal candidates for use in photocatalytic systems [24]. However, to maximize their efficiency in water treatment applications, these materials must be carefully tailored to optimize their photocatalytic activity, stability, and selectivity [25].

The process of tailoring conducting polymers for photocatalysis involves adjusting their electronic properties, morphology, and surface characteristics. This can be achieved through a variety of strategies, including doping with metals or non-metals, incorporating nanoparticles, and functionalizing the polymer surface. Such modifications can improve the polymer's light absorption, charge carrier mobility, and catalytic performance under visible light, all of which are crucial for efficient water purification [26, 27].

Through careful design and optimization, tailored conducting polymers can play a crucial role in developing next-generation technologies for clean water, contributing to global efforts to ensure access to safe and clean drinking water for all.

## 16.6 Tailored Conducting Polymers for Adsorption of Dyes and Heavy Metals

Industrial processes, particularly in textiles, mining, and manufacturing, release a wide range of pollutants into the environment, including toxic dyes and heavy metals [28]. The removal of such pollutants from wastewater is a critical challenge, and traditional treatment methods, including chemical precipitation and filtration, often struggle with the efficiency and selectivity required to address these complex contaminants [29, 30].

Conducting polymers have emerged as promising materials for this purpose, owing to their unique combination of high surface area, tunable chemical composition, and excellent adsorption properties [31]. These polymers, which exhibit electrical conductivity as well as the ability to interact with a variety of chemical species can be tailored for targeted adsorption of pollutants like toxic dyes and heavy metals [32].

The tuning of conducting polymers typically involves doping with various metal or non-metal species, functionalizing the polymer surface with specific groups, or incorporating nanoparticles to increase the material's surface reactivity and affinity for contaminants [33]. These modifications can improve the polymer's ability to adsorb specific pollutants, increase the polymer's surface area, and enhance its mechanical stability under practical conditions [34].

The ability to design conducting polymers with specific properties for targeted adsorption opens up exciting new possibilities for advanced materials in environmental remediation, providing a sustainable pathway for cleaner water and healthier ecosystems.

## 16.7 Tailored Conducting Polymers for Surface Protection

Surface protection is a critical concern across a wide range of industries, including electronics, automotive, construction, and marine sectors, where materials are frequently exposed to harsh environmental conditions. In these settings, surfaces are subjected to a variety of challenges such as corrosion, wear, environmental degradation, and fouling, all of which can compromise the longevity, functionality, and performance of materials [35, 36].

Conducting polymers, with their unique combination of electrical conductivity, chemical versatility, and processability, have gained significant attention as promising candidates for surface protection applications [7]. By tuning their chemical composition, morphology, and electronic properties, conducting polymers can be engineered to deliver enhanced corrosion resistance, abrasion resistance, and even active functionalities such as anti-fouling or self-healing properties [37].

Various strategies are used to design conducting polymers with specific surface protection capabilities [38]. For example, doping with metal or non-metal elements can improve the corrosion resistance of conducting polymers, while functionalization with hydrophobic or biocompatible groups can impart water resistance or antifouling properties [39].

There are challenges and opportunities in developing conducting polymers for surface protection, including issues related to long-term stability, adhesion to substrates, and the durability of the protective films under dynamic environmental conditions [40].

This research area represents a promising frontier for the development of next-generation materials capable of meeting the evolving demands of industries where surface integrity and longevity are paramount.

## 16.8 Modification of Conducting Polymers for Electronics and Optoelectronics

The rapid advancement of electronics and optoelectronics has revolutionized numerous industries, ranging from consumer electronics to energy harvesting, medical devices, and telecommunications. As electronic devices become smaller, more efficient, and more versatile, there is an increasing need for novel materials that can meet the complex demands of these high-performance applications [41]. Conducting polymers, which combine the electrical properties of metals with the flexibility and processability of organic materials, have emerged as promising candidates for next-generation electronic and optoelectronic devices [42].

Various techniques are used to modify conducting polymers to optimize their performance for use in electronics and optoelectronics. These modification strategies can include doping with electron-rich or electron-deficient materials, altering the polymer's molecular structure to enhance conjugation length, blending with other functional materials, and surface treatments to improve interface properties [43]. The modification of conducting polymers is crucial to improving their electrical conductivity, charge transport, stability, and optical characteristics, all of which are critical for devices such as organic light-emitting diodes (OLEDs), organic solar cells (OPVs), transistors, sensors, and flexible displays [44].

By understanding the different methods of tuning conducting polymers, researchers and engineers can push the boundaries of what is possible with organic-based devices, paving the way for more efficient, flexible, and cost-effective electronic and optoelectronic technologies in the future.

## 16.9 Tailored Conducting Polymers as Energy Storage Devices

As the global demand for sustainable and renewable energy sources continues to rise, the development of efficient energy storage systems becomes increasingly vital. Energy storage devices, such as batteries and supercapacitors, play a crucial role in facilitating the widespread adoption of renewable energy by ensuring stable power supply, managing energy fluctuations, and enabling energy conservation. Traditional energy storage materials, including metals and inorganic compounds, have shown significant limitations in terms of cost, environmental impact, and performance in some areas [45]. In the recent years, conducting polymers have emerged as promising candidates for next-generation energy storage devices [46].

Conducting polymers offer numerous advantages as energy storage materials, such as lightweight design, high surface area, and the ability to undergo redox reactions, which are essential for both charge storage and release [47]. Modifying the structure, doping levels, and morphology of conducting polymers enables their properties to be adjusted to meet the specific requirements of energy storage applications, including enhancing charge capacity, cycling stability, and rate performance [48].

Various strategies are used to tailor conducting polymers for energy storage applications. These strategies include altering the polymer's chemical composition, adjusting doping levels, and combining conducting polymers with other materials to enhance their electrochemical performance [49]. By fine-tuning factors such as the polymer's electroactivity, porosity, and mechanical properties, researchers can design materials that maximize energy density, improve efficiency, and extend the lifespan of energy storage devices [50].

In supercapacitors, conducting polymers can serve as efficient electrodes by providing high capacitance and fast charge-discharge cycles. In batteries, they can function as electrodes or electrode additives, enhancing energy storage capacity while offering advantages in terms of flexibility and lightweight design. Moreover, hybrid systems that combine conducting polymers with other nanomaterials or inorganic components, such as carbon-based materials and metal oxides, are becoming increasingly popular as they offer the potential for even greater performance improvements [51].

The ability to tailor their properties opens up exciting possibilities for more efficient, sustainable, and flexible energy storage solutions.

## 16.10 Hybrid Conducting Polymers as Sensors and Biosensors

The need for advanced sensors and biosensors has surged in recent years, driven by applications across a wide range of fields, including environmental monitoring, healthcare, food safety, and industrial process control. Sensors that can detect minute changes in chemical, biological, or physical conditions are crucial for real-time diagnostics, early disease detection, pollution monitoring, and the prevention of hazardous events [52]. Conducting polymers form hybrid systems that significantly enhance sensor performance, enabling highly sensitive, selective, and versatile detection platforms [53].

Hybrid conducting polymers leverage the strengths of both organic conducting polymers and inorganic materials, offering superior properties that can be tailored for specific sensing applications [54]. These materials allow for enhanced charge transport, increased surface area, improved stability, and the ability to interface with biological systems. The incorporation of functional materials, such as metal nanoparticles, carbon-based nanomaterials, or biological recognition elements (e.g., enzymes, antibodies, and nucleic acids), allows for a more effective response to target analytes, leading to sensors that are more sensitive and selective than traditional materials [55]. Furthermore, these hybrids often exhibit the ability to operate in diverse environments, from aqueous solutions to harsh industrial settings [56].

Through careful design and optimization, hybrid materials offer exciting opportunities to create more efficient, cost-effective, and flexible sensors capable of addressing some of the most critical challenges in health, environment, and industry.

## 16.11 Tuning of Conducting Polymers for Applications in Materials Science

Conducting polymers have become a cornerstone of modern materials science, offering a unique blend of properties that make them suitable for a diverse range of applications. Unlike traditional metals or ceramics, conducting polymers combine the flexibility, lightweight nature, and processability of organic materials with the electrical conductivity typically associated with inorganic semiconductors and metals [57].

Tuning process allows researchers to achieve a desired balance of conductivity, mechanical properties, optical characteristics, and environmental stability, making them ideal candidates for applications in materials science [58]. The ability to control these properties at the molecular level provides unprecedented flexibility in designing materials for a wide array of cutting-edge technologies, ranging from flexible electronics and energy storage devices to sensors, actuators, and advanced coatings [59].

In coatings and corrosion protection, tailored conducting polymers can offer greater durability and resistance to environmental degradation. Additionally, the incorporation of conducting polymers into composite materials has opened new possibilities in areas like smart materials, sensors, and wearable electronics [60].

With the ability to precisely control their properties, conducting polymers are poised to enable a wide range of innovations in the development of advanced materials, offering exciting prospects for the future of technology and industry.

## 16.12 Biodegradable Conducting Polymers for Smart Drug Delivery

The field of drug delivery has undergone a dramatic transformation over the past few decades, driven by the desire to improve the efficacy, targeting, and controlled release of therapeutic agents while minimizing side effects [61]. Among the most promising developments in drug delivery technology are smart drug delivery systems, which are designed to release drugs in response to specific environmental stimuli (e.g., pH, temperature, and enzymes) or biological markers. These systems have the potential to provide more precise and controlled treatment, especially for complex conditions like cancer, chronic diseases, and infections [62].

Biodegradable conducting polymers have emerged as a powerful class of materials in this context, offering unique advantages for the development of smart drug delivery systems. Conducting polymers can respond to environmental stimuli, such as pH or electrical signals, enabling the controlled release of drugs, while their biodegradable nature ensures that they do not accumulate in tissues or cause long-term toxicity [63].

By combining conducting polymers with other materials, such as nanoparticles, lipids, or targeting ligands, researchers have developed hybrid systems that not only offer controlled drug release but also enhance targeting, improve bioavailability, and facilitate cellular uptake [64].

There are mechanisms by which biodegradable conducting polymers enable smart drug delivery, focusing on their ability to respond to external stimuli such as changes in pH (which occurs in different parts of the body, such as the stomach or tumors), electrical fields, or the presence of specific enzymes [65]. These properties can be leveraged to trigger drug release at specific sites in the body, ensuring that the drug is delivered in the right amount, at the right time, and to the right location [66].

Through their tunable properties, biodegradability, and responsiveness to biological stimuli, these materials hold immense potential for improving the specificity, efficiency, and safety of drug therapies, paving the way for more effective treatments with fewer side effects [67]. As research continues to advance, biodegradable conducting polymers are poised to play a pivotal role in the future of nanomedicine and personalized drug delivery.

## 16.13 Tailored Conducting Polymers for Biomedical Engineering

Biomedical engineering, which lies at the intersection of engineering, biology, and medicine, has seen significant advancements with the development of novel materials that can interact with biological systems in a controlled and dynamic manner [68]. These organic materials combine the flexibility, processability, and biocompatibility of polymers with the electrical conductivity typically associated with metals and semiconductors. This unique combination allows conducting polymers to function as bridges between biological systems and electronic devices, making them invaluable for a wide range of applications in biomedical engineering, from biosensors and tissue engineering to drug delivery systems and neural interfaces [69].

Tailoring conducting polymers involves modifying their chemical structure, doping them with specific ions or molecules, or blending them with other materials to optimize their performance for a particular biomedical application. This enables the design of materials that can respond to external stimuli, interact with cells and tissues in specific ways, and integrate seamlessly with living systems [70].

By tuning the molecular structure, doping levels, and morphology, conducting polymers can be designed to achieve specific functionalities, such as enhanced electrical conductivity for neural interfaces, improved mechanical properties for scaffolding in tissue engineering, or controlled drug release for therapeutic applications

[71]. There are specific applications of tailored conducting polymers in biomedical engineering, such as their role in electroactive scaffolds for tissue regeneration, biosensors for detecting biomarkers, and bioelectronic devices like neural stimulators and electrochemical sensors [72].

The ability to modify these materials at the molecular level opens up new opportunities to develop more effective, responsive, and adaptable devices for a variety of medical purposes, ultimately improving the quality of care and enabling personalized treatments [73]. As research in this area continues to evolve, tailored conducting polymers will play a key role in shaping the future of biomedical technologies, paving the way for smarter, more efficient, and more sustainable medical devices and therapies.

## 16.14 Modification of Conducting Polymers for Smart Textiles and Wearables

The emergence of smart textiles and wearable technologies has revolutionized the world around us, enabling a wide range of applications from healthcare monitoring and fitness tracking to environmental sensing and interactive fashion [74]. Conducting polymers, with their unique combination of electrical conductivity, flexibility, and processability, have become essential components in the development of these smart materials. Their ability to be integrated into fabrics and textiles, coupled with their tunable properties, makes them particularly attractive for use in wearable devices and smart textiles [75, 76].

By altering the polymer's composition and structure, including the introduction of functional groups, doping with different materials, or combining conducting polymers with other flexible materials, their properties can be tuned to provide high conductivity, durability, and comfort in wearable applications. These modifications also ensure that conducting polymers retain their functionality under the mechanical stresses and environmental conditions typical of textiles and wearables, such as washing, stretching, or exposure to moisture [77].

As research continues to advance in this area, conducting polymers will remain at the forefront of innovations that enhance the capabilities, comfort, and efficiency of wearable devices in everyday life.

## 16.15 Emerging Applications of Conducting Polymers in Architecture

The field of architecture has long been driven by the pursuit of innovative materials that can enhance the functionality, esthetics, and sustainability of buildings and infrastructure [78]. In the recent years, the integration of advanced materials such as conducting polymers has opened up new possibilities for transforming the built environment [23, 79].

Conducting polymers are particularly suited for integration into architectural systems due to their ability to be tuned for specific properties such as conductivity, mechanical strength, transparency, and environmental responsiveness [80]. The ability to control their electronic properties through doping, blending, and surface modification makes them ideal for use in applications such as smart windows, energy-harvesting façades, self-healing coatings, and interactive building elements [81]. Additionally, their flexibility and lightweight nature offer significant advantages in the design of flexible and wearable building components, as well as in retrofitting existing structures with advanced functionality [82].

One of the most exciting developments in the use of conducting polymers in architecture is their potential to create multifunctional building materials. For example, conducting polymers can be incorporated into transparent conductive films, allowing windows to act as both passive elements for light transmission and active components for energy generation or sensing. Similarly, they can be used in self-cleaning or self-healing coatings that respond to environmental stimuli, reducing maintenance costs and extending the lifespan of building materials [83]. In addition, the integration of conducting polymers into sensor networks within buildings offers the potential for real-time monitoring of environmental conditions such as humidity, temperature, air quality, and structural health [84].

The versatility of these materials, combined with ongoing innovations in material science, will undoubtedly play a pivotal role in shaping the next generation of buildings that are not only more functional and energy-efficient but also more interactive, responsive, and environmentally conscious. Conducting polymers are helping to pave the way toward a future where architecture is not just a passive environment, but an active, dynamic component of the built world.

## 16.16 Conclusion and Future Aspects

The study and development of conducting polymers continue to offer new opportunities for innovation across a variety of sectors. Through careful tuning of their properties, conducting polymers are poised to make significant contributions in both environmental sustainability and healthcare technologies.

This book underscores the importance of interdisciplinary approaches, blending material science, chemistry, physics, and engineering, to optimize the performance of conducting polymers. It is clear that, as research and development continue, conducting polymers will increasingly be incorporated into real-world solutions for pressing environmental and biomedical challenges. However, further exploration into their long-term stability, cost-effectiveness, and scalability will be essential for maximizing their practical applications.

In conclusion, conducting polymers represent a dynamic and evolving field of research that holds immense potential. Their versatility and adaptability make them key materials for the future of both environmental and biomedical innovations. Continued research, innovation, and collaboration across scientific disciplines will be crucial in unlocking the full potential of these remarkable materials.

## References


1. Namsheer K, Rout CS (2021) Conducting polymers: a comprehensive review on recent advances in synthesis, properties and applications. RSC Adv 11(10):5659–5697. https://doi.org/10.1039/d0ra07800j
2. Ding H, Hussein AM, Ahmad I, Latef R, Abbas JK, Ali ATA, Saeed SM, Abdulwahid AS, Ramadan MF, Rasool HA, Elawady A (2024) Conducting polymers in industry: a comprehensive review on the characterization, synthesis and application. Alex Eng J 88:253–267. https://doi.org/10.1016/j.aej.2024.01.029
3. Guo X, Facchetti A (2020) The journey of conducting polymers from discovery to application. Nat Mater 19(9):922–928. https://doi.org/10.1038/s41563-020-0778-5
4. Dubey N, Kushwaha CS, Shukla SK (2019) A review on electrically conducting polymer bionanocomposites for biomedical and other applications. Int J Polym Mater 69(11):709–727. https://doi.org/10.1080/00914037.2019.1605513
5. Nezakati T, Seifalian A, Tan A, Seifalian AM (2018) Conductive polymers: opportunities and challenges in biomedical applications. Chem Rev 118(14):6766–6843. https://doi.org/10.1021/acs.chemrev.6b00275
6. Wang W, Li Z, Li M, Fang L, Chen F, Han S, Lan L, Chen J, Chen Q, Wang H, Liu C, Yang Y, Yue W, Xie Z (2022) High-transconductance, highly elastic, durable and recyclable all-polymer electrochemical transistors with 3D micro-engineered interfaces. Nano-Micro Lett 14(1). https://doi.org/10.1007/s40820-022-00930-5
7. Inoue A, Yuk H, Lu B, Zhao X (2020) Strong adhesion of wet conducting polymers on diverse substrates. Sci Adv 6(12). https://doi.org/10.1126/sciadv.aay5394
8. Yang J, Du M, Wang L, Li S, Wang G, Yang X, Zhang L, Fang Y, Zheng W, Yang G, Jiang X (2018) Bacterial cellulose as a supersoft neural interfacing substrate. ACS Appl Mater Interfaces 10(39):33049–33059. https://doi.org/10.1021/acsami.8b12083
9. Maria IP, Paulsen BD, Savva A, Ohayon D, Wu R, Hallani R, Basu A, Du W, Anthopoulos TD, Inal S, Rivnay J, McCulloch I, Giovannitti A (2021) The effect of alkyl spacers on the mixed ionic-electronic conduction properties of N-type polymers. Adv Funct Mater 31(14). https://doi.org/10.1002/adfm.202008718
10. Besford QA, Yong H, Merlitz H, Christofferson AJ, Sommer J, Uhlmann P, Fery A (2021) FRET-integrated polymer brushes for spatially resolved sensing of changes in polymer conformation. Angew Chem Int Ed 60(30):16600–16606. https://doi.org/10.1002/anie.202104204
11. Park J, Staiger A, Mecking S, Winey KI (2022) Ordered nanostructures in thin films of precise ion-containing multiblock copolymers. ACS Central Sci 8(3):388–393. https://doi.org/10.1021/acscentsci.1c01594
12. Cavassin P, Holzer I, Tsokkou D, Bardagot O, Réhault J, Banerji N (2022) Electrochemical doping in ordered and disordered domains of conjugated polymers. arXiv (Cornell University). https://doi.org/10.48550/arxiv.2212.09468
13. Zhang S, Alesadi A, Mason GT, Chen K, Freychet G, Galuska L, Cheng Y, St Onge PBJ, Ocheje MU, Ma G, Qian Z, Dhakal S, Ahmad Z, Wang C, Chiu Y, Rondeau-Gagné S, Xia W, Gu X (2021) Molecular origin of strain-induced chain alignment in PDPP-based semiconducting polymeric thin films. Adv Funct Mater 31(21). https://doi.org/10.1002/adfm.202100161
14. Gregory SA, Li Y, Monroe TD, Li J, Yee SK, Losego MD (2021) Vapor phase infiltration doping of the semiconducting polymer poly(aniline) with TiCl4 + H2O: mechanisms, reaction kinetics, and electrical and optical properties. ACS Appl Polym Mater 3(2):720–729. https://doi.org/10.1021/acsapm.0c01014

15. Snyder BL, Mohammed HS, Samways DSK, Shipp DA (2020) Drug delivery and drug efficacy from amorphous poly(thioether anhydrides). Macromol Biosci 20(5). https://doi.org/10.1002/mabi.201900377
16. Criado-Gonzalez M, Dominguez-Alfaro A, Lopez-Larrea N, Alegret N, Mecerreyes D (2021) Additive manufacturing of conducting polymers: recent advances, challenges, and opportunities. ACS Appl Polym Mater 3(6):2865–2883. https://doi.org/10.1021/acsapm.1c00252
17. Tamburrino F, Barone S, Paoli A, Razionale AV (2021) Post-processing treatments to enhance additively manufactured polymeric parts: a review. Virtual Phys Prototyp 16(2):221–254. https://doi.org/10.1080/17452759.2021.1917039
18. Li L, Chen X, Torkelson JM (2020) Covalent adaptive networks for enhanced adhesion: exploiting disulfide dynamic chemistry and annealing during application. ACS Appl Polym Mater 2(11):4658–4665. https://doi.org/10.1021/acsapm.0c00720
19. Yoon SE, Kang Y, Jeon GG, Jeon D, Lee SY, Ko S, Kim T, Seo H, Kim B, Kim JH (2020) Exploring wholly doped conjugated polymer films based on hybrid doping: strategic approach for optimizing electrical conductivity and related thermoelectric properties. Adv Funct Mater 30(42). https://doi.org/10.1002/adfm.202004598
20. Makvandi P, Iftekhar S, Pizzetti F, Zarepour A, Zare EN, Ashrafizadeh M, Agarwal T, Padil VVT, Mohammadinejad R, Sillanpaa M, Maiti TK, Perale G, Zarrabi A, Rossi F (2020) Functionalization of polymers and nanomaterials for water treatment, food packaging, textile and biomedical applications: a review. Environ Chem Lett 19(1):583–611. https://doi.org/10.1007/s10311-020-01089-4
21. Gan D, Huang Z, Wang X, Jiang L, Wang C, Zhu M, Ren F, Fang L, Wang K, Xie C, Lu X (2019) Graphene oxide-templated conductive and redox-active nanosheets incorporated hydrogels for adhesive bioelectronics. Adv Funct Mater 30(5). https://doi.org/10.1002/adfm.201907678
22. Mao L, Hu S, Gao Y, Wang L, Zhao W, Fu L, Cheng H, Xia L, Xie S, Ye W, Shi Z, Yang G (2020) Biodegradable and electroactive regenerated bacterial cellulose/MXene (Ti3C2Tx) composite hydrogel as wound dressing for accelerating skin wound healing under electrical stimulation. Adv Healthc Mater 9(19). https://doi.org/10.1002/adhm.202000872
23. Luo H, Kaneti YV, Ai Y, Wu Y, Wei F, Fu J, Cheng J, Jing C, Yuliarto B, Eguchi M, Na J, Yamauchi Y, Liu S (2021) Nanoarchitectured porous conducting polymers: from controlled synthesis to advanced applications. Adv Mater 33(29). https://doi.org/10.1002/adma.202007318
24. Lee C, Javed H, Zhang D, Kim J, Westerhoff P, Li Q, Alvarez PJJ (2018) Porous electrospun fibers embedding TIO2 for adsorption and photocatalytic degradation of water pollutants. Environ Sci Technol 52(7):4285–4293. https://doi.org/10.1021/acs.est.7b06508
25. Marques AC, Vale M, Vicente D, Schreck M, Tervoort E, Niederberger M (2021) Porous silica microspheres with immobilized Titania nanoparticles for in-flow solar-driven purification of wastewater. Glob Chall 5(5). https://doi.org/10.1002/gch2.202000116
26. Shu C, Han C, Yang X, Zhang C, Chen Y, Ren S, Wang F, Huang F, Jiang J (2021) Boosting the photocatalytic hydrogen evolution activity for D–Π–A conjugated microporous polymers by statistical copolymerization. Adv Mater 33(26). https://doi.org/10.1002/adma.202008498
27. Zhang Y, Lv H, Zhang Z, Wang L, Wu X, Xu H (2021) Stable unbiased photo-electrochemical overall water splitting exceeding 3% efficiency via covalent triazine framework/metal oxide hybrid photoelectrodes. Adv Mater 33(15). https://doi.org/10.1002/adma.202008264
28. Yaseen DA, Scholz M (2018) Textile dye wastewater characteristics and constituents of synthetic effluents: a critical review. Int J Environ Sci Technol 16(2):1193–1226. https://doi.org/10.1007/s13762-018-2130-z
29. Ren L, Zhao G, Pan L, Chen B, Chen Y, Zhang Q, Xiao X, Xu W (2021) Efficient removal of dye from wastewater without selectivity using activated carbon-Juncus effusus porous fibril composites. ACS Appl Mater Interfaces 13(16):19176–19186. https://doi.org/10.1021/acsami.0c22104
30. Li C, Guggenberger P, Han SW, Ding W, Kleitz F (2022) Ultrathin covalent organic framework anchored on graphene for enhanced organic pollutant removal. Angew Chem Int Ed 61(35). https://doi.org/10.1002/anie.202206564

31. Feng D, Liu P, Wang Q (2020) Selective microwave sintering to prepare multifunctional poly(ether imide) bead foams based on segregated carbon nanotube conductive network. Ind Eng Chem Res 59(13):5838–5847. https://doi.org/10.1021/acs.iecr.0c00090
32. Hong Y, Thirion D, Subramanian S, Yoo M, Choi H, Kim HY, Stoddart JF, Yavuz CT (2020) Precious metal recovery from electronic waste by a porous porphyrin polymer. Proc Natl Acad Sci 117(28):16174–16180. https://doi.org/10.1073/pnas.2000606117
33. Klemes MJ, Ling Y, Ching C, Wu C, Xiao L, Helbling DE, Dichtel WR (2019) Reduction of a Tetrafluoroterephthalonitrile-β-Cyclodextrin polymer to remove anionic micropollutants and perfluorinated alkyl substances from water. Angew Chem Int Ed 58(35):12049–12053. https://doi.org/10.1002/anie.201905142
34. Yang S, Yong X, Zhao B, Wu Y, Deng J (2021) Biomass-derived acetylenic polymer monoliths prepared by high internal phase emulsion template method and used for adsorbing cationic pollutants. Macromol Chem Phys 222(6). https://doi.org/10.1002/macp.202000448
35. Kulyk B, Freitas MA, Santos NF, Mohseni F, Carvalho AF, Yasakau K, Fernandes AJS, Bernardes A, Figueiredo B, Silva R, Tedim J, Costa FM (2021) A critical review on the production and application of graphene and graphene-based materials in anti-corrosion coatings. Crit Rev Solid State Mater Sci 47(3):309–355. https://doi.org/10.1080/10408436.2021.1886046
36. Ma Y, Yang Y, Cao D, Fang J, Zhu C, Yang M, Fang L, Lu C, Xu Z (2021) Facile repair of anti-corrosion polymeric composite coatings based on light triggered self-healing. Macromol Mater Eng 306(9). https://doi.org/10.1002/mame.202100106
37. Peng T, Xiao R, Rong Z, Liu H, Hu Q, Wang S, Li X, Zhang J (2020) Polymer nanocomposite-based coatings for corrosion protection. Chem Asian J 15(23):3915–3941. https://doi.org/10.1002/asia.202000943
38. Jin Z, Mei H, Pan L, Liu H, Cheng L (2021) Superhydrophobic self-cleaning hierarchical micro-/nanocomposite coating with high corrosion resistance and durability. ACS Sustain Chem Eng 9(11):4111–4121. https://doi.org/10.1021/acssuschemeng.0c09059
39. Xu H, Fan S, Lu Y, Feng H, Qiu J (2020) Proposal and verification of a novel superhydrophobic-conductive anti-corrosion polyaniline-silica coating. Bull Chem Soc Jpn 93(9):1114–1120. https://doi.org/10.1246/bcsj.20200051
40. Brito J, Hlushko H, Abbott A, Aliakseyeu A, Hlushko R, Sukhishvili SA (2021) Integrating antioxidant functionality into polymer materials: fundamentals, strategies, and applications. ACS Appl Mater Interfaces 13(35):41372–41395. https://doi.org/10.1021/acsami.1c08061
41. Khokhar D, Jadoun S, Arif R, Jabin S (2020) Functionalization of conducting polymers and their applications in optoelectronics. Polym-Plast Technol Mater 60(5):465–487. https://doi.org/10.1080/25740881.2020.1819312
42. Murad AR, Iraqi A, Aziz SB, Abdullah SN, Brza MA (2020) Conducting polymers for optoelectronic devices and organic solar cells: a review. Polymers 12(11):2627. https://doi.org/10.3390/polym12112627
43. Wu X, Fu W, Chen H (2022) Conductive polymers for flexible and stretchable organic optoelectronic applications. ACS Appl Polym Mater 4(7):4609–4623. https://doi.org/10.1021/acsapm.2c00519
44. Thapa YN, Kafle BP, Adhikari R (2023) Properties and applications of conjugated polymers for flexible electronics: current trends and perspectives. Flex Wear Sens:97–114
45. Magu TO, Agobi AU, Hitler L, Dass PM (2019) A review on conducting polymers-based composites for energy storage application. J Chem Rev 1(1):19–34. https://doi.org/10.33945/sami/jcr.2019.1.1934
46. Khan J, Khan A, Rubab B, Jamshaid F, Al-Kahtani AA, Dahshan A (2023) Exploring the progression of energy storage toward flexibility: metal-organic framework and conducting polymer aspects. Appl Mater Today 34:101906. https://doi.org/10.1016/j.apmt.2023.101906
47. Li C, Zhang K, Cheng X, Li J, Jiang Y, Li P, Wang B, Peng H (2023) Polymers for flexible energy storage devices. Prog Polym Sci 143:101714. https://doi.org/10.1016/j.progpolymsci.2023.101714
48. Pati SK, Patra D, Muduli S, Mishra S, Park S (2023) Energy storage application of conducting polymers featuring dual acceptors: exploring conjugation and flexible chain length effects. Small 19(26). https://doi.org/10.1002/smll.202300689

49. Kim J, Lee J, You J, Park M, Hossain MSA, Yamauchi Y, Kim JH (2016) Conductive polymers for next-generation energy storage systems: recent progress and new functions. Mater Horiz 3(6):517–535. https://doi.org/10.1039/c6mh00165c
50. Shi Y, Peng L, Ding Y, Zhao Y, Yu G (2015) Nanostructured conductive polymers for advanced energy storage. Chem Soc Rev 44(19):6684–6696. https://doi.org/10.1039/c5cs00362h
51. Dhandapani E, Thangarasu S, Ramesh S, Ramesh K, Vasudevan R, Duraisamy N (2022) Recent development and prospective of carbonaceous material, conducting polymer and their composite electrode materials for supercapacitor—a review. J Energy Storage 52:104937. https://doi.org/10.1016/j.est.2022.104937
52. Dakshayini B, Reddy KR, Mishra A, Shetti NP, Malode SJ, Basu S, Naveen S, Raghu AV (2019) Role of conducting polymer and metal oxide-based hybrids for applications in ampereometric sensors and biosensors. Microchem J 147:7–24. https://doi.org/10.1016/j.microc.2019.02.061
53. Iqbal S, Ahmad S (2017) Recent development in hybrid conducting polymers: synthesis, applications and future prospects. J Ind Eng Chem 60:53–84. https://doi.org/10.1016/j.jiec.2017.09.038
54. Abhishek N, Verma A, Singh A, Vandana N, Kumar T (2023) Metal-conducting polymer hybrid composites: a promising platform for electrochemical sensing. Inorg Chem Commun 157:111334. https://doi.org/10.1016/j.inoche.2023.111334
55. Lakard B (2020) Electrochemical biosensors based on conducting polymers: a review. Appl Sci 10(18):6614. https://doi.org/10.3390/app10186614
56. Ramanaviciene A, Ramanavicius A (2004) Towards the hybrid biosensors based on biocompatible conducting polymers. In: UV solid-state light emitters and detectors, pp 287–296. https://doi.org/10.1007/978-1-4020-2103-9_25
57. Tajik S, Beitollahi H, Nejad FG, Shoaie IS, Khalilzadeh MA, Asl MS, Van Le Q, Zhang K, Jang HW, Shokouhimehr M (2020) Recent developments in conducting polymers: applications for electrochemistry. RSC Adv 10(62):37834–37856. https://doi.org/10.1039/d0ra06160c
58. Sumdani MG, Islam MR, Yahaya ANA, Safie SI (2021) Recent advancements in synthesis, properties, and applications of conductive polymers for electrochemical energy storage devices: a review. Polym Eng Sci 62(2):269–303. https://doi.org/10.1002/pen.25859
59. Salim E, Abdelghany A, Tarabiah A (2023) Ameliorating and tuning the optical, dielectric, and electrical properties of hybrid conducting polymers/metal oxide nanocomposite for optoelectronic applications. Mater Chem Phys 313:128788. https://doi.org/10.1016/j.matchemphys.2023.128788
60. Yan Y, Han M, Jiang Y, Ng ELL, Zhang Y, Owh C, Song Q, Li P, Loh XJ, Chan BQY, Chan SY (2024) Electrically conductive polymers for additive manufacturing. ACS Appl Mater Interfaces 16(5):5337–5354. https://doi.org/10.1021/acsami.3c13258
61. Paramshetti S, Angolkar M, Fatease AA, Alshahrani SM, Hani U, Garg A, Ravi G, Osmani R, a. M. (2023) Revolutionizing drug delivery and therapeutics: the biomedical applications of conductive polymers and composites-based systems. Pharmaceutics 15(4):1204. https://doi.org/10.3390/pharmaceutics15041204
62. Puiggalí-Jou A, Del Valle LJ, Alemán C (2019) Drug delivery systems based on intrinsically conducting polymers. J Control Release 309:244–264. https://doi.org/10.1016/j.jconrel.2019.07.035
63. Raj GK, Singh E, Hani U, Ramesh K, Talath S, Garg A, Savadatti K, Bhatt T, Madhuchandra K, Osmani R, a. M. (2023) Conductive polymers and composites-based systems: an incipient stride in drug delivery and therapeutics realm. J Control Release 355:709–729. https://doi.org/10.1016/j.jconrel.2023.02.017
64. Alkahtani ME, Elbadawi M, Chapman CAR, Green RA, Gaisford S, Orlu M, Basit AW (2023) Electroactive polymers for on-demand drug release. Adv Healthc Mater 13(3). https://doi.org/10.1002/adhm.202301759
65. Aragón-Navas A, López-Cano JJ, Johnson M, Sigen A, Vicario-De-La-Torre M, Andrés-Guerrero V, Tai H, Wang W, Bravo-Osuna I, Herrero-Vanrell R (2023) Smart biodegradable hydrogels: drug-delivery platforms for treatment of chronic ophthalmic diseases affecting the back of the eye. Int J Pharm 649:123653. https://doi.org/10.1016/j.ijpharm.2023.123653

66. Wong WF, Ang KP, Sethi G, Looi CY (2023) Recent advancement of medical patch for transdermal drug delivery. Medicina 59(4):778. https://doi.org/10.3390/medicina59040778
67. Geszke-Moritz M, Moritz M (2024) Biodegradable polymeric nanoparticle-based drug delivery systems: comprehensive overview, perspectives and challenges. Polymers 16(17):2536. https://doi.org/10.3390/polym16172536
68. Nasser RA, Arya SS, Alshehhi KH, Teo JC, Pitsalidis C (2024) Conducting polymer scaffolds: a new frontier in bioelectronics and bioengineering. Trends Biotechnol 42(6):760–779. https://doi.org/10.1016/j.tibtech.2023.11.017
69. Maziz A, Özgür E, Bergaud C, Uzun L (2021) Progress in conducting polymers for biointerfacing and biorecognition applications. Sens Actuators Rep 3:100035. https://doi.org/10.1016/j.snr.2021.100035
70. Van Le C, Yoon H (2024) Advances in the use of conducting polymers for healthcare monitoring. Int J Mol Sci 25(3):1564. https://doi.org/10.3390/ijms25031564
71. Khan T, Vadivel G, Ramasamy B, Murugesan G, Sebaey TA (2024) Biodegradable conducting polymer-based composites for biomedical applications—a review. Polymers 16(11):1533. https://doi.org/10.3390/polym16111533
72. Guo B, Ma XP (2018) Conducting polymers for tissue engineering. Biomacromolecules 19(6):1764–1782. https://doi.org/10.1021/acs.biomac.8b00276
73. Joy N, Gopalan GP, Eldho J, Francis R (2016) Conducting polymers: biomedical applications. Biomed Appl Polym Mater Compos:37–89. https://doi.org/10.1002/9783527690916.ch3
74. Grancarić AM, Jerković I, Koncar V, Cochrane C, Kelly FM, Soulat D, Legrand X (2017) Conductive polymers for smart textile applications. J Ind Text 48(3):612–642. https://doi.org/10.1177/1528083717699368
75. Ghazal H, Nasser A, Maraae A, Allam L, Beltagy Z, Abd-Elaal L, Elshamy M (2024) Conductive polymer coated textile and their applications. J Text Color Polym Sci 0(0):0. https://doi.org/10.21608/jtcps.2024.259584.1305
76. Mohamed AL, El-Naggar ME, Hassabo AG (2021) Preparation of hybrid nanoparticles to enhance the electrical conductivity and performance properties of cotton fabrics. J Mater Res Technol 12:542–554. https://doi.org/10.1016/j.jmrt.2021.02.035
77. Ding Y, Jiang J, Wu Y, Zhang Y, Zhou J, Zhang Y, Huang Q, Zheng Z (2024) Porous conductive textiles for wearable electronics. Chem Rev 124(4):1535–1648. https://doi.org/10.1021/acs.chemrev.3c00507
78. Umoh AA, Adefemi A, Ibekwe KI, Etukudoh EA, Ilojianya VI, Nwokediegwu ZQS (2024) Green architecture and energy efficiency: a review of innovative design and construction techniques. Eng Sci Technol J 5(1):185–200. https://doi.org/10.51594/estj.v5i1.743
79. Zhang C, Li Y, Kang W, Liu X, Wang Q (2021) Current advances and future perspectives of additive manufacturing for functional polymeric materials and devices. SusMat 1(1):127–147. https://doi.org/10.1002/sus2.11
80. Hofmann AI, Östergren I, Kim Y, Fauth S, Craighero M, Yoon M, Lund A, Müller C (2020) All-polymer conducting fibers and 3D prints via melt processing and templated polymerization. ACS Appl Mater Interfaces 12(7):8713–8721. https://doi.org/10.1021/acsami.9b20615
81. Khandelwal H, Schenning APHJ, Debije MG (2017) Infrared regulating smart window based on organic materials. Adv Energy Mater 7(14). https://doi.org/10.1002/aenm.201602209
82. Globa A, Costin G, Tokede O, Wang R, Khoo CK, Moloney J (2021) Hybrid kinetic facade: fabrication and feasibility evaluation of full-scale prototypes. Archit Eng Des Manag 18(6):791–811. https://doi.org/10.1080/17452007.2021.1941739
83. Chen J, Zhu Y, Huang J, Zhang J, Pan D, Zhou J, Ryu JE, Umar A, Guo Z (2020) Advances in responsively conductive polymer composites and sensing applications. Polym Rev 61(1):157–193. https://doi.org/10.1080/15583724.2020.1734818
84. Jiang J, Jung N, Boor BE (2021) Using building energy and smart thermostat data to evaluate indoor ultrafine particle source and loss processes in a Net-Zero energy house. ACS ES&T Eng 1(4):780–793. https://doi.org/10.1021/acsestengg.1c00002

GPSR Compliance

*The European Union's (EU) General Product Safety Regulation (GPSR) is a set of rules that requires consumer products to be safe and our obligations to ensure this.*

*If you have any concerns about our products, you can contact us on ProductSafety@springernature.com*

In case Publisher is established outside the EU, the EU authorized representative is:

Springer Nature Customer Service Center GmbH
Europaplatz 3
69115 Heidelberg, Germany

**Batch number: 10755329**

Printed by Printforce, the Netherlands